INFORMATION RETRIEVAL

INFORMATION RETRIEVAL: SEARCHING IN THE 21ST CENTURY

Ayşe Göker
City University London, UK

John Davies
BT, UK

A John Wiley and Sons, Ltd., Publication

Library of Congress Cataloging-in-Publication Data
Information retrieval : searching in the 21st century / [edited by] Ayşe Göker, John Davies.
        p. cm.
    Includes bibliographical references and index.
    ISBN 978-0-470-02762-2
    1.   Information retrieval. I. Göker, Ayşe.   II. Davies, J. (N. John)
    ZA3075.I55 2009
    025.5′24 – dc22
                                                          2009025912

A catalogue record for this book is available from the British Library.

ISBN: 978-0-470-02762-2 (H/B)

Set in 9/11 Times New Roman by Laserwords Private Ltd, Chennai, India.
Printed and bound in Great Britain by CPI Antony Rowe, Chippenham, Wiltshire.

Dedication

I would like to thank my husband, parents and wider family for all their support to my efforts to reach my full potential. This book is one of the tangible outcomes of this process. My father, in particular, would have loved to have seen it.

Ayşe Göker

Contents

Foreword

In the forty years since I started working in the field, and indeed for some years before that (almost since Calvin Mooers coined the term *information storage and retrieval* in the 1950s), there have been a significant number of books on information retrieval. Even if we ignore the more specialist research monographs and the 'readers' of previously published papers, I can find on my shelves or in my mental library many books that attempt (probably with the IR student in mind) to construct a coherent and systematic way of defining and presenting information retrieval as a field of study and of application.

Often such a book is the work of a single author, or perhaps a pair working together. Such works can clearly have an advantage in respect of coherence; the field is necessarily presented from a single viewpoint. On the other hand, they can also suffer for the same reason. The IR field is rich (more so now than it has ever been), and it is difficult within a single viewpoint to do justice to this richness. Readers, on the other hand, have to be constructed out of the materials to hand: the published papers, each of which has taken its own view, probably with a much narrower field of vision, and different from that of the other chosen papers.

The present book attempts the tricky task of combining the breadth of vision of multiple authors with the coherence of a single integrated work. The richness of the field is apparent in the range of chapters: from formal mathematical modelling to user context, from parallel computation to semantic search.

The topics covered also vary greatly in their historical association with the field. *Categorisation*, for example, has been around as an IR technique for quite a long time – though Stuart Watt brings a new perspective. *Mobile search* (David Mountain, Hans Myrhaug and Ayşe Göker), however, is a relatively recent development. The use of formal models (*information retrieval models*, Djoerd Hiemstra) goes back almost to the beginning, as does experimental evaluation (*user-centred evaluation of information retrieval systems*, Pia Borlund), though in both cases there have been huge changes in the past decade.

This same decade has witnessed the huge growth of the World Wide Web, and the developing dominance of web search engines (*web information retrieval*, Nick Craswell and David Hawking) as the glue which holds the web together. For many people today, IR *is* web search. It is true that there has been a huge amount of influence in both directions: search engines are largely based on techniques from both the IR research community and from previous operational systems, while IR research and practice in other environments has learnt a great deal from the forcing-house that is the web search space. This dominance of the web as *the* domain of interest is well reflected in many of the chapters in the present volume.

It is important, however, to remember that IR is not all about web search, and that the web space presents both problems and opportunities which differ from those in other domains. The desktop, the enterprise, specialist collections such as scientific papers are all examples of different domains for which search functionality is a fundamental requirement. There are references to several of these throughout the book, but specific domains with their own chapters are *multimedia resource discovery* (Stefan Rüger) and *image users' needs and searching behaviour* (Stina Westman). The user theme is taken further in the *context and information retrieval* (Ayşe Göker, Hans Myrhaug and Ralf Bierig). More generic problem areas are addressed in *cross-language information retrieval* (Daqing He and Jianqiang Wang), in *semantic search* (John Davies, Alistair Duke and Atanas Kiryakov) and in the chapter on *natural language processing* (Tony Russell-Rose and Mark Stevenson). Finally, a chapter on *performance issues and parallelism* (Andrew MacFarlane) addresses more technical computing concerns.

Information retrieval, from being the rather arcane subject in which I did my masters degree forty years ago, has become one of the defining technologies of the twenty first century. I believe the present book does justice to this status.

Stephen Robertson
2008

Preface

This project originated during Ayşe Göker's time as Chair of the British Computer Society Information Retrieval Specialist Group (BCS IRSG) when John Davies presented the initial opportunity. The IRSG Committee took this up and as editors we broadened the contributors list.

Each project has its challenges and this book is no exception. We thank Andrew MacFarlane of City University London first and foremost for his collegiate support and superb organisational skills combined with subject knowledge. He gave us a great boost, particularly in the latter part of the project.

We would also like to thank Margaret Graham who provided initial input on the project.

We would like to thank colleagues at City University London who each contributed in significant ways: David Bawden, Jonathan Raper, Jo Wood, Jason Dykes, and Tamara Eisenschitz. Likewise, Stuart Watt, previously of Robert Gordon University and now a consultant. Last but not least, we thank Hans Myrhaug of AmbieSense for his wholehearted commitment and support. He has been willing to carry out any task required to move the project forward. We also appreciate his fantastic eye for detail, and his skills in graphic design.

Each chapter was reviewed by at least four referees looking at the subject matter, coverage, and readability. We thank them accordingly. The referees were as follows:

Leif Azzopardi, University of Glasgow; Ralf Bierig, Rutgers University; Malcolm Clark, Robert Gordon University; John Davies, BT Research; Tamara Eisenschitz, City University London; Ayşe Göker, City University London; Sarah Hinton, Wiley; Fang Huang, Robert Gordon University; Rowan January, Wiley; Paul Lewis, Southampton University; Andrew MacFarlane, City University London; Carol Peters, Italian National Research Council; Stefan Rüger, The Open University; Ian Ruthven, Strathclyde University; Tamar Sadah, Ex Libris Ltd; Aidan Slingsby, City University London; Alan Smeaton, Dublin City University; Sarah Tilley, Wiley; Fiona Walsh, Robert Gordon University.

About the Editors

Ayşe Göker

Dr. Ayşe Göker is a senior academic at City University London. Her research since the early 90s has focused on developing novel search techniques and environments, with an emphasis on personalised and context-sensitive information retrieval and management systems. These occur particularly within mobile and wireless computing, and also in bibliographic and web environments. Her skills are in identifying user needs and developing innovative systems that meet them. In international collaborations she has also been successful, with extensive experience in designing projects and managing teams to implement them. On the teaching side, Ayşe has developed course modules in information systems on several degree programmes at both postgraduate and undergraduate levels.

Ayşe is also a company co-founder of AmbieSense Ltd, a mobile information system company. This project began as the AmbieSense EU-IST project at Robert Gordon University, Aberdeen where she was a Reader and project leader. She holds a lifetime Enterprise Fellowship from the Royal Society of Edinburgh and Scottish Enterprise. More recently she was selected for and completed the Massachussetts Institute of Technology (MIT) Entrepreneurship Development Program in Boston, USA.

In her profession, she has been the Chair of the British Computing Society's Specialist Group in Information Retrieval (BCS IRSG) (2000-2005). She was recognised for the totality of her endeavours by becoming a finalist in the Blackberry Women & Technology Awards (2005) for Best Woman in Technology (Academia).

John Davies

Dr John Davies leads the Semantic Technology research group at BT. Current interests centre around the application of semantic web technology to business intelligence, information integration, knowledge management and service-oriented environments. He is Project Director of the €12m ACTIVE EU integrated project. He co-founded the European Semantic Web conference series. He is also chairman of the European Semantic Technology Conference and a Vice-President of the Semantic Technology Institute. He chairs the NESSI Semantic Technology working group. He has written and edited many papers and books in the areas of the semantic technology, web-based information management and knowledge management; and has served on the program committee of numerous conferences in these and related areas. He is a Fellow of the British Computer Society and a Chartered Engineer. Earlier research at BT led to the development of a set of knowledge management tools which are the subject of a number of patents. These tools were spun out of BT and are now marketed by Infonic Ltd, of which Dr Davies is Group Technical Advisor. Dr Davies received the BT Award for Technology Entrepreneurship for his contribution to the creation of Infonic.

List of Contributors

Ralf Bierig
Rutgers University, USA

Pia Borlund
Royal School of Library and Information
Science, Denmark

Nick Craswell
Microsoft, UK

John Davies
BT, UK

Alistair Duke
BT, UK

Ayşe Göker
City University London, UK

David Hawking
Funnelback, Australia; and
Australian National University

Daqing He
University of Pittsburgh, USA

Djoerd Hiemstra
University of Twente, The Netherlands

Atanas Kiryakov
Ontotext, Bulgaria

Andrew MacFarlane
City University London, UK

David Mountain
City University London, UK

Hans Myrhaug
AmbieSense, UK

Stephen Robertson
Microsoft Research Cambridge, UK; and
City University London, UK

Stefan Rüger
The Open University, UK;
Imperial College London, UK; and
University of Waikato, New Zealand

Tony Russell-Rose
Endeca, UK

Mark Stevenson
Sheffield University, UK

Jianqiang Wang
State University of New York at Buffalo,
USA

Stuart Watt
Information Balance, Canada

Stina Westman
Helsinki University of Technology,
Finland

Introduction

This book covers a diverse range of important themes in modern information retrieval. We first provide a general overview of information retrieval and then describe the chapters with a roadmap for guidance on the topics covered.

What is Information Retrieval?

Information retrieval, simply put, is about finding information. It encompasses the topics of *information need, search, retrieval*, and *access*. It can involve a range of content and media. Work on the storage and retrieval of data began very soon after the invention of the digital computer. The term *information retrieval* was coined around 1952, and as early as 1945, Bush (1945) was talking about a 'device in which an individual stores all his books, records, and communications, and which is mechanized so that it may be consulted with exceeding speed and flexibility'. The increasing amount and diversity of information that is available to us now, along with the challenges these bring, make this field a continually evolving and exciting area.

Information retrieval (IR) can be explained with the following typical problem (Figure 1): On the one hand, we have a person with an information need. This need, which can initially be vague, somehow needs to be articulated into a request which describes this. The request is then converted into a *search statement* or as is now more often referred to as a *query*. On the other hand, we have information stored in collections. They sit as a potentially valuable resource, but in order to be found they need to be represented somehow and then subsequently indexed. The challenge is to provide a good match between these two in order to ensure the information presented is of relevance to the person with the original query.

More specifically, information retrieval is the process of matching the *query* against the *information objects* that are *indexed*. An index is an optimised data structure that is built on top of the information objects, allowing faster access for the search process. The indexer tokenises the text (parsing), removes words with little semantic value (so-called *stop-words*), and unifies word families (so-called *stemming*). The same is done for the query as well. *Users* express their information need as a *request (search terms)* and it is formulised as a *query* for the retrieval system. The information system responds by *matching information objects*, which are *relevant* to this query. Information retrieval focuses on finding relevant information rather than simple pattern matching. It is also important to note that *relevance* is a subjective notion, since different users may make various judgments about the relevance or non-relevance of particular documents or information objects to given questions.

A *retrieval strategy* (model) is an algorithm and related structures that takes a query and a set of documents and assigns a *similarity measure* between the query and each document. This similarity represents relevance to the user query. Documents are then *ranked* on the basis of their similarity to the query and presented to the user. This process can be repeated and the query can be modified.

Figure 2 depicts the concepts that are usually provided by an information retrieval system.

Information retrieval has a number of models (as will be discussed in Chapter 1). Three main models are the *Boolean model*, the *Vector Space model*, and the *Probabilistic model*. The Boolean model is built on binary relevance (documents are treated either as relevant or non-relevant). It uses *exact match search strategy* based on Boolean expressions (such as AND, OR, NOT). The Vector Space model maps both query and documents in a vector space and uses spatial distance as a similarity measure.

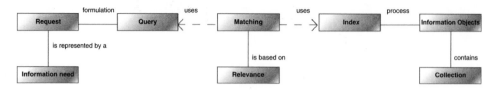

Figure 1   Information retrieval in general

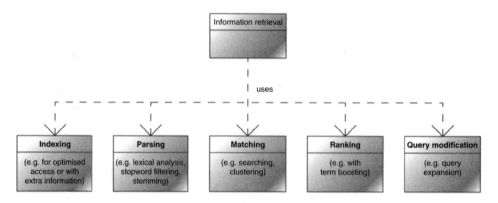

Figure 2   Information retrieval processes

On the other hand, the probabilistic model estimates document relevance as a probability using user feedback for iterative improvement. Both the vector space and the probabilistic model support ranking. There are other models that are also available, but these are a common start to the field.

Information retrieval systems tend to incorporate several activities that are also shown in Figure 2. Broadly speaking, there is *indexing, parsing, matching, ranking*, and *query modification*. These processes are commonly found in search engines. In practice, some of these can be brought together. For example, parsing can be part of an indexing process, and likewise the matching and ranking can be grouped under one umbrella – though not all methods of matching need to have a ranking.

Much of the research and development in information retrieval is aimed at improving retrieval efficiency (see also Chapter 2). Measures referred to as *precision* and *recall* are usually used to measure effectiveness. Precision is the ratio of the number of relevant documents retrieved in relation to the total number of documents retrieved. Recall is the ratio of the number of relevant documents retrieved in relation to the total number of relevant documents. There are other measures too, but these provide a core foundation to the field. If we are to improve information retrieval systems, then it is imperative we have some means of evaluating them against each other.

Overview of the Book

In selecting the chapters for this volume, we have attempted to offer comprehensive coverage of the main themes in modern information retrieval.

We begin with a look at the theoretical underpinnings of the field of information retrieval. As the first chapter explains, there are multiple formal IR models and each may be suited for some IR tasks and less well suited for others, and the chapter offers an excellent introduction to the principal ones in use today. With all IT systems, it is critical to know whether the system performs as intended. Believing therefore that evaluation is an important aspect of information retrieval, and further that the user view of the system and their satisfaction with the retrieved information objects are paramount, we proceed with Chapter 2, on user-centred evaluation.

An increasing proportion of information is non-textual in format: consider, for example, the emergence of *YouTube* and *Flickr* and the many similar systems as web-based sources of video and audio information. With this in mind, the next two chapters are devoted to retrieval of non-textual information resources. In a comprehensive overview of the area, Chapter 3 introduces the basic concepts of multimedia resource discovery, and looks at challenges of multimedia indexing, before discussing novel services which can be offered in the multimedia context. This chapter is followed by a more detailed look at the particular issues raised by image retrieval with a particular focus on user needs analysis (Chapter 4).

Web-based IR has of course emerged as a large area of research and development in its own right over the last 15 years, and the next chapter is devoted to this important topic. The distinctive characteristics of web IR are discussed, and the way in which IR on the web has been developed to deal with these characteristics is described. Like the web, the availability and use of mobile information environments have expanded at a dramatic rate in recent years, and mobile search is the subject of Chapter 6. The diversity and modes of use of mobile information environments globally are discussed and the distinctive features of mobile search identified. Two compelling mobile search applications from very different environments are then described. A key aspect of mobile environments is the requirement for context sensitivity, and a wider discussion on the role of context in IR is the subject of Chapter 7. The importance and role of context information in handling challenges of information search and retrieval are discussed. The capture and representation of context information is described along with examples and guidelines on how a context-aware information system or application can be developed. Consideration is then given to system evaluation in such context-aware information applications.

Unlike the topics covered by some other chapters, categorisation is an area which has been researched since the early days of IR and it continues to be an important topic. A fresh perspective is offered in Chapter 8 by Stuart Watt who interestingly focuses on the purposes of categorisation from the human problem-solving perspective and provides a framework of approaches to further work on categorisation in IR.

Chapter 9 looks at the role of semantic web technology in information retrieval. Central to this approach is the semantic analysis of the target document collection using techniques such as named entity recognition and information extraction, on the hypothesis that this deeper analysis of the target texts may offer improved IR. These semantic analysis techniques are described, followed by a survey of systems using such an approach. Following this chapter, a related topic is covered: the role of natural language technology in IR (Chapter 10). A range of natural language processing (NLP) techniques relevant to IR are described, along with how they can be applied to IR. The usefulness of their contribution to date and future prospects for NLP in IR are discussed. Cross-language IR (CLIR) is an area of increasing interest driven partly by the growing availability of non-English resources (e.g. on the web). Chapter 11 provides a comprehensive overview of the state of the art. We finish with a chapter of central importance to all types of IR: the use of parallel computing to improve the performance of what is by nature a computationally intensive task (Chapter 12). We look at different methods for distributing inverted file data on a parallel computer and the challenges of developing a model by comparing the output of a predictive model against empirical data.

The book is intended for graduate students, later stage undergraduate students, researchers and information/computer scientists, and in particular the growing number of industrial IT professionals whose role requires them to have an understanding of this increasingly important topic. Exercises have been included at the end of each chapter to help readers test and further develop their understanding of the material presented.

Reference

Bush, V. (1945). As we may think. *Atlantic Monthly* **176**(1): 101–108.

1

Information Retrieval Models

Djoerd Hiemstra

1.1 Introduction

Many applications that handle information on the internet would be completely inadequate without the support of information retrieval technology. How would we find information on the world wide web if there were no web search engines? How would we manage our email without spam filtering? Much of the development of information retrieval technology, such as web search engines and spam filters, requires a combination of *experimentation* and *theory*. Experimentation and rigorous empirical testing are needed to keep up with increasing volumes of web pages and emails. Furthermore, experimentation and constant adaptation of technology is needed in practice to counteract the effects of people who deliberately try to manipulate the technology, such as email spammers. However, if experimentation is not guided by theory, engineering becomes trial and error. New problems and challenges for information retrieval come up constantly. They cannot possibly be solved by trial and error alone. So, what is the theory of information retrieval?

There is not one convincing answer to this question. There are many theories, here called *formal models,* and each model is helpful for the development of some information retrieval tools, but not so helpful for the development of others. In order to understand information retrieval, it is essential to learn about these retrieval models. In this chapter, some of the most important retrieval models are gathered and explained in a tutorial style. But first, we will describe what exactly it is that these models model.

1.1.1 Terminology

An information retrieval system is a software programme that stores and manages information on documents, often textual documents, but possibly multimedia. The system assists users in finding the information they need. It does not explicitly return information or answer questions. Instead, it informs on the existence and location of documents that might contain the desired information. Some suggested documents will, hopefully, satisfy the user's information need. These documents are called *relevant* documents. A perfect retrieval system would retrieve only the relevant documents and no irrelevant documents. However, perfect retrieval systems do not exist and will not exist, because search statements are necessarily incomplete and relevance depends on the subjective opinion of the user. In practice, two users may pose the same query to an information retrieval system and judge the relevance of the retrieved documents differently. Some users will like the results, others will not.

Information Retrieval: Searching in the 21st Century   edited by A. Göker & J. Davies
© 2009 John Wiley & Sons, Ltd

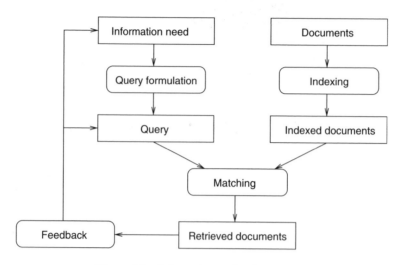

Figure 1.1   Information retrieval processes

There are three basic processes an information retrieval system has to support: the representation of the content of the documents, the representation of the user's information need, and the comparison of the two representations. The processes are visualised in Figure 1.1. In the figure, squared boxes represent data and rounded boxes represent processes.

Representing the documents is usually called the *indexing* process. The process takes place offline, that is, the end user of the information retrieval system is not directly involved. The indexing process results in a representation of the document. Often, full-text retrieval systems use a rather trivial algorithm to derive the index representations, for instance an algorithm that identifies words in an English text and puts them to lower case. The indexing process may include the actual storage of the document in the system, but often documents are only stored partly, for instance only the title and the abstract, plus information about the actual location of the document.

Users do not search just for fun, they have a need for information. The process of representing their *information need* is often referred to as the *query formulation* process. The resulting representation is the query. In a broad sense, query formulation might denote the complete interactive dialogue between system and user, leading not only to a suitable query, but possibly also to the user better understanding his/her information need: This is denoted by the feedback process in Figure 1.1.

The comparison of the query against the document representations is called the *matching* process. The matching process usually results in a ranked list of documents. Users will walk down this document list in search of the information they need. Ranked retrieval will hopefully put the relevant documents towards the top of the ranked list, minimising the time the user has to invest in reading the documents. Simple, but effective ranking algorithms use the frequency distribution of terms over documents, but also statistics over other information, such as the number of hyperlinks that point to the document. Ranking algorithms based on statistical approaches easily halve the time the user has to spend on reading documents. The theory behind ranking algorithms is a crucial part of information retrieval and the major theme of this chapter.

1.1.2   What is a model?

There are two good reasons for having models of information retrieval. The first is that models guide research and provide the means for academic discussion. The second reason is that models can serve as a blueprint to implement an actual retrieval system.

Mathematical models are used in many scientific areas with the objective to understand and reason about some behaviour or phenomenon in the real world. One might for instance think of a model of our solar system that predicts the position of the planets on a particular date, or one might think of a model of the world climate that predicts the temperature, given the atmospheric emissions of greenhouse gases. A model of information retrieval predicts and explains what a user will find relevant, given the user query. The correctness of the model's predictions can be tested in a controlled experiment. In order to do predictions and reach a better understanding of information retrieval, models should be firmly grounded in intuitions, metaphors and some branch of mathematics. Intuitions are important because they help to get a model accepted as reasonable by the research community. Metaphors are important because they help to explain the implications of a model to a bigger audience. For instance, by comparing the earth's atmosphere with a greenhouse, non-experts will understand the implications of certain models of the atmosphere. Mathematics is essential to formalise a model, to ensure consistency, and to make sure that it can be implemented in a real system. As such, a model of information retrieval serves as a blueprint which is used to implement an actual information retrieval system.

1.1.3   Outline

The following sections will describe a total of eight models of information retrieval rather extensively. Many more models have been suggested in the information retrieval literature, but the selection made in this chapter gives a comprehensive overview of the different types of modelling approaches. We start out with two models that provide structured query languages, but no means to rank the results in Section 1.2. Section 1.3 describes vector space approaches, Section 1.4 describes probabilistic approaches, and Section 1.5 concludes this chapter.

1.2   Exact Match Models

In this section, we will address two models of information retrieval that provide exact matching, i.e. documents are either retrieved or not, but the retrieved documents are not ranked.

1.2.1   The Boolean model

The Boolean model is the first model of information retrieval and probably also the most criticised model. The model can be explained by thinking of a query term as an unambiguous definition of a set of documents. For instance, the query term economic simply defines the set of all documents that are indexed with the term economic. Using the operators of George Boole's mathematical logic, query terms and their corresponding sets of documents can be combined to form new sets of documents. Boole defined three basic operators, the logical product called AND, the logical sum called OR and the logical difference called NOT. Combining terms with the AND operator will define a document set that is smaller than or equal to the document sets of any of the single terms. For instance, the query social AND economic will produce the set of documents that are indexed both with the term social and the term economic, i.e. the intersection of both sets. Combining terms with the OR operator will define a document set that is bigger than or equal to the document sets of any of the single terms. So, the query social OR political will produce the set of documents that are indexed with either the term social or the term political, or both, i.e. the union of both sets. This is visualised in the Venn diagrams of Figure 1.2 in which each set of documents is visualised by a disc. The intersections of these discs and their complements divide the document collection into 8 non-overlapping regions, the unions of which give 256 different Boolean combinations of 'social, political and economic documents'. In Figure 1.2, the retrieved sets are visualised by the shaded areas.

An advantage of the Boolean model is that it gives (expert) users a sense of control over the system. It is immediately clear why a document has been retrieved, given a query. If the resulting document set is either too small or too big, it is directly clear which operators will produce respectively a bigger or smaller set. For untrained users, the model has a number of clear disadvantages. Its main

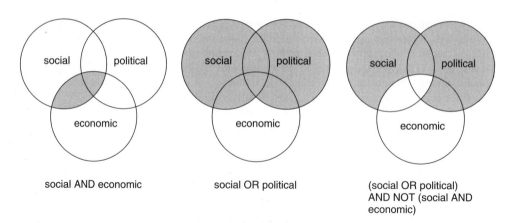

social AND economic   social OR political   (social OR political)
                                                                                AND NOT (social AND
                                                                                economic)

Figure 1.2   Boolean combinations of sets visualised as Venn diagrams

disadvantage is that it does not provide a ranking of retrieved documents. The model either retrieves a document or not, which might lead to the system making rather frustrating decisions. For instance, the query `social AND worker AND union` will of course not retrieve a document indexed with `party`, `birthday` and `cake`, but will likewise not retrieve a document indexed with `social` and `worker` that lacks the term `union`. Clearly, it is likely that the latter document is more useful than the former, but the model has no means to make the distinction.

1.2.2   *Region models*

Region models (Burkowski 1992; Clarke *et al.* 1995; Navarro and Baeza-Yates 1997; Jaakkola and Kilpelainen 1999) are extensions of the Boolean model that reason about arbitrary parts of textual data, called segments, extents or *regions*. Region models model a document collection as a linearised string of words. Any sequence of consecutive words is called a region. Regions are identified by a start position and an end position. Figure 1.3 shows a fragment from Shakespeare's *Hamlet* for which we numbered the word positions. The figure shows the region that starts at word 103 and ends at word 131. The phrase 'stand, and unfold yourself' is defined by the region that starts on position 128 in the text, and ends on position 131. Some regions might be predefined because they represent a logical element in the text, for instance the line spoken by Bernardo which is defined by region (122, 123).

Region systems are not restricted to retrieving documents. Depending on the application, we might want to search for complete plays using some textual queries, we might want to search for scenes referring to speakers, we might want to retrieve speeches by some speaker, we might want to search for single lines using quotations and referring to speakers, etc. When we think of the Boolean model as operating on sets of documents where a document is represented by a nominal identifier, we could think of a region model as operating on sets of regions, where a region is represented by two ordinal identifiers: the start position and the end position in the document collection. The Boolean operators AND, OR and NOT might be defined on sets of regions in a straightforward way as set intersection, set union and set complement. Region models use at least two more operators: CONTAINING and CONTAINED_BY. Systems that supports region queries can process complex queries, such as the following that retrieves all lines in which Hamlet says 'farewell': (`<LINE> CONTAINING farewell`) `CONTAINED_BY` (`<SPEECH> CONTAINING` (`<SPEAKER> CONTAINING Hamlet`)).

There are several proposals of region models that differ slightly. For instance, the model proposed by Burkowski (1992) implicitly distinguishes mark-up from content. As above, the query `<SPEECH> CONTAINING Hamlet` retrieves all speeches that contain the word 'Hamlet'. In later publications Clarke *et al.* (1995) and Jaakkola and Kilpelainen (1999) describe region models that do *not* distinguish

```
    ⋮
<ACT>
 <TITLE>ACT¹⁰³ I¹⁰⁴</TITLE>
 <SCENE>
  <TITLE>SCENE¹⁰⁵ I!¹⁰⁶ Elsinore!¹⁰⁷ A¹⁰⁸ platform¹⁰⁹ before¹¹⁰ the¹¹¹
         castle!¹¹²</TITLE>
  <STGDIR>FRANCISCO¹¹³ at¹¹⁴ his¹¹⁵ post!¹¹⁶ Enter¹¹⁷ to¹¹⁸ him¹¹⁹
         BERNARDO¹²⁰</STGDIR>
  <SPEECH>
   <SPEAKER>BERNARDO¹²¹</SPEAKER>
   <LINE>Who's¹²² there?¹²³</LINE>
  </SPEECH>
  <SPEECH>
   <SPEAKER>FRANCISCO¹²⁴</SPEAKER>
   <LINE>Nay!¹²⁵ answer¹²⁶ me!¹²⁷ stand!¹²⁸ and¹²⁹ unfold¹³⁰
         yourself!¹³¹</LINE>
    ⋮
```

Figure 1.3   Position numbering of example data

mark-up from content. In their system, the operator FOLLOWED_BY is needed to match opening and closing tags, so the query would be somewhat more verbose: (<speech> FOLLOWED_BY </speech>) CONTAINING Hamlet. In some region models, such as the model by Clarke *et al.* (1995), the query A AND B does not retrieve the intersection of sets A and B, but instead retrieves the smallest regions that contain a region from both set A and set B.

1.2.3   Discussion

The Boolean model is firmly grounded in mathematics and its intuitive use of document sets provides a powerful way of reasoning about information retrieval. The main disadvantage of the Boolean model and the region models is their inability to rank documents. For most retrieval applications, ranking is of the utmost importance and ranking extensions have been proposed of the Boolean model (Salton *et al.* 1983) as well as of region models (Mihajlovic 2006). These extensions are based on models that take the need for ranking as their starting point. The remaining sections of this chapter discuss these models of ranked retrieval.

1.3   Vector Space Approaches

Hans Peter Luhn was the first to suggest a statistical approach to searching information (Luhn 1957). He suggested that in order to search a document collection, the user should first prepare a document that is similar to the documents needed. The degree of similarity between the representation of the prepared document and the representations of the documents in the collection is used to rank the search results. Luhn formulated his similarity criterion as follows:

> The more two representations agreed in given elements and their distribution, the higher would be the probability of their representing similar information.

Following Luhn's similarity criterion, a promising first step is to count the number of elements that the query and the index representation of the document share. If the document's index representation is a vector $\vec{d} = (d_1, d_2, \cdots, d_m)$ of which each component $d_k (1 \le k \le m)$ is associated with an index

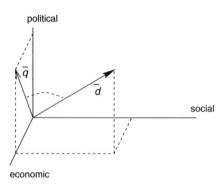

Figure 1.4   A query and document representation in the vector space model

term; and if the query is a similar vector $\vec{q} = (q_1, q_2, \cdots, q_m)$ of which the components are associated with the same terms, then a straightforward similarity measure is the vector inner product:

$$\text{score}(\vec{d}, \vec{q}) = \sum_{k=1}^{m} d_k \cdot q_k \tag{1.1}$$

If the vector has binary components, i.e. the value of the component is 1 if the term occurs in the document or query and 0 if not, then the vector product measures the number of shared terms. A more general representation would use natural numbers or real numbers for the components of the vectors $\vec{d}$ and $\vec{q}$.

1.3.1   The vector space model

Gerard Salton and his colleagues suggested a model based on Luhn's similarity criterion that has a stronger theoretical motivation (Salton and McGill 1983). They considered the index representations and the query as vectors embedded in a high-dimensional Euclidean space, where each term is assigned a separate dimension. The similarity measure is usually the cosine of the angle that separates the two vectors $\vec{d}$ and $\vec{q}$. The cosine of an angle is 0 if the vectors are orthogonal in the multidimensional space and 1 if the angle is $0°$. The cosine formula is given by:

$$\text{score}(\vec{d}, \vec{q}) = \frac{\sum_{k=1}^{m} d_k \cdot q_k}{\sqrt{\sum_{k=1}^{m} (d_k)^2} \cdot \sqrt{\sum_{k=1}^{m} (q_k)^2}} \tag{1.2}$$

The metaphor of angles between vectors in a multidimensional space makes it easy to explain the implications of the model to non-experts. Up to three dimensions, one can easily visualise the document and query vectors. Figure 1.4 visualises an example document vector and an example query vector in the space that is spanned by the three terms `social`, `economic` and `political`. The intuitive geometric interpretation makes it relatively easy to apply the model to new information retrieval problems. The vector space model guided research in, for instance, automatic text categorisation and document clustering.

Measuring the cosine of the angle between vectors is equivalent to normalising the vectors to unit length and taking the vector inner product. If index representations and queries are properly normalised, then the vector product measure of Equation (1.1) does have a strong theoretical motivation. The formula then becomes:

$$\text{score}(\vec{d}, \vec{q}) = \sum_{k=1}^{m} n(d_k) \cdot n(q_k)$$

where

$$n(v_k) = \frac{v_k}{\sqrt{\sum_{k=1}^{m}(v_k)^2}}$$ (1.3)

1.3.2   Positioning the query in vector space

Some rather *ad hoc*, but quite successful retrieval algorithms are nicely grounded in the vector space model if the vector lengths are normalised. An example is the relevance feedback algorithm by Joseph Rocchio (Rocchio 1971). Rocchio suggested the following algorithm for relevance feedback, where $\vec{q}_{old}$ is the original query, $\vec{q}_{new}$ is the revised query, $\vec{d}_{rel}^{(i)}$ ($1 \leq i \leq r$) is one of the r documents the user selected as relevant, and $\vec{d}_{nonrel}^{(i)}$ ($1 \leq i \leq n$) is one of the n documents the user selected as non-relevant.

$$\vec{q}_{new} = \vec{q}_{old} + \frac{1}{r}\sum_{i=1}^{r}\vec{d}_{rel}^{(i)} - \frac{1}{n}\sum_{i=1}^{n}\vec{d}_{nonrel}^{(i)}$$ (1.4)

The normalised vectors of documents and queries can be viewed as points on a hypersphere at unit length from the origin. In Equation (1.4), the first sum calculates the centroid of the points of the known relevant documents on the hypersphere. In the centroid, the angle with the known relevant documents is minimised. The second sum calculates the centroid of the points of the known non-relevant documents. Moving the query towards the centroid of the known relevant documents and away from the centroid of the known non-relevant documents is guaranteed to improve retrieval performance.

1.3.3   Term weighting and other caveats

The main disadvantage of the vector space model is that it does not in any way define what the values of the vector components should be. The problem of assigning appropriate values to the vec-tor components is known as *term weighting*. Early experiments by Salton (1971) and Salton and Yang (1973) showed that term weighting is not a trivial problem at all. They suggested so-called *tf.idf* weights, a combination of term frequency *tf*, which is the number of occurrences of a term in a document, and *idf*, the inverse document frequency, which is a value inversely related to the document frequency *df*, which is the number of documents that contain the term. Many modern weighting algorithms are versions of the family of *tf.idf* weighting algorithms. Salton's original *tf.idf* weights perform relatively poorly, in some cases worse than simple *idf* weighting. They are defined as:

$$d_k = q_k = tf(k, d) \cdot \log \frac{N}{df(k)}$$ (1.5)

where $tf(k, d)$ is the number of occurrences of the term k in the document d, $df(k)$ is the number of documents containing k, and N is the total number of documents in the collection. Another problem with the vector space model is its implementation. The calculation of the cosine measure needs the values of all vector components, but these are not available in an inverted file. In practice, the normalised values and the vector product algorithm have to be used. Either the normalised weights have to be stored in the inverted file, or the normalisation values have to be stored separately. Both are problematic in case of incremental updates of the index. Adding a single new document changes the document frequencies of terms that occur in the document, which changes the vector lengths of every document that contains one or more of these terms.

1.4   Probabilistic Approaches

Several approaches that try to define term weighting more formally are based on *probability theory*. The notion of the probability of something, for instance the probability of relevance notated as $P(R)$, is usually formalised through the concept of an experiment, where an experiment is the process by which an observation is made. The set of all possible outcomes of the experiment is called the sample space. In the case of $P(R)$ the sample space might be {*relevant*, *irrelevant*}, and we might define the random variable R to take the values {0, 1}, where $0 =$ irrelevant and $1 =$ relevant.

Let's define an experiment for which we take one document from the collection at random. If we know the number of relevant documents in the collection, say 100 documents are relevant, and we know the total number of documents in the collection, say 1 million, then the quotient of those two defines the probability of relevance $P(R=1) = 100/1000\,000 = 0.0001$. Suppose furthermore that $P(D_k)$ is the probability that a document contains the term k with the sample space {0, 1}, (0 = the document does not contain term k, 1 = the document contains term k), then we will use $P(R, D_k)$ to denote the *joint probability distribution* with outcomes {(0, 0), (0, 1), (1, 0) and (1, 1)}, and we will use $P(R|D_k)$ to denote the *conditional probability distribution* with outcomes {0, 1}. So, $P(R=1|D_k=1)$ is the probability of relevance if we consider documents that contain the term k.

Note that the notation $P(\ldots)$ is overloaded. Whenever we are talking about a different random variable or sample space, we are also talking about a different measure P. So, one equation might refer to several probability measures, all ambiguously referred to as P. Also note that random variables such as D and T might have different sample spaces in different models. For instance, D in the probabilistic indexing model is a random variable denoting '*this* is the relevant document', that has as possible outcomes the identifiers of the documents in the collection. However, D in the probabilistic retrieval model is a random variable that has as possible outcomes all possible document descriptions, which in this case are vectors with binary components d_k that denote whether a document is indexed by term k or not.

1.4.1   The probabilistic indexing model

As early as 1960, Bill Maron and Larry Kuhns (Maron and Kuhns 1960) defined their probabilistic indexing model. Unlike Luhn, they did not target automatic indexing by information retrieval systems. Manual indexing was still guiding the field, so they suggested that a human indexer, who runs through the various index terms T that possibly apply to a document D, assigns a probability $P(T|D)$ to a term given a document instead of making a yes/no decision for each term. So, every document ends up with a set of possible index terms, weighted by $P(T|D)$, where $P(T|D)$ is the probability that, if a user wants information of the kind contained in document D, he/she will formulate a query by using T. Using Bayes' rule, i.e.

$$P(D|T) = \frac{P(T|D)P(D)}{P(T)} \tag{1.6}$$

they then suggest to rank the documents by $P(D|T)$, that is, the probability that the document D is relevant, given that the user formulated a query by using the term T. Note that $P(T)$ in the denominator of the right-hand side is constant for any given query term T, and consequently documents might be ranked by $P(T|D)P(D)$ which is a quantity proportional to the value of $P(D|T)$. In the formula, $P(D)$ is the *a priori* probability of relevance of document D.

Whereas $P(T|D)$ is defined by the human indexer, Maron and Kuhns suggest that $P(D)$ can be defined by statistics on document usage, i.e. by the quotient of the number of uses of document D by the total number of document uses. So, their usage of the document prior $P(D)$ can be seen as the very first description of popularity ranking, which became important for internet search (see Section 1.4.6). Interestingly, an estimate of $P(T|D)$ might be obtained in a similar way by storing, for each use of a document, also the query term that was entered to retrieve the document in the first place. Maron and Kuhns state that 'such a procedure would of course be extremely impractical', but

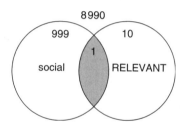

Figure 1.5 Venn diagram of the collection given the query term `social`

in fact, such techniques – rank optimization using so-called click-through rates – are now common in web search engines as well (Joachims *et al.* 2005). Probabilistic indexing models were also studied by Fuhr (1989).

1.4.2 The probabilistic retrieval model

Whereas Maron and Kuhns introduced ranking by the probability of relevance, it was Stephen Robertson who turned the idea into a principle. He formulated the probability ranking principle, which he attributed to William Cooper, as follows (Robertson 1977).

> If a reference retrieval system's response to each request is a ranking of the documents in the collections in order of decreasing probability of usefulness to the user who submitted the request, where the probabilities are estimated as accurately as possible on the basis of whatever data has been made available to the system for this purpose, then the overall effectiveness of the system to its users will be the best that is obtainable on the basis of that data.

This seems a rather trivial requirement indeed, since the objective of information retrieval systems is defined in Section 1.1 as to 'assist users in finding the information they need', but its implications might be very different from Luhn's similarity principle. Suppose a user enters a query containing a single term, for instance the term `social`. If all documents that fulfil the user's need were known, it would be possible to divide the collection into four non-overlapping document sets as visualised in the Venn diagram of Figure 1.5. The figure contains additional information about the size of each of the non-overlapping sets. Suppose the collection in question has 10 000 documents, of which 1000 contain the word 'social'. Furthermore, suppose that only 11 documents are relevant to the query of which 1 contains the word 'social'. If a document is taken at random from the set of documents that are indexed with `social`, then the probability of picking a relevant document is $1/1000 = 0.0010$. If a document is taken at random from the set of documents that are *not* indexed with `social`, then the probability of relevance is bigger: $10/9000 = 0.0011$. Based on this evidence, the best performance is achieved if the system returns documents that are not indexed with the query term `social`, that is, to present first the documents that are *dissimilar* to the query. Clearly, such a strategy violates Luhn's similarity criterion.

Stephen Robertson and Karen Spärck-Jones based their probabilistic retrieval model on this line of reasoning (Robertson and Spärck-Jones 1976). They suggested to rank documents by $P(R|D)$, that is the probability of relevance R given the document's content description D. Note that D is here a vector of binary components, each component typically representing a term, whereas in the previous section D was the 'relevant document'. In the probabilistic retrieval model the probability $P(R|D)$ has to be interpreted as follows: there might be several, say 10, documents that are represented by the same D. If 9 of them are relevant, then $P(R|D) = 0.9$. To make this work in practice, we use Bayes' rule on the probability odds $P(R|D)/P(\overline{R}|D)$, where $\overline{R}$ denotes irrelevance. The odds allow us to ignore $P(D)$ in the computation while still providing a ranking by the probability of relevance.

Additionally, we assume independence between terms given relevance.

$$\frac{P(R|D)}{P(\overline{R}|D)} = \frac{P(D|R)P(R)}{P(D|\overline{R})P(\overline{R})} = \frac{\prod_k P(D_k|R)P(R)}{\prod_k P(D_k|\overline{R})P(\overline{R})} \tag{1.7}$$

Here, D_k denotes the kth component (term) in the document vector. The probabilities of the terms are defined as above from examples of relevant documents, that is, in Figure 1.5, the probability of social given relevance is $1/11$. A more convenient implementation of probabilistic retrieval uses the following three order-preserving transformations. First, the documents are ranked by sums of logarithmic odds, instead of the odds themselves. Second, the *a priori* odds of relevance $P(R)/P(\overline{R})$ is ignored. Third, we subtract $\sum_k \log(P(D_k = 0|R)/P(D_k = 0|\overline{R}))$, i.e. the score of the empty document, from all document scores. This way, the sum over all terms, which might be millions of terms, only includes non-zero values for terms that are present in the document.

$$\text{matching-score}(D) = \sum_{k \in \text{matching terms}} \log \frac{P(D_k = 1|R)\, P(D_k = 0|\overline{R})}{P(D_k = 1|\overline{R})P(D_k = 0|R)} \tag{1.8}$$

In practice, terms that are not in the query are also ignored in Equation (1.8). Making full use of the probabilistic retrieval model requires two things: examples of relevant documents and long queries. Relevant documents are needed to compute $P(D_k|R)$, that is, the probability that the document contains the term k given relevance. Long queries are needed because the model only distinguishes term presence and term absence in documents and as a consequence, the number of distinct values of document scores is low for short queries. For a one-word query, the number of distinct probabilities is two (either a document contains the word or not), for a two-word query it is four (the document contains both terms, or only the first term, or only the second, or neither), for a three-word query it is eight, etc. Obviously, this makes the model inadequate for web search, for which no relevant documents are known beforehand and for which queries are typically short. However, the model is helpful in, for instance, spam filters. Spam filters accumulate many examples of relevant (no spam or 'ham') and irrelevant (spam) documents over time. To decide if an incoming email is spam or ham, the full text of the email can be used instead of a just few query terms.

1.4.3   The 2-Poisson model

Bookstein and Swanson (1974) studied the problem of developing a set of statistical rules for the purpose of identifying the index terms of a document. They suggested that the number of occurrences *tf* of terms in documents could be modelled by a mixture of two Poisson distributions as follows, where X is a random variable for the number of occurrences.

$$P(X = tf) = \lambda \frac{e^{-\mu_1}(\mu_1)^{tf}}{tf!} + (1 - \lambda)\frac{e^{-\mu_2}(\mu_2)^{tf}}{tf!} \tag{1.9}$$

The model assumes that the documents were created by a random stream of term occurrences. For each term, the collection can be divided into two subsets. Documents in subset one treat a subject referred to by a term to a greater extent than documents in subset two. This is represented by λ which is the proportion of the documents that belong to subset one and by the Poisson means μ_1 and $\mu_2(\mu_1 \geq \mu_2)$ which can be estimated from the mean number of occurrences of the term in the respective subsets. For each term, the model needs these three parameters, but unfortunately, it is unknown to which subset each document belongs. The estimation of the three parameters should therefore be done iteratively by applying, e.g. the expectation maximisation algorithm (Dempster *et al.* 1977) or alternatively by the method of moments, as done by Harter (1975).

If a document is taken at random from subset one, then the probability of relevance of this document is assumed to be equal to, or higher than, the probability of relevance of a document from subset two;

because the probability of relevance is assumed to be correlated with the extent to which a subject referred to by a term is treated, and because $\mu_1 \geq \mu_2$. Useful terms will make a good distinction between relevant and non-relevant documents, that is, both subsets will have very different Poisson means μ_1 and μ_2. Therefore, Harter (1975) suggests the following measure of effectiveness of an index term that can be used to rank the documents given a query.

$$ z = \frac{\mu_1 - \mu_2}{\sqrt{\mu_1 + \mu_2}} \tag{1.10} $$

The 2-Poisson model's main advantage is that it does not need an additional term weighting algorithm to be implemented. In this respect, the model contributed to the understanding of information retrieval and inspired some researchers in developing new models, as shown in the next paragraph. The model's biggest problem, however, is the estimation of the parameters. For each term there are three unknown parameters that cannot be estimated directly from the observed data. Furthermore, despite the model's complexity, it still might not fit the actual data if the term frequencies differ very much per document. Some studies therefore examine the use of more than two Poisson functions, but this makes the estimation problem even more intractable (Margulis 1993).

Robertson *et al.* (1981) proposed to use the 2-Poisson model to include the frequency of terms within documents in the probabilistic model. Although the actual implementation of this model is cumbersome, it inspired Stephen Robertson and Stephen Walker in developing the Okapi BM25 term weighting algorithm, which is still one of the best performing term weighting algorithms (Robertson and Walker 1994; Spärck-Jones *et al.* 2000).

1.4.4 Bayesian network models

In 1991, Howard Turtle proposed the *inference network model* (Turtle and Croft 1991) which is formal in the sense that it is based on the Bayesian network mechanism (Metzler and Croft 2004). A Bayesian network is an acyclic directed graph (a directed graph is acyclic if there is no directed path $A \to \cdots \to Z$ such that $A = Z$) that encodes probabilistic dependency relationships between random variables. The presentation of probability distributions as directed graphs, makes it possible to analyse complex conditional independence assumptions by following a graph theoretic approach. In practice, the inference network model is comprised of four layers of nodes: document nodes, representation nodes, query nodes and the information need node. Figure 1.6 shows a simplified inference network model. All nodes in the network represent binary random variables with values $\{0, 1\}$. To see how the model works in theory, it is instructive to look at a subset of the nodes, for instance the nodes r_2, q_1, q_3 and I, and ignore the other nodes for the moment. By the chain rule of probability, the joint probability of the nodes r_2, q_1, q_3 and I is:

$$ P(r_2, q_1, q_3, I) = P(r_2)P(q_1|r_2)P(q_3|r_2, q_1)P(I|r_2, q_1, q_3) \tag{1.11} $$

The directions of the arcs suggest the dependence relations between the random variables. The event 'information need is fulfilled' ($I = 1$) has two possible causes: query node q_1 is true, or query node q_3 is true (remember we are ignoring q_2). The two query nodes in turn depend on the representation node r_2. So, the model makes the following conditional independence assumptions.

$$ P(r_2, q_1, q_3, I) = P(r_2)\, P(q_1|r_2)\, P(q_3|r_2)\, P(I|q_1, q_3) \tag{1.12} $$

On the right-hand side, the third probability measure is simplified because q_1 and q_3 are independent, given their parent r_2. The last part $P(I|q_1, q_3)$ is simplified because I is independent of r_2, given its parents q_1 and q_3.

Straightforward use of the network is impractical if there are a large number of query nodes. The number of probabilities that have to be specified for a node grows exponentially with its number

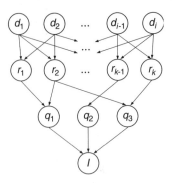

Figure 1.6   A simplified inference network

of parents. For example, a network with n query nodes requires the specification of 2^{n+1} possible values of $P(I|q_1, q_2, \cdots, q_n)$ for the information need node. For this reason, all network layers need some form of approximation. Metzler and Croft (2004) describe the following approximations. For the document layer they assume that only a single document is observed at a time, and for every single document a separate network is constructed for which the document layer is ignored. For the representation layer of every network, the probability of the representation nodes (which effectively are priors now, because the document layer is ignored) are estimated by some retrieval model. Note that a representation node is usually a single term, but it might also be a phrase. Finally, the query nodes and the information need node are approximated by standard probability distributions defined by so-called believe operators. These operators combine probability values from representation nodes and other query nodes in a fixed manner. If the values of $P(q_1|r_2)$, and $P(q_3|r_2)$ are given by p_1 and p_2, then the calculation of $P(I|r_2)$ might be done by operators such as 'and', 'or', 'sum', and 'wsum'.

$$P_{\text{and}}(I|r_2) = p_1 \cdot p_2$$
$$P_{\text{or}}(I|r_2) = 1 - ((1 - p_1)(1 - p_2)) \tag{1.13}$$
$$P_{\text{sum}}(I|r_2) = (p_1 + p_2)/2$$
$$P_{\text{wsum}}(I|r_2) = w_1\, p_1 + w_2\, p_2$$

It can be shown that, for these operators, so-called link matrices exists, that is, for each operator there exists a definition of for instance $P(I|q_1, q_3)$ that can be computed, as shown in Equation (1.13). So, although the link matrix that belongs to the operator may be huge, it does not exist in practice and its result can be computed in linear time. One might argue though, that the approximations on each network layer make it questionable if the approach still deserves to be called a 'Bayesian network model'.

1.4.5   Language models

Language models were applied to information retrieval by a number of researchers in the late 1990s (Ponte and Croft 1998; Hiemstra and Kraaij 1998; Miller *et al.* 1999). They originate from probabilistic models of language generation developed for automatic speech recognition systems in the early 1980s (e.g. Rabiner 1990). Automatic speech recognition systems combine probabilities of two distinct models: the acoustic model, and the language model. The acoustic model might for instance produce the following candidate texts in decreasing order of probability: 'food born thing', 'good corn sing', 'mood morning', and 'good morning'. Now, the language model would determine that the phrase 'good morning' is much more probable, i.e. it occurs more frequently in English than the other phrases. When combined with the acoustic model, the system is able to decide that 'good morning' was the most likely utterance, thereby increasing the system's performance.

For information retrieval, language models are built for each document. By following this approach, the language model of the book you are reading now would assign an exceptionally high probability to the word 'retrieval', indicating that this book would be a good candidate for retrieval if the query contains this word. Language models take the same starting point as the probabilistic indexing model by Maron and Kuhns described in Section 1.4.1. That is, given D – the document is relevant – the user will formulate a query by using a term T with some probability $P(T|D)$. The probability is defined by the text of the documents. If a certain document consists of 100 words, and of those the word 'good' occurs twice, then the probability of 'good', given that the document is relevant, is simply defined as 0.02. For queries with multiple words, we assume that query words are generated independently from each other, i.e. the conditional probabilities of the terms $T_1, T_2, \cdots$ given the document are multiplied:

$$P(T_1, T_2, \cdots | D) = \prod_i P(T_i|D) \tag{1.14}$$

As a motivation for using the probability of the query given the document, one might think of the following experiment. Suppose we ask one million monkeys to pick a good three-word query for several documents. Each monkey will point three times at random to each document. Whatever word the monkey points to, will be the (next) word in the query. Suppose that seven monkeys accidentally pointed to the words 'information', 'retrieval' and 'model' for document 1, and only two monkeys accidentally pointed to these words for document 2. Then, document 1 would be a better document for the query 'information retrieval model' than document 2.

The above experiment assigns zero probability to words that do not occur anywhere in the document, and because we multiply the probabilities of the single words, it assigns zero probability to documents that do not contain all of the words. For some applications this is not a problem. For instance for a web search engine, queries are usually short and it will rarely happen that no web page contains all query terms. For many other applications empty results happen much more often, which might be problematic for the user. Therefore, a technique called *smoothing* is applied. Smoothing assigns some non-zero probability to unseen events. One approach to smoothing takes a linear combination of $P(T_i|D)$ and a background model $P(T_i)$ as follows.

$$P(T_1, \cdots, T_n | D) = \prod_{i=1}^{n} (\lambda P(T_i|D) + (1 - \lambda)P(T_i)) \tag{1.15}$$

The background model $P(T_i)$ might be defined by the probability of term occurrence in the collection, i.e. by the quotient of the total number of occurrences in the collection divided by the length of the collection. In the equation, λ is an unknown parameter that has to be set empirically. Linear interpolation smoothing accounts for the fact that some query words do not seem to be related to the relevance of documents at all. For instance in the query 'capital of the Netherlands', the words 'of' and 'the' might be seen as words from the user's general English vocabulary, and not as words from the relevant document he/she is looking for. In terms of the experiment above, a monkey would either pick a word at random from the document with probability λ or the monkey would pick a word at random from the entire collection. A more convenient implementation of the linear interpolation models can be achieved with order-preserving transformations that are similar to those for the probabilistic retrieval model (see Equation 1.8). We multiply both sides of the equation by $\prod_i (1 - \lambda)P(T_i)$ and take the logarithm, which leads to:

$$\text{matching score}(d) = \sum_{k \in \text{ matching terms}} \log \left(1 + \frac{tf(k, d)}{\sum_t tf(t, d)} \cdot \frac{\sum_t cf(t)}{cf(k)} \cdot \frac{\lambda}{1 - \lambda} \right) \tag{1.16}$$

where, $P(T_i = t_i) = cf(t_i)/\sum_t cf(t)$, and $cf(t) = \sum_d tf(t, d)$.

There are many approaches to smoothing, most pioneered for automatic speech recognition (Chen and Goodman 1996). Another approach to smoothing that is often used for information retrieval is so-called Dirichlet smoothing, which is defined as (Zhai and Lafferty 2004):

$$P(T_1 = t_1, \cdots, T_n = t_n | D = d) = \prod_{i=1}^{n} \frac{tf(t_i, d) + \mu P(T_i = t_i)}{(\sum_t tf(t, d)) + \mu} \tag{1.17}$$

Here, μ is a real number $\mu \geq 0$. Dirichlet smoothing accounts for the fact that documents are too small to reliably estimate a language model. Smoothing by Equation (1.17) has a relatively big effect on small documents, but relatively small effect on bigger documents.

The equations above define the probability of a query given a document, but obviously, the system should rank by the probability of the documents given the query. These two probabilities are related by Bayes' rule as follows.

$$P(D | T_1, T_2, \cdots, T_n) = \frac{P(T_1, T_2, \cdots, T_n | D) P(D)}{P(T_1, T_2, \cdots, T_n)} \tag{1.18}$$

The left-hand side of Equation (1.18) cannot be used directly because the independence assumption presented above assumes terms are independent, given the document. So, in order to compute the probability of the document D given the query, we need to multiply Equation (1.15) by $P(D)$ and divide it by $P(T_1, \cdots, T_n)$. Again, as stated in the previous paragraph, the probabilities themselves are of no interest, only the ranking of the document by the probabilities is. And since $P(T_1, \cdots, T_n)$ does not depend on the document, ranking the documents by the numerator of the right-hand side of Equation (1.18) will rank them by the probability given the query. This shows the importance of $P(D)$, the marginal probability, or prior probability of the document, i.e. it is the probability that the document is relevant if we do not know the query (yet). For instance, we might assume that long documents are more likely to be useful than short documents. In web search, such so-called static rankings (see Section 1.4.6), are commonly used. For instance, documents with many links pointing to them are more likely to be relevant, as shown in the next section.

1.4.6 Google's PageRank model

When Sergey Brin and Lawrence Page launched the web search engine Google in 1998 (Brin and Page 1998), it had two features that distinguished it from other web search engines. It had a simple no-nonsense search interface, and, it used a radically different approach to rank the search results. Instead of returning documents that closely match the query terms (i.e. by using any of the models in the preceding sections), they aimed at returning high-quality documents, i.e. documents from trusted sites. Google uses the hyperlink structure of the web to determine the quality of a page, called *PageRank*. Web pages that are linked at from many places around the web are probably worth looking at: they must be *high-quality* pages. If pages that have links from other high-quality web pages, for instance DMOZ or Wikipedia[1], then that is a further indication that they are likely to be worth looking at. The PageRank of a page d is defined as $P(D = d)$, i.e. the probability that d is relevant as used in the probabilistic indexing model in Section 1.4.1 and as used in the language modelling approach in Section 1.4.5. It is defined as:

$$P(D = d) = (1 - \lambda) \frac{1}{\#\text{pages}} + \lambda \sum_{i | i \text{ links to } d} P(D = i) P(D = d | D = i) \tag{1.19}$$

If we ignore $(1 - \lambda)/\#\text{pages}$ for the moment, then the PageRank $P(D = d)$ is recursively defined as the sum of the PageRanks $P(D = i)$ of all pages i that link to d, multiplied by the probability $P(D = d | D = i)$ of following a link from i to d. One might think of the PageRank as the probability that a random surfer visits a page. Suppose we ask the monkeys from the previous section to surf the

---

[1] See http://dmoz.org and http://wikipedia.org

web from a randomly chosen starting point i. Each monkey will now click on a random hyperlink with the probability $P(D = d|D = i)$ which is defined as one divided by the number of links on page i. This monkey will end up in d. But other monkeys might end up in d as well: those that started on another page that happens to link to d. After letting the monkeys surf a while, the highest-quality pages, i.e. the best-connected pages, will have most monkeys that look at it.

The above experiment has a similar problem with zero probabilities as the language modelling approach. Some pages might have no links pointing to them, so they will get a zero PageRank. Others might not link to any other page, so you cannot leave the page by following hyperlinks. The solution is also similar to the zero probability problem in the language modelling approach: we smooth the model by some background model, in this case the background is uniformly distributed over all pages. With some unknown probability λ a link is followed, but with probability $1 - \lambda$ a random page is selected, which is like a monkey typing in a random (but valid) URL.

PageRank is a so-called static ranking function, that is, it does not depend on the query. It is computed once off-line at indexing time by iteratively calculating the page rank of pages at time $t + 1$ from the page ranks as calculated in a previous interations at time t until they do not change significantly anymore. Once the PageRank of every page is calculated it can be used during querying. One possible way to use PageRank during querying is as follows: select the documents that contain all query terms (i.e. a Boolean AND query) and rank those documents by their page rank. Interestingly, a simple algorithm such as this would not only be effective for web search, it can also be implemented very efficiently (Richardson *et al.* 2006). In practice, web search engines such as Google use many more factors in their ranking than just page rank alone. In terms of the probabilistic indexing model and the language modelling approaches, static rankings are simply document priors, i.e. the *a priori* probability of the document being relevant, that should be combined with the probability of terms given the document. Document priors can be easily combined with standard language modelling probabilities and are as such powerful means to improve the effectiveness of, for instance, queries for home pages in web search (Kraaij *et al.* 2002).

1.5   Summary and Further Reading

There is no such thing as a dominating model or theory of information retrieval, unlike the situation in, for instance, the area of databases where the relational model is *the* dominating database model. In information retrieval, some models work for some applications, whereas others work for other applications. For instance, the region models introduced in Section 1.2.2 have been designed to search in semi-structured data; the vector space models in Section 1.3 are well suited for similarity search and relevance feedback in many (also non-textual) situations if a good weighting function is available; the probabilistic retrieval model of Section 1.4.2 might be a good choice if examples of relevant and non-relevant documents are available; language models in Section 1.4.5 are helpful in situations that require models of language similarity or document priors; and the PageRank model of Section 1.4.6 is often used in situations that need modelling of more of less static relations between documents. This chapter describes these and other information retrieval models in a tutorial style in order to explain the consequences of modelling assumptions. Once the reader is aware of the consequences of modelling assumptions, he or she will be able to choose a model of information retrieval that is adequate in new situations.

Whereas the citations in the text are helpful for background information and for putting things in a historical context, we recommend the following publications for people interested in a more in-depth treatment of information retrieval models. A good starting point for the region models of Section 1.2.2 would be the overview article of Hiemstra and Baeza-Yates (2009). An in-depth description of the vector space approaches of Section 1.3, including, for instance, latent semantic indexing is given by Berry *et al.* (1999). The probabilistic retrieval model of Section 1.4.2 and developments based on the model are well described by Spärck-Jones *et al.* (2000); de Campos *et al.* (2004) edited a special issue on Bayesian networks for information retrieval as described in Section 1.4.4. An excellent overview of statistical language models for information retrieval is given by Zhai (2008). Finally, a good follow-up article on Google's PageRank is given by Henzinger (2001).

Exercises

1.1 In the Boolean model of Section 1.2.1, there are a large number of queries that can be formulated with three query terms, for instance `one OR two OR three`, or `(one OR two) AND three`, or possibly `(one AND three) OR (two AND three)`. Some of these queries, however, for instance the last two queries, return the same set of documents. *How many different sets of documents can be specified given three query terms?* Explain your answer.

(a) 8
(b) 9
(c) 256
(d) unlimited

1.2 Given a query $\vec{q}$ and a document $\vec{d}$ in the vector space model of Section 1.3.1. Suppose the similarity between $\vec{q}$ and $\vec{d}$ is 0.08. Suppose we interchange the full contents of the document with the query, that is, all words from $\vec{q}$ go to $\vec{d}$ and all words from $\vec{d}$ go to $\vec{q}$. *What will now be the similarity between $\vec{q}$ and $\vec{d}$?* Explain your answer.

(a) smaller than 0.08
(b) equal: 0.08
(c) bigger than 0.08
(d) it depends on the term weighting algorithm

1.3 In the vector approach of Section 1.3 that uses Equation (1.1) and *tf.idf* term weighting, suppose we add some documents to the collection. *Do the weights of the terms in the documents that were indexed before change?* Explain your answer.

(a) no
(b) yes, it affects the *tf*'s of terms in other documents
(c) yes, it affects the *idf*'s of terms in other documents
(d) yes, it affects the *tf*'s and the *idf*'s of terms in other documents

1.4 In the vector space model using the cosine similarity of Equation (1.2) and *tf.idf* term weighting, suppose we add some documents to the collection. *Do the weights of the terms in the documents that were indexed before change?* Explain your answer.

(a) no, other documents are unaffected
(b) yes, the same weights as in Exercise 1.3
(c) yes, more weights change than in Exercise 1.3, but not all
(d) yes, all weights in the index change

1.5 In the probabilistic model of Section 1.4.2, two documents might get the same score. *How many different scores do we expect to get if we enter three query terms?* Explain your answer.

(a) 8
(b) 9
(c) 256
(d) unlimited

1.6 For the probabilistic retrieval model of Section 1.4.2, suppose we query for the word *retrieval*, and document D has more occurrences of *retrieval* than document E. *Which document will be ranked first?* Explain your answer.

(a) D will be ranked before E
(b) E will be ranked before D

(c) it depends on the model's implementation

(d) it depends on the lengths of D and E

1.7 In the language modeling approach of Section 1.4.5, suppose the model does not use smoothing. In the case we query for the word *retrieval*, and document D consisting of 100 words in total, contains the word *retrieval* 4 times. *What is $P(T = retrieval|D)$?*

(a) smaller than $4/100 = 0.04$

(b) equal to $4/100 = 0.04$

(c) bigger than $4/100 = 0.04$

(d) it depends on the term weighting algorithm

1.8 In the language modeling approach of Section 1.4.5, suppose we use a linear combination of a document model and a collection model as in Equation (1.15). *What happens if we take $\lambda = 1$?* Explain your answer.

(a) all docucments get a probability > 0

(b) documents that contain at least one query term get a probability > 0

(c) only documents that contain all query terms get a probability > 0

(d) the system returns a randomly ranked list

References

Berry, M. W, Z. Drmac and E. R. Jessup (1999). Matrices, vector spaces, and information retrieval. *SIAM Review* **41**(2) 335–362.

Bookstein, A. and D. Swanson (1974). Probabilistic models for automatic indexing. *Journal of the American Society for Information Science* **25**(5) 313–318.

Brin, S. and L. Page (1998). The anatomy of a large-scale hypertextual Web search engine. *Computer Networks and ISDN Systems* **30**(1–7) 107–117.

Burkowski, F. (1992). Retrieval activities in a database consisting of heterogeneous collections of structured texts. In *Proceedings of the 15th ACM SIGIR Conference on Research and Development in Information Retrieval (SIGIR'92)* pp. 112–125.

Campos, L. M. de, J. M. Fernandez-Luna and J. F. Huete (eds) (2004). Special issue on Bayesian networks and information retrieval. *Information Processing and Management* **40**(5).

Chen, S. and J. Goodman (1996). An empirical study of smoothing techniques for language modeling. In *Proceedings of the Annual Meeting of the Association for Computational Linguistics*, pp. 310–318.

Clarke, C., G. Cormack, and F. Burkowski (1995). Algebra for structured text search and a framework for its implementation. *The Computer Journal* **38**(1) 43–56.

Dempster, A., N. Laird, and D. Rubin (1977). Maximum likelihood from incomplete data via the em-algorithm plus discussions on the paper. *Journal of the Royal Statistical Society* **39**(B) 1–38.

Fuhr, N. (1989). Models for retrieval with probabilistic indexing. *Information processing and management* **25**(1) 55–72.

Harter, S. (1975). An algorithm for probabilistic indexing. *Journal of the American Society for Information Science* **26**(4) 280–289.

Henzinger, M. R. (2001) Hyperlink Analysis for the Web. *IEEE Internet Computing* **5**(1) 45–50.

Hiemstra, D. and R. Baeza-Yates (2009). Structured text retrieval models. In M. Tamer zsu and Ling Liu (eds), *Encyclopedia of Database Systems*, Springer.

Hiemstra, D. and W. Kraaij (1998). Twenty-One at TREC-7: Ad-hoc and cross-language track. In *Proceedings of the 7th Text Retrieval Conference TREC-7*, pp. 227–238. NIST Special Publication 500-242.

Jaakkola, J. and P. Kilpelainen (1999). Nested text-region algebra. *Technical Report CR-1999-2*, Department of Computer Science, University of Helsinki.

Joachims, T., L. Granka, B. Pan, H. Hembrooke and G. Gay (2005). Accurately interpreting click-through data as implicit feedback. In *Proceedings of the 28th ACM SIGIR Conference on Research and Development in Information Retrieval (SIGIR'05)*, pp. 154–161.

Kraaij, W., T. Westerveld, and D. Hiemstra (2002). The importance of prior probabilities for entry page search. In *Proceedings of the 25th ACM Conference on Research and Development in Information Retrieval (SIGIR'02)*, pp. 27–34.

Luhn, H. (1957). A statistical approach to mechanised encoding and searching of litary information. *IBM Journal of Research and Development* **1**(4) 309–317.

Margulis, E. (1993). Modelling documents with multiple Poisson distributions. *Information Processing and Management* **29** 215–227.

Maron, M. and J. Kuhns (1960). On relevance, probabilistic indexing and information retrieval. *Journal of the Association for Computing Machinery* **7** 216–244.

Metzler, D. and W. Croft (2004). Combining the language model and inference network approaches to retrieval. *Information Processing and Management* **40**(5) 735–750.

Mihajlovic, V. (2006). *Score Region Algebra: A flexible framework for structured information retrieval*. PhD Thesis, University of Twente.

Miller, D., T. Leek and R. Schwartz (1999). A hidden Markov model information retrieval system. In *Proceedings of the 22nd ACM Conference on Research and Development in Information Retrieval (SIGIR'99)*, pp. 214–221.

Navarro, G., and R. Baeza-Yates (1997). Proximal nodes: A model to query document databases by content and structure. *ACM Transactions on Information Systems* **15** 400–435.

Ponte, J. and W. Croft (1998). A language modeling approach to information retrieval. In *Proceedings of the 21st ACM Conference on Research and Development in Information Retrieval (SIGIR'98)* pp. 275–281.

Rabiner, L. (1990). A tutorial on hidden markov models and selected applications in speech recognition. In A. Waibel and K. Lee (eds), *Readings in speech recognition*, pp. 267–296. Morgan Kaufmann.

Richardson, M., A. Prakash, and E. Brill (2006). Beyond PageRank: machine learning for static ranking. In *Proceedings of the 15th International Conference on World Wide Web*, pp. 707715–. ACM Press.

Robertson, S. (1977). The probability ranking principle in ir. *Journal of Documentation* **33**(4) 294–304.

Robertson, S. and K. Spärck-Jones (1976). Relevance weighting of search terms. *Journal of the American Society for Information Science* **27** 129–146.

Robertson, S. and S. Walker (1994). Some simple effective approximations to the 2-poisson model for probabilistic weighted retrieval. In *Proceedings of the 17th ACM Conference on Research and Development in Information Retrieval (SIGIR'94)*, pp. 232–241.

Robertson, S. E., C. J. van Rijsbergen, and M. F. Porter (1981). Probabilistic models of indexing and searching. In R. Oddy *et al.* (eds), *Information Retrieval Research*, pp. 35–56. Butterworths.

Rocchio, J. (1971). Relevance feedback in information retrieval. In G. Salton (ed.), *The Smart Retrieval System: Experiments in Automatic Document Processing*, pp. 313–323. Prentice Hall.

Salton, G. (1971). The SMART retrieval system: *Experiments in automatic document processing*. Prentice-Hall.

Salton, G. and E. A. Fox and H. Wu (1983). *Extended Boolean information retrieval*. *Communications of the ACM* **26**(11) 1022–1036.

Salton, G. and M. McGill (1983). *Introduction to Modern Information Retrieval*. McGraw-Hill.

Salton, G. and C. Yang (1973). On the specification of term values in automatic indexing. *Jounral of Documentation* **29**(4) 351–372.

Spärck-Jones, K., S. Walker, and S. Robertson (2000). A probabilistic model of information retrieval: Development and comparative experiments (part 1 and 2). *Information Processing and Management* **36**(6) 779–840.

Turtle, H. and W. Croft (1991). Evaluation of an inference network-based retrieval model. *ACM Transactions on Information Systems* **9**(3) 187–222.

Zhai, C. and J. Lafferty (2004). A study of smoothing methods for language models applied to information retrieval. *ACM Transactions on Information Systems* **22**(2) 179–214.

Zhai, C. (2008). Statistical Language Models for Information Retrieval. *Foundations and Trends in Information Retrieval* **2**(3) 137–213.

2

User-centred Evaluation of Information Retrieval Systems

Pia Borlund

2.1 Introduction

Evaluation of Information Retrieval (IR) systems is important because we need to know whether the designed and developed system performs as intended. Similarly, we need to know how the users use and perceive the system, that is, the users' intuitive use, the application of interactive IR (IIR) search facilities, and how the users apply and assess relevance in the IR process. Or in other words, as the overall objective of IR systems is to retrieve relevant information, we do of course need to know whether the systems and users, in part and in combination, succeed in doing so. This leaves us with three approaches to IR systems evaluation: the *system-driven approach*, which focuses on the system side; the classical *user-oriented approach* that concentrates on the users and the users' use of the system and their satisfaction with the retrieved information objects; and the *cognitive IR approach* that views and treats the system and users as a whole and conducts the evaluation accordingly by use of, e.g. the IIR evaluation model. Of the three evaluation approaches, the *system-driven approach*, the *user-oriented approach*, and the *cognitive IR approach*, the present chapter focuses mainly on the two user-involving approaches, which as one is denoted *user-centred evaluation*.

The objective of the chapter is to outline the characteristics of user-centred IR systems evaluation in order to highlight the complexity and level of variables to consider in this type of evaluation. This is achieved by an insightful overview of the landmarks in user-centred evaluation and its historical development: the MEDLARS test (Lancaster 1969); the Okapi project (e.g. Walker 1989; Robertson 1997a; Beaulieu and Jones 1998); and the IIR evaluation model (Borlund 2000a; 2000b; 2003a). Because the predominant approach to IR systems evaluation is the system-driven approach, the point of departure for introducing the characteristics of user-centred characteristics is a brief introduction of the Cranfield model that signifies the system-driven approach to IR systems evaluation.

2.1.1 Background

User-centred IR systems evaluation is complex to create and assess due to the increased number of variables to incorporate within the evaluation, and the difficulty of selecting appropriate evaluation criteria for study. In spite of the complicated nature of user-centred evaluation, this form of evaluation

Information Retrieval: Searching in the 21st Century edited by A. Göker & J. Davies
© 2009 John Wiley & Sons, Ltd

is used to understand the effectiveness of individual IR systems and user search interactions. The growing incorporation of users into the evaluation process also reflects the changing nature of IR within society; for example, more and more people have access to IR systems through Internet search engines, but have little training or guidance in how to use these systems effectively. Similarly, new types of search systems such as recommender systems and IIR facilities are becoming available to wide groups of end-users. Hence user-centred IR systems evaluation becomes more essential and important than ever, as the end-user information searcher of today no longer necessarily is an information search specialist like the librarians who in the past conducted information searching on behalf of the user.

It is, however, interesting to note, that though IR systems have developed and become more interactive and end-user-oriented, the evaluation most used still is the system-driven approach that builds upon the Cranfield model (Cleverdon *et al.* 1966; Cleverdon and Keen 1966). The Cranfield model is based on the principle of test collections, that is: a collection of documents, a collection of queries, and a collection of independent (pre-assessed) relevance assessments[1], and the measurement of recall[2] and precision[3] ratio. The series of tests carried out in Cranfield, UK that resulted in the Cranfield model constitutes the empirical research tradition on the development and testing of IR systems (Gull 1956; Thorne 1955; Cleverdon 1960; 1962; Cleverdon *et al.* 1966; Cleverdon and Keen 1966). The Cranfield model represents the archetype of the system-driven approach to IR evaluation, which is viewed as a laboratory type of test, carried out as a 'black box experiment' based on batch retrieval with the purpose of observing inputs and outputs, and measuring the outcome (Robertson and Hancock-Beaulieu 1992). In comparison, the classical user-oriented approach to IR evaluation is characterised as a diagnostic type of investigation. This type of evaluation takes place and involves users within the operational environment of the test and, depending on the test purpose, allows for as many search runs necessary to satisfy the information needs (Robertson and Hancock-Beaulieu 1992). The objective of the Cranfield model is to keep all variables controlled and to obtain results that can state conclusions about the effectiveness of retrieval performance in general. From a system-driven viewpoint, a system is defined as the retrieval mechanism and databases of information objects. The queries are held to be identical to static information needs, and relevance judgements are assumed to be an objective indication of whether or not the document is about the query, and can as such be made by any knowledgeable person. Consequently, the Cranfield model suffers from limitation due to its restricted assumptions on the cognitive and behavioural features of the environment in which IR systems function.

The user-oriented evaluation approach defines the IR system in a much broader way and views the searching and retrieval processes as a whole. The main purpose of this type of evaluation is concerned with how well the user, the retrieval mechanism, and the database interact – extracting information – under real-life operational conditions. In this approach the relevance judgements have to be given by the original user in relation to his or her personal information need, which may change over session time. The assumption is that the relevance judgements represent the value of the documents for a particular user at a particular moment, hence the assessment can only be made by the user at that time.

The core of the discussion between the two major approaches to evaluation of IR systems, the system-driven approach *versus* the classical user-oriented approach, is clearly outlined by Robertson and Hancock-Beaulieu (1992, p. 460) who state that '[t]he conflict between laboratory and operational experiments is essentially a conflict between, on the one hand, control over experimental variables, observability, and repeatability, and on the other hand, realism'. In summary, the two main approaches differ from each other in respect to the issues of experimental control *versus* realism. They further

---

[1] A relevance assessment is the decision about whether a given document is relevant, or not, to a given query. Thus relevance is a performance criterion for IR systems evaluation, that is, the testing of the system's ability to retrieve relevant documents and hold back non-relevant documents.

[2] Recall is the ratio of the number of relevant documents retrieved divided with the number of relevant documents in the collection.

[3] Precision is the ratio of the number of relevant documents retrieved divided by the number of documents retrieved.

differ in their attitude towards: 1) the nature of the information need; 2) the research environment used for experimentation; 3) the roles of the interface as an intermediary, the user, and the information retrieval systems; and 4) the question as to where the retrieval systems ends and the automatic interface intermediary begins (Ingwersen 1992, p. 87).

2.1.2   Chapter outline

This chapter focuses on the need for, and qualities of, user-centred IR systems evaluation. The objective of the chapter is to present the characteristics of user-centred evaluation of (I)IR systems. The underlying view of the chapter is that the system-oriented and user-centred approaches to IR systems evaluation complement each other, and do not contradict each other. Where the system-driven approach is easily introduced due to the Cranfield model, the user-oriented approach is less easily explained; partly, because the user-oriented approach is not framed within a model like that of the Cranfield model, and partly because the focus of user-oriented IR systems evaluation can be much broader than the narrower IR effectiveness focus of the system-driven approach. Hence, Section 2.2 presents the MEDLARS[4] test by Lancaster (1969) as a classical example of user-oriented IR systems evaluation, in order to illustrate the characteristics and broader focus of this type of evaluation. This is followed by an introduction to Okapi[5] in Section 2.3. Okapi is a British research project founded on the approach to user-oriented IR systems research (e.g. Walker 1989; Robertson 1997a; Beaulieu and Jones 1998). The purpose of introducing Okapi is partly to point out how Okapi, as a representative of the user-oriented research tradition, is actively involved in TREC[6], and hence employs the two evaluation approaches according to test purpose. Partly, to illustrate how the Okapi research on relevance feedback leads to the development of IIR facilities, which emphasise the call for alternative evaluation approaches as dealt with in Section 2.4 on the cognitive IR approach to IR systems evaluation. Section 2.4 introduces the third evaluation approach, the cognitive IR approach that focuses specifically on IIR systems and applies the *IIR evaluation model*. This approach is anchored in the holistic nature of the cognitive viewpoint by being a hybrid of the system-driven and the user-oriented approaches, building upon each of their central characteristics of control and realism. Section 2.5 closes the chapter with a summary of the main characteristics and the strengths and weaknesses of the three approaches to IR systems evaluation: the system-driven; the classical user-oriented; and the cognitive IR approach as illustrated with the IIR evaluation model.

2.2   The MEDLARS Test

The present MEDLARS review is a description of the evaluation carried out and reported on by Lancaster (1969). The MEDLARS system (now known as Medline[7] and PubMed[8]) has been evaluated by others than Lancaster (1969), for instance by Salton (1972) in his comparison of the conventional indexing of MEDLARS and automatic indexing by the Smart[9] system. The reader should note that the MEDLARS system was not an end-user system at the time of testing, instead information search specialists searched on behalf of the user. Lancaster describes MEDLARS as '. . . a multipurpose system, a prime purpose being the production of *Index Medicus* and other recurring bibliographies'(Lancaster 1969, p. 119). The aim of Lancaster's MEDLARS test was to evaluate the existing system and to find out ways in which the performance of the system could be improved.

---

[4] MEDLARS is an acronym for: MEDical Literature Analysis and Retrieval System.

[5] Okapi is an acronym for: Online Keyword Access to Public Information.

[6] TREC is an acronym for: Text REtrieval Conference. In brief, TREC is an annual research workshop and provides for large-scale test collections and methodologies, and builds on the Cranfield model (Voorhees and Harman 2005a). TREC is hosted by the US Government's National Institute of Standards and Technology (NIST) (http://trec.nist.gov/).

[7] Medline URL: http://medline.cos.com/

[8] PubMed URL: http://www.ncbi.nlm.nih.gov/pubmed/

[9] Smart is an acronym for: System for the Mechanical Analysis and Retrieval of Text.

2.2.1   Description of Lancaster's test of MEDLARS

The planning of the MEDLARS evaluation began in December 1965 and was carried out during 1966 and 1967. Lancaster (1969, p. 120) outlines the principal objectives of the evaluation to be: 1) to study the demanded types of search requirements of MEDLARS users; 2) to determine how effectively and efficiently the present MEDLARS service meets these requirements; 3) to recognise factors adversely affecting performance; and 4) to disclose ways in which the requirements of MEDLARS users may be satisfied more efficiently and/or economically. Lancaster further explains how the users were to relate the prime search requirements to the following factors:

1. The *coverage* of MEDLARS (i.e., the proportion of the useful literature on a particular topic, within the time limits imposed, that is indexed into the system).
2. Its *recall* power (i.e., its ability to retrieve 'relevant' documents, which within the context of this evaluation, means documents of value in relation to an information need that prompted a request to MEDLARS).
3. Its *precision* power (i.e., its ability to hold back 'non-relevant' documents).
4. The *response time* of the system (i.e., the time elapsing between receipt of a request at a MEDLARS center and delivery to the user of a printed bibliography).
5. The *format* in which search results are presented.
6. The amount of *effort* the user must personally expend in order to achieve a satisfactory response from the system... (Lancaster 1969, p. 120).

The document collection used was the one available on the MEDLARS service at the time, which according to Robertson (1981, p. 20) had the size of approximately 700 000 items. 302 real information requests were used to search the database, and the users requesting the information made the relevance assessments. A recall base was generated based on a number of documents judged relevant by the users in response to their own requests, but found by means outside MEDLARS. The sources for these documents were: (a) those already known to the user; and (b) documents found by MEDLARS staff through sources other than MEDLARS or *Index Medicus*. Subsequently, each user was asked to assess relevance of a sample of the output from the MEDLARS search, together with selected documents from other sources (Robertson 1981, p. 20). The relevance assessments were carried out with reference to three categories of relevance: major, minor, or no value. In addition, the users explained their relevance judgements by indicating why particular items were of major, minor, or no value to their information need (e.g. see Martyn and Lancaster (1981, p. 165) for an example of the relevance questionnaire used in the MEDLARS test). Precision and relative recall ratios were calculated and compared to the degree of exhaustivity[10] and specificity[11] of the requests as expressed by use of the controlled entry vocabulary. Analysis of failures was carried out, and a classification of reasons for failure was devised (Lancaster 1969, p. 125, Table 5).

Variables were built into the test. The primary one concerned the level of interaction between the user and the system. The interaction could take place at three different levels, which refers to, i.e., the originality or level of influence on the expression of the user's information request (Lancaster 1969, p. 120). At one level the interaction takes place as *personal interaction*, that is, the user makes a visit to a MEDLARS centre and negotiates his or her requirements directly with an information search specialist. A second level of interaction is identified as *local interaction*, which means that the user's information request comes by mail to the centre, but is submitted by a librarian or information specialist on behalf of the requester. The third interaction level concerns *no local interaction*, which implies that no interference by librarian or information specialist has taken place in the process of interpreting or translating the request into the controlled vocabulary used by MEDLARS. In other

---

[10] Exhaustivity and specificity are well-established concepts within indexing. Exhaustivity refers to the comprehensiveness of the indexing in question, which consequently affects retrieval results.

[11] Specificity refers to the accuracy of the indexing. Specificity of indexing makes for high precision and low recall.

words, the request comes directly by mail from the requester. The different levels of interaction enabled a comparison of the results obtained. The main product of the test was a detailed analysis of the reasons for failure. The result is interesting, because '… it appears that the best request statements (i.e. those that most closely reflect the actual area of information need) are those written down by the requester in his own natural-language narrative terms' (Lancaster 1969, p. 138). For the user-oriented approach to IR systems evaluation, this result is seen as a strong piece of evidence and motivation for exactly this type of evaluation. That only the user who owns the need for information, can represent and assess that information need, and hereby give insight as to how well the system works under real-life operational conditions in a given situation and context at that time.

The results, the focus, and the approach of the MEDLARS test helped the contemporary system-driven researchers to understand problems of IR as well as how to design future systems and tests. In the following quotation, Spärck Jones (1981b, p. 230) comments on the effect of Lancaster's test by describing the situation prior to the MEDLARS test: '[i]t was thus not at all obvious how systems should be designed to perform well, modulo a performance for recall or precision, in particular environments, especially outside established frameworks like those represented by the MEDLARS system, or for situations and needs clearly resembling those of existing systems'. Robertson (1981, pp. 20–21) descri-bes how Cranfield II and MEDLARS are two of the classical tests, with Cranfield II being a highly controlled and artificial experiment, and MEDLARS being an investigation of an operational system, as far as possible under realistic conditions. They both play a significant part in creating the archetypes of tests, by each representing the opposite poles of the system-driven – user-oriented spectrum.

2.2.2 Evaluation characteristics

A number of characteristics about user-centred IR systems evaluation can be derived from Lancaster's MEDLARS test, e.g. the evaluation takes place in an operational environment, and employs a large-scale, dynamic document collection. The evaluation involves real users and their genuine information needs. The users judge relevance of the retrieved information objects for their own information needs. Relevance is (often) judged according to several degrees of relevance, that is, as non-binary relevance. In the case of MEDLARS, relevance is judged according to three degrees of relevance: major, minor, or not relevant. The evaluation is 'diagnostic' in nature. The objective of this type of test is to identify and understand the causes of search failures, and/or users' system use and search behaviour, thus to allow corrective actions to be taken to improve the existing system and advance knowledge about user IR system interaction. Hence, this type of evaluation often includes a combination of effectiveness and efficiency aspects as well as interface evaluation issues, compared with the system-driven approach that focuses on the aspect of effectiveness only.

Another interesting point to make relates to performance effectiveness. Usually, the user-oriented approach collects relevance assessments in a non-binary form (that is, as several degrees of rele-vance) and calculates performance effectiveness as recall and precision. But the formulas of recall and precision allow only for a binary relevance representation, and do not (traditionally) provide for a non-binary indication of how relevant the relevant information objects are. Thus, it is habitually the case in tests where non-binary relevance judgements are applied that two or more relevance categories get merged into the binary scale of relevant and non-relevant in order to facilitate the calculation of precision (and recall) values (e.g. Su 1992). According to Schamber (1994, p. 18), the relevance cat-egories are merged because it is assumed that no information is being lost in the merging process. A consequence of this is also seen in how precision is calculated as the mean of the relevance values in cases where the users' relevance assessments are indicated as numerical relevance values (e.g. Borlund and Ingwersen 1998; Reid 2000). It should also be noted that it is primarily precision that is computed, as knowledge about the number of relevant documents in the collection is required to compute recall. This information is not available when the document collection used is very big and continually updated with more documents, e.g. the web. The solution to this is to do what Lancaster did, namely to construct a recall base and compute relative recall, and not exact recall, where all relevant documents in the collection are to be known. Despite the deficiencies of recall and precision (from a user-oriented point of view), they are the most used measures of IR effectiveness. The use of precision and recall can be viewed as an attempt by the user-oriented approach to compute and present results on conditions of the system-driven approach in order to make the results 'comparable' to those of the system-driven. Another simple explanation is that the user-oriented approach has no other effectiveness measures to employ, and hence uses precision (and recall) regardless of the deficiencies.

Yet another characteristic of the tests within the user-oriented approach is the uniqueness of the tests due to different operational test settings, different users with individual information needs, the employment of subjective relevance assessments, different research foci, and research questions from test to test leading to different research designs, etc. Therefore the results from one test are not necessarily directly comparable to the results of another test. The issue of incomparability of test results is often put forward as a disadvantage of the user-oriented approach to IR systems evaluation. It is, however, an expected consequence and limitation rather than a disadvantage of this form of evaluation, which wishes to understand the effectiveness and use of individual IR systems and user search interactions.

What is considered a disadvantage of this approach is the fact that it is rather time and resource demanding to plan and carry out this type of evaluation, and furthermore to analyse and correlate the collected data. Again, Lancaster's MEDLARS test is an illustrative example. The planning of the MEDLARS test started in December 1965, the evaluation was conducted in 1966 and 1967, and the results were published in 1969 (Lancaster 1969, p. 120). Bearing this in mind makes the contribution of the Okapi project even more impressive, which will be discussed next.

2.3   The Okapi Project

Okapi is the name of a series of generations of test retrieval systems. Okapi can be viewed as a test bed including an experimental test system – or versions of systems – and can as such be considered the user-oriented IR approach's answer to Salton's Smart project within the system-driven IR evaluation approach (e.g. Salton 1981). The foci of the system tests and the different installations of the systems are reported on in a number of publications (e.g. Walker 1989; Walker and De Vere 1990; Robertson and Hancock-Beaulieu 1992; Beaulieu *et al.* 1996; Robertson 1997a; Beaulieu and Jones 1998). Even an entire issue of the *Journal of Documentation* (Robertson 1997a) has had the theme of Okapi and IR research. The Okapi systems have had their homes in various places. From 1982 to 1989 Okapi was based at the Polytechnic of Central London (now the University of Westminster). Since 1989 Okapi has been located at City University London.

2.3.1   The objectives of Okapi

Okapi is a test search system designed for end-users who are not expert searchers. Throughout its history Okapi has been used as the basis for real services to groups of users. The initial Okapi investigations started out with Okapi functioning as an OPAC[12] that was made available in a number of libraries to the actual users of those libraries. Later, various Okapi systems have been made available in similar ways to groups of researchers over a network with a database of scientific abstracts (Robertson 1997b, p. 5). The operational set up of the Okapi was made specifically with the purpose of creating an environment in which ideas could be subject to user trials. The operational set up of the Okapi investigations has allowed for end-users to be observed in a large variety of ways, from straight transaction logs to questionnaires, interviews, and direct observation (Robertson 1997b, p. 5). This means that the Okapi team gets additional qualitatively based information on the functionality and performance of, e.g. relevance feedback. Robertson (1997b, p. 5) explains how the quality of user-oriented Okapi investigations supply evidence concerning relevance feedback, not just on its retrieval effectiveness in the usual sense, but also on the ways in which users actually make use of it, and how useful they find it. The Okapi project has contributed to the understanding that users can be given access to sophisticated IR and feedback techniques and that users are capable of using these techniques effectively. At the same time the research shows that there are difficulties involved

---

[12] OPAC is an acronym for: Online Public Access Catalogue.

in providing these techniques, and that users have to be given guidance in the use of these techniques, as pointed out by Robertson (1997b, p. 6).

Robertson (1997b, pp. 3–4) elegantly describes the typical Okapi system and its underlying probabilistically based retrieval and relevance feedback mechanisms:

> 'What the user sees first is an invitation to enter a query in free-text form. This free-text query is then parsed into (generally) a list of single word-stems; each stem is given a weight based on its collection frequency. The system then produces a ranked list of documents according to a best-match function based on the term weights, and shows the user titles of the top few items in the list. The user can scroll the list and select any title for viewing of the full record. Having seen the full record, he or she is asked to make relevance judgement ('Is this the kind of thing you want?') in a yes/no term. Once the user has marked a few items as relevant, he or she has the opportunity to perform relevance feedback search ('More like the ones you have chosen?'). For this purpose, the system extracts terms from the chosen documents and makes up a new query from these terms. This is normally referred to as query expansion, although the new query may not necessarily contain all the original terms entered by the user. The new query is run and produces a ranked list in the usual fashion, and the process can iterate'.

The quotation by Robertson of a typical Okapi system session reveals the multi-layer of the system architecture and use of the system, which is in accordance with the broad system definition in the present evaluation approach. The broad definition is further reflected in the investigative and experimental division of the focus of Okapi tests on: IR system techniques, information searching behaviour, and interface issues. The inclusion of research on interface issues emphasises the point of difference between the two main approaches to IR systems evaluation made by Ingwersen (1992, p. 87) about the question as to where IR systems ends and the automatic interface intermediary begins. The interface research of Okapi concerns interaction issues such as the user's perception of system functions in relation to information searching tasks (e.g. Beaulieu 1997; Beaulieu and Jones 1998). Via Okapi, research attention is given to the issue of 'cognitive load' that IR systems put on the users. Cognitive load refers to the intellectual effort required from the user in order to work the system. As such the research areas of IR and human computer interaction (HCI), especially with the introduction of OPACs and IIR systems, become overlapping. Beaulieu and Jones (1998, pp. 246–247) explain that cognitive load '... involves not only the requirement to understand conceptual elements of the system [...] but to make meaningful decisions based on that understanding. When searchers have to make decisions without adequate understanding, their efforts can be counter-productive for overall system effectiveness'. This they illustrate with an example of how users who do not realise how query expansion operates may avoid the trouble of making relevance judgements if they can, and lose one important advantage of probabilistic retrieval. So the ideal interface for an IIR system is one that meets the (inexperienced) end-user as transparent and self-explanatory so that no extra cognitive burden is put on the user.

The Okapi project further serves as a very good example of how the two main approaches to IR systems evaluation complement each other. In that, Okapi has taken part in every round and often in several tracks of TREC since the beginning of TREC[13] in 1992.

2.3.2 Okapi at TREC

In brief, TREC is an annual research workshop and provides for large-scale test collections and methodologies, and builds on the Cranfield model (Voorhees and Harman 2005a). TREC is hosted by the US Government's National Institute of Standards and Technology (NIST). Since 1994 TREC has been extended with more tracks than *ad hoc* and routing text retrieval,

---

[13] TREC is an acronym for: Text REtrieval Conference. For more information about TREC, the reader is directed to the book *TREC: Experiments and Evaluation in Information Retrieval* by Voorhees and Harman (2005a), as well as to the following URL: http://trec.nist.gov/

examples of additional TREC tracks are: Interactive (1994–2003); Spanish (1994–1996); Confusion (1995–1996); Database merging (1995–1996); Filtering (1995–2001); Chinese (1996–1997); Natural Language Processing (NLP) (1996–1997); Speech (1997–2000); Cross-language (1997–2002); Web (1999–2004); Genomics (2004–); Blog (2006–) (Voorhees and Harman 2005b, p. 8, Table 1.1); (URL: http://trec.nist.gov/tracks.html).

Okapi has taken part in several TREC tracks over the years, e.g. *ad hoc*, routing, interactive, filtering, and web tracks. According to Robertson *et al.* (1997, p. 20) the Okapi team uses TREC to improve some of the automatic techniques used in Okapi, especially the term-weighting function and the algorithms for term selection for query expansion. Prior to TREC, the Okapi team had worked only with operational systems or with small-scale partially controlled experiments with real collections (Robertson *et al.* 1997, p. 23). As such, TREC is a change of test culture and environment to the Okapi test tradition. However, the Okapi team maintain their interest in real users and information needs because to them the most critical and most difficult areas of IR system design are currently in the area of interaction and user interfaces (Robertson *et al.* 1997, p. 32). It is clear that the Okapi team sees their participation in the TREC experiments as complementary to the real-user experiments (Robertson 1997b, p. 6). Further, Okapi's participation in TREC illustrates how systems, or elements of systems, at some point may be tested according to system-driven principles, as well as how the interactive functionality of parts or more 'finished' and complete IIR systems are best evaluated and validated by involvement of end-users and potentially dynamic information needs. Okapi's participation in TREC also shows that the boundaries between the two evaluation approaches are not clear-cut, and the two approaches supplement each other.

2.3.3   The impact of Okapi

The Okapi project represents a strong and illustrative case of user-oriented IR systems evaluations (e.g. Walker 1989; Walker and De Vere 1990; Robertson and Hancock-Beaulieu 1992; Beaulieu *et al.* 1996; Robertson 1997a; Beaulieu and Jones 1998). Further, Okapi demonstrates in relation to TREC the complementary nature of the two main approaches to IR systems evaluation. In addition, Okapi and the series of tests in which probability-based IR techniques such as relevance feedback have been tested, emphasise the need for alternative approaches to evaluation of IIR techniques and search facilities, which we focus on in the following section. Though the present introduction of Okapi is made from a historical viewpoint, it is appropriate to point out that Okapi is still active and maintained (personal communication with Dr A. MacFarlane 2007)[14]. Recently, Okapi has been extended to handle XML documents and element retrieval for INEX[15] (Lu *et al.* 2006; Robertson *et al.* 2006; Lu *et al.* 2007). The Okapi-Pack, which is a complete implementation of the Okapi system, is available from the Centre for Interactive Systems Research (CISR) at City University for a nominal fee when used for research purposes only (URL: http://www.soi.city.ac.uk/~andym/OKAPI-PACK/). According to MacFarlane (personal communication 2007) the long-term plan is to replace Okapi with an open source version, Okapi++.

2.4   The Interactive IR Evaluation Model

IIR systems are defined as systems where the user dynamically conducts searching tasks and correspondingly reacts to systems responses over session time. By incorporating the communicative behaviour by users, IIR systems are defined just as broadly as in the user-oriented approach, and the focus of the evaluation is similarly wider than in non-interactive IR. The focus of IIR evaluation includes all the user's activities of interaction with the retrieval and feedback mechanisms as well as

---

[14] Personal communication with Dr. Andrew Macfarlane, Department of Information Science, City University, London. November 2007.

[15] INEX is an acronym for: INitiative for the Evaluation of XML retrieval. For more information about INEX the reader is directed to: URL: http://inex.is.informatik.uni-duisburg.de/

the retrieval outcome itself. The overall purpose of the evaluation of IIR systems is to evaluate the system in a way which reflects the interactive information searching and retrieval processes of system and user, and takes into account the potential dynamic natures of information needs and relevance. This can be achieved by the alternative approach to IIR systems evaluation, referred to as the IIR evaluation model (Borlund 2000b; 2003a). The model provides a framework for the collection and analysis of IR interaction data. The key element of the IIR evaluation model is the use of realistic scenarios, known as simulated work task situations. A simulated work task situation is a short 'cover story' that describes the context and situation for the task at hand, leading to IR. A simulated work task situation accomplishes two functions central to IR: 1) it prompts and develops a simulated information need by allowing for user interpretations of the simulated work task situation, leading to cognitively individual information need interpretations as in real life; and 2) it is the platform against which situational relevance[16] is judged. Further, by being the same for all test persons, experimental control is provided. Hence, the concept of a simulated work task situation ensures the test has both realism and control.

2.4.1   The cognitive IR approach

The IIR evaluation model is founded on the cognitive approach to IR (e.g. Belkin 1980; Ingwersen 1992; 1996). The cognitive viewpoint is concerned with humans' dynamic and interactive processing and use of information, and its holistic way of viewing the IR system, the user, the information need, and the underlying reason for the information need, as well as the context for the search interaction as a whole. The breakthrough of this viewpoint within IR happened with Belkin's introduction of the concept of an 'anomalous state of knowledge', shortened to the acronym ASK, and commonly known as the ASK hypothesis (Belkin 1980; Belkin *et al.* 1982). An ASK is the user's recognition of an insufficient knowledge model due to an external situation, e.g. a given work task situation, which results in an information need, for instance, in order to reduce uncertainty. The result of the ASK hypothesis is that the user's information need is seen as a reflection of an anomalous state of knowledge. This is a further development of the scientific perception of the information need from a static concept (as viewed and applied in the system-driven approach to IR) to a user-individual and potentially dynamic concept (as employed by the user-oriented approaches to IR). The user's perception of an information need is thus triggered by the interpretation of a given situation, a problem to be solved or a state of interest to be fulfilled, under influence of the user's current cognitive and emotional state, which again is affected by the cultural and social context within which the user acts. The main issue is how to operationalise this in an evaluation setting, which brings us back to the aforementioned key element: the simulated work task situation of the IIR evaluation model.

2.4.2   The three parts of the IIR evaluation model

Basically, the IIR evaluation model consists of three parts:

Part 1. A set of components which aims at ensuring a functional, valid, and realistic setting for the evaluation of IIR systems;

Part 2. Empirically based recommendations for the application of the concept of a simulated work task situation; *and*

Part 3. Alternative performance measures[17] capable of managing non-binary based relevance assessments.

---

[16] Situational relevance is a user-subjective type of relevance that considers the usefulness of the assessed information objects according to the task at hand. Thus this type of relevance is user-individual and situation, task, and context dependent. Further, the outcome of the users' relevance judgement of situational relevance might change over time, and be different from user to user. For more information about the concept of relevance in IR – the multidimensionality and potentially dynamic nature of relevance – the reader is directed to, e.g. Borlund (2003b).

[17] With respect to the performance measures of recall and precision traditionally employed.

Parts 1 and 2 concern the collection of data, whereas part 3 concerns data analysis. As such the IIR evaluation model is comparable to the Cranfield model's two main parts: the principle of test collections, and the employment of recall and precision. The first part of the model deals with the experimental setting. This part of the model is identical to the traditional user-oriented approach in that it involves potential users as test persons, applies the test persons' individual and potentially dynamic information need interpretations, and supports assignment of multidimensional relevance (that is, relevance assessment according to various types of relevance, several degrees of relevance, and multiple relevance criteria) and dynamic relevance assessments (allowing that the users' perception of relevance can change over time). The present approach differs from the traditional user-oriented approach with the introduction of simulated work task situations as a tool for the creation of simulated, but realistic, information need interpretations. Hence, part 2 outlines recommendations for how to create and use simulated work task situations. The major challenge within this approach lies in the design of realistic and applicable simulated work task situations. Previous research (Borlund 2000a; b) shows that a well-designed situation is one: which the test persons can relate to and in which they can identify themselves; that the test persons find topically interesting; and which provides enough imaginative context in order for the test persons to be able to relate and apply the situation. The ongoing PIAN project by Ruthven and Borlund has set out to further verify recommendations for simulated work task-based evaluation, and hereby emphasise the strength of the recommendations and provide clearer guidelines for the employment of this approach. One of the issues they will look into is the effect that topical interest of users might have on search behaviour (Lei *et al.* 2006). It is furthermore recommendable to employ a combination of simulated and test persons' genuine information needs. Hence, genuine information needs function as a baseline against the simulated information needs, both at a specific test person level and at a more general level. In addition, they provide information about the systems' effect on this type of information needs. The inclusion of genuine information needs is also useful in the pilot test of the test setting, and the test persons' perceptions of the simulated work task situations, as they can inspire to 'realistic' and user-adaptable simulated work task situations.

The set of components combined with the second part of the model, recommendations for the application of simulated work task situations, provides an experimental setting[18] that enables the facilitation of evaluation of IIR systems as realistically as possible with reference to actual information seeking and retrieval processes, though still in a relatively controlled evaluation environment. The third and final part of the model is a call for alternative performance measures that are capable of managing non-binary-based relevance assessments, as a result of the application of the model parts 1 and 2. The dominating use of the ratios of recall and precision for the measurement of the effectiveness of IR performance, also within the traditional user-oriented approach to IR systems evaluation, have forced researchers to reconsider whether these measures are sufficient in relation to the effectiveness evaluation of IIR systems. Spink and colleagues comment on the situation in the following way: '[t]he current IR evaluation measures are. . .not designed to assist end-users in evaluation of their information seeking behavior (and an information problem) in relation to their use of an IR system. Thus, these measures have limitations for IR system users and researchers' (Spink *et al.* 1998, p. 604). Nevertheless, the reason for the measures' well-established positions is due to the clear and intuitively understandable definitions, combined with the fact that they represent important aspects of IR, are easy to use, and the results are comparable (Spärck Jones 1971, p. 97). However, the measures view relevance as a binary concept and do not allow for the often non-binary approach taken in, e.g. the user-centred approaches. Further, the measures do not distinguish between the different types of relevance involved, but treat them as the one and same type. In order to illustrate the need for alternative performance measures, the measures of relative relevance (RR) and ranked half-life (RHL) (Borlund and Ingwersen 1998) were introduced, followed up by the stronger measures of cumulative gain (CG) with and without discount by Järvelin and Kekäläinen (2000). The RR measure is intended to satisfy the need for correlating the various types of relevance applied in the evaluation of IR and

---

[18] An experimental setting, in this context, necessarily includes a database of information objects as well as the system(s) under investigation. However, these components are not explicitly dealt with here. It is assumed that the system to be tested, other technical facilities, and the facilities of where to carry out the experiment are already arranged for.

specifically IIR systems. The RHL informs about the position of the assessed information objects as to how well the system is capable of satisfying a user's need for information at a given level of perceived relevance. In line with RHL, the CG measures are position measures. The assumptions of the CG measures are that:

- Highly relevant documents are more valuable than marginal ones; and
- The lower the ranked position of a retrieved document, the less valuable it is for the user, because the less likely it is that the user will ever examine it (Järvelin & Kekäläinen 2000, p. 42).

The recommendation is to apply these measures in combination with the traditional performance measures of recall and precision. Ingwersen and Järvelin (2005) further discuss the strengths and weaknesses of the RR, RHL, and CG measures.

It seems the IIR evaluation model, and especially the use of simulated work task situations, satisfies a demand within the research area of information seeking behaviour, as of how to conduct comparative studies of IR behavioural patterns. Though the model is developed for IIR performance evaluation, it is also suitable for research in information seeking behaviour, which is due to the realism inherent in the use of simulated work task situations.

Section 2.5 summarises the characteristics of the three approaches to IR systems evaluation, the system-driven approach that builds upon the Cranfield model, and the two user-centred approaches, the classical user-oriented approach and the cognitive IR approach, respectively.

2.5 Summary

The majority of characteristics of the three IR evaluation approaches presented throughout the chapter are depicted in Table 2.1, and hence the table and this section of the chapter provide a summary of the characteristics, of which some have been discussed directly and others only briefly or indirectly touched upon. Table 2.1 consists of four columns, of which the first column lists the characteristics in question, which are hereafter presented according to each of the three approaches: the system-driven approach; the user-oriented approach; and the cognitive IR approach, exemplified by the IIR evaluation model. As with the present chapter, Table 2.1 starts out with the system-driven approach because this approach is well known and easily introduced and characterised due to the Cranfield model, and hence provides a comparable basis for the characterisations of the two user-centred approaches.

When listed in this schematic form the approaches appear very '*black and white*' as '*either-or*' options, but may be considered as '*both-and*' options. The present test situation reveals that this is not the case. Take for example the previously mentioned INEX project (Section 2.3.3). It builds upon the principle of test collections of the Cranfield model, but employs a nuanced view of relevance, treating relevance as a non-binary concept. In addition, it considers the retrieved structured document elements in respect to exhaustivity and specificity, like Lancaster (1969). In reality it might be less clear to distinguish between the various approaches than Table 2.1 indicates. Therefore the main issue when evaluating IR systems is to define and maintain the focus of the evaluation in the formulation of research questions and the design of the test. On one hand the evaluation deals with the function of the system therefore the focus on system performance. On the other hand, IR is about human needs for, and reactions to, information, hence IR system evaluation cannot take place without an interpretation of these underlying concepts of information need and relevance. Again, it is evident how the approaches, the system-driven and the two user-centred approaches, balance each other.

As pointed out by Robertson and Hancock-Beaulieu (1992, p. 460) the central issues we are dealing with are those of realism and experimental control. The IIR evaluation model of the cognitive IR evaluation approach aims at meeting both by being a hybrid of the two major evaluation approaches.

In summary, Table 2.1 depicts how the shortcomings of the two main evaluation approaches, the lack of realism (system-driven approach) and the lack of experimental control (user-oriented approach), both illustrate the considered qualities of each of the approaches: the element of realism of the user-oriented approach due to the user-involvement, and the element of experimental control of the system-driven approach. The two approaches aim at the same goal: reliability of the test results,

Table 2.1 Summary of three approaches to IR systems evaluation: the system-driven approach; the user-oriented approach; and the cognitive IR approach, exemplified by the IIR evaluation model

Aspect	System-driven approach	User-oriented approach	Cognitive IR approach
The user	• Is not present	• Is present, and is perceived as part of the system	• Is present, and is perceived as part of the system
Information need	• Predesigned query = information need • Static	• The user's genuine information need • Potentially dynamic	• Simulated information needs + the user's genuine information need • Potentially dynamic
Relevance	• Objective • Static • Binary	• Subjective • Potentially dynamic • Non-binary (but often treated in a binary way: recall/precision)	• Subjective • Potentially dynamic • Non-binary (but might also be treated in a binary way: recall/precision)
Model	• The Cranfield model • Laboratory environment • Batch retrieval • Experimental control • Focus: system performance	• No formalized model – depends on research focus • Real-life operational environment • Allows for successive search sessions, depending on research focus • Realism • Focus: the user and system use	• The IIR evaluation model – allows for different research foci • Semi-laboratory/semi-real-life • Allows for successive search sessions, depending on research focus • Realism and experimental control • Focus: the user and the system in interaction
Performance measures	• Recall/precision • Effectiveness	• Recall/precision • Effectiveness/efficiency/information searching behaviour	• Alternative performance measures + recall/precision • Effectiveness/efficiency/information searching behaviour
Strengths	• Experimental control • Quick and cheap to carry out • Applicable for IR techniques under development	• Realism • Users form part of the test • Genuine, potentially dynamic information needs • Subjective, non-binary, potentially dynamic relevance • Applicable for developed and operational IR systems	• Realism • Experimental control • Users form part of the test • Genuine and simulated potentially dynamic information needs • Subjective, non-binary, potentially dynamic relevance • Applicable for developed and operational IR systems
Weaknesses	• Lacks realism • No users are involved • Artificial information needs • Objective, binary and static relevance	• No experimental control • Resource demanding	• Resource demanding

but reach it differently. By reliability we mean trustworthiness of the experimental results. With the system-driven approach reliability of the test results is achieved through control over the experimental variables and the repeatability of the test results. The user-oriented approach puts the user (and his or her context/situation) in focus with reference to system development, design, and evaluation, which is carried out according to the (potential) end-user's information use, retrieval, and searching behaviour with the objective of obtaining realistic results. In other words, with the user-oriented approach the test results become reliable, though not repeatable, by being loyal to the IR and searching processes. This is, however, not the same as saying that user-oriented tests cannot be repeated; they can be carried out again but do not necessarily lead to results of precise numerical level, owing to the human factors. That is, the test persons' individual information needs, the test persons' capability of expressing and treating those needs, e.g. reflected in the numbers of search formulations, search runs, the search facilities and strategies used, and the relevance assessment behaviour (e.g. relevance criteria, degrees of relevance) necessary in the process of reaching satisfaction of the information needs. With this listing of human factors and characteristics of elements of user-centred IR systems evaluation, we are back to the starting point and objective of the chapter: namely to bring attention to these characteristics.

As this chapter is an introduction to the characteristics of user-centred IR systems evaluation only, it closes with recommendations for further reading. But before doing so, a fundamental piece of advice is rightfully given: in user-centred evaluation the design of the test is everything; and in order to design well, the research focus must be clear – preferably formulated in precise and exact research questions.

For further reading on IR systems evaluation the reader is directed to the book in memory of Cleverdon edited by Spärck Jones (1981a) titled *Information Retrieval Experiments*. Or the hands-on paper by Tague-Sutcliffe (1992) which gives ten decisions to make when conducting empirical IR research. Also the ARIST chapters by Harter and Hert (1997) and Wang (1999) are recommendable when considering approaches, issues, and methods for IR systems evaluation and evaluation of information user behaviour. The recent book by Ingwersen and Järvelin (2005) that aims at integrating research in information seeking and IR deserves attention, too. Most recently, a special issue on IR user evaluation has been published in the international journal *Information Processing and Management* (Borlund and Ruthven 2008), which presents a variety of user-centred IR systems evaluations that might serve as inspiration.

Exercises

2.1 Write a simulated work task situation that involves a topic with search for some information on two countries.

2.2 When you have done this, pick out some search terms and try them on the following search engines and meta-search engines e.g.:

- Google: www.google.com
- AltaVista: www.altavista.com
- Microsoft Live Search: www.live.com
- Clusty: http://clusty.com/
- Metacrawler: www.metacrawler.com/

You may find other search engines via Search Engine Watch on the following URL http://searchenginewatch.com/showPage.html?page=Links

Write both natural language queries and Boolean queries.

2.3 Using the topic and criteria you have defined in your simulated work task situation, evaluate your searches with the following metrics:

- Precision at 5 documents retrieved (P @ 5).
- Precision at 10 documents retrieved (P @ 10).
- Consider possibilities for calculating recall.
- Rate of repeated documents (RT): Record the number of duplicates per search. This figure should be in the range 0–10.
- Link broken (LB): Record the number of broken links per search. This figure should be in the range 0–10.
- Not retrieved (NT): Record the total number of documents not retrieved by that search. This figure should be in the range 0–10.
- Spam: Record the number of Spam documents per search. This figure should be in the range 0–10.

Tabulate the results and evaluate your searches.

2.4 Following on from this:

- Reflect upon the criteria and factors affecting your relevance judgements in obtaining the precision values.
- Reflect upon the need for expressing more degrees of relevance than 'relevant', 'non-relevant' when judging relevance of retrieved documents.
- Consider the ways in which you could modify your queries given the documents that you think match your information need.

References

Beaulieu, M. (1997). Experiments on interfaces to support query expansion. *Journal of Documentation*, **53**(1), 8–19.

Beaulieu, M. and Jones, S. (1998). Interactive Searching and Interface Issues in the Okapi Best Match Probabilistic Retrieval System. *Interacting with Computers*, **10**, 237–248.

Beaulieu, M., Robertson, S. and Rasmussen, E. (1996). Evaluating Interactive Systems in TREC. *Journal of the American Society for Information Science*, **47**(1), 85–94.

Belkin, N. J. (1980). Anomalous states of knowledge as a basis for information retrieval. *Canadian Journal of Information Science*, **5**, 133–143.

Belkin, N. J., Oddy, R. and Brooks, H. (1982). ASK for information retrieval: part I. Background and theory. *Journal of Documentation*, **38**(2), 61–71.

Borlund, P. (2000a). *Evaluation of Interactive Information Retrieval Systems*. Åbo: Åbo Akademi University Press. Doctoral Thesis, Åbo Akademi University.

Borlund, P. (2000b). Experimental Components for the Evaluation of Interactive Information Retrieval Systems. *Journal of Documentation*, **56**(1), 71–90.

Borlund, P. (2003a). The IIR Evaluation Model: a Framework for Evaluation of Interactive Information Retrieval Systems. *Information Research*, **8**(3). Available at: http://informationr.net/ir/8-3/paper152.html

Borlund, P. (2003b). The Concept of Relevance in IR. In: *Journal of the American Society for Information Science and Technology*, **54**(10), 913–925.

Borlund, P. and Ingwersen, P. (1998). Measures of Relative Relevance and Ranked Half-life: Performance Indicators for Interactive IR. In: Croft, B. W., Moffat, A., van Rijsbergen, C. J, Wilkinson, R. and Zobel, J. (Eds.) *Proceedings of the 21st ACM Sigir Conference on Research and Development of Information Retrieval*. Melbourne, 1998. Australia: ACM Press/York Press, 324–331.

Borlund, P. and Ruthven, I. (eds). (2008). Special Issue: Evaluation of Interactive Information Retrieval Systems. In: *Information Processing & Management*, **44**(1).

Cleverdon, C. W. (1960). *Aslib Cranfield Research Project: Report on the First Stage of an Investigation into the Comparative Efficiency of Indexing Systems*. Cranfield: the College of Aeronautics.

Cleverdon, C. W. (1962). *Aslib Cranfield Research Project: Report on the Testing and Analysis of an Investigation into the Comparative Efficiency of Indexing Systems*. Cranfield.

Cleverdon, C. W. and Keen, E. M. (1966). *Aslib Cranfield Research Project: Factors Determining the Performance of Indexing Systems*. Vol. 2: Results. Cranfield.

Cleverdon, C. W., Mills, J. and Keen, E. M. (1966). *Aslib Cranfield Research Project: Factors Determining the Performance of Indexing Systems*. Vol. 1: Design.

Gull, C. D. (1956). Seven Years of Work on the Organization of Materials in the Special Library. *American Documentation*, **7**, 320–329.

Harter, S. P. and Hert, C. A. (1997). Evaluation of Information Retrieval Systems: Approaches, Issues, and Methods. In: Williams, M. E., ed., *Annual Review of Information Science and Technology*, **32**, 3–94.

Ingwersen, P. (1992). *Information retrieval interaction*. London: Taylor Graham.

Ingwersen, P. (1996). Cognitive perspectives of information retrieval interaction: elements of a cognitive IR theory. *Journal of Documentation*, **52**(1), 3–50.

Ingwersen, P. and Järvelin, K. (2005). *The Turn: Integration of Information Seeking Retrieval in Context*. Dordrecht, Netherlands: Springer Verlag.

Järvelin, K. and Kekäläinen, J. (2000). IR Evaluation Methods for Retrieving Highly Relevant Documents. In: Belkin, N. J., Ingwersen, P. and Leong, M-K. (eds) *Proceedings of the 23rd ACM Sigir Conference on Research and Development of Information Retrieval*. Athens, Greece, 2000. New York, N.Y.: ACM Press, 2000, 41–48.

Lancaster, W. F. (1969). Medlars: Report on the Evaluation of its Operating Efficiency. *American Documentation*, **20**, 119–142.

Lei, W., Ruthven, I. and Borlund, P. (2006). The Effects on Topic Familiarity on Online Search Behaviour and Use of Relevance Criteria. In: Lalmas, M., MacFarlane, A., Rüger, S., Tombros, A., Tsikrika, T., and Yavlinsky, A. (eds). *Advances in Information Retrieval. 28th European Conference on IR Research, ECIR 2006*, London, UK. Berlin/Heidelberg: Springer Verlag. 2006. 456–459.

Lu, W., Robertson, S. E. and Macfarlane, A. (2006). Field-Weighted XML Retrieval Based on BM25. In: Fuhr, N., Lalmas, M., Malik, S. and Kazai, G. (eds): *Advances in XML Information Retrieval and Evaluation: Fourth Workshop of the INitiative for the Evaluation of XML Retrieval (INEX 2005)*, Dagstuhl, 28–30 November 2005, Lecture Notes in Computer Science, Vol 3977, Springer-Verlag,

Lu, W., Robertson, S. E. and Macfarlane, A. (2007). CISR at INEX 2006. In: Fuhr, N. Lalmas, M., and Trotman, A. (eds), *Comparative Evaluation of XML Information Retrieval Systems: 5th International Workshop of the Initiative for the Evaluation of XML Retrieva (INEX 2006)*, Dagstuhl, Germany, LNCS 4518, Springer-Verlag, (2007), 57–63.

Martyn, J. and Lancaster, F. W. (1981). *Investigative Methods in Library and Information Science: an Introduction*. Virginia: Information Resources Press. 1981. (2nd impression September 1991).

Reid, J. (2000). A Task-oriented Non-interactive Evaluation Methodology for Information Retrieval Systems. *Information Retrieval*, **2**(1), 113–127.

Robertson, S. E. (1981). The Methodology of Information Retrieval Experiment. In: Spärck Jones, K. (Ed.) *Information Retrieval Experiments*. London: Butterworths, 9–31.

Robertson, S. E. (1997a). (ed.) Special Issue on Okapi. *Journal of Documentation*, **53**(1).

Robertson, S. E. (1997b). Overview of the Okapi Projects. *Journal of Documentation*, **53**(1), 3–7.

Robertson, S. E. and Hancock-Beaulieu, M. M. (1992). On the Evaluation of IR Systems. *Information Processing and Management*, **28**(4), 457–466.

Robertson, S. E., Lu, W. and MacFarlane, A. (2006). XML-structured documents: retrievable units and inheritance. In: Legind Larsen, H., Pasi, G., Ortiz-Arroyo, D., Andreasen, T., Christiansen, H. (Eds.) *Proceedings of Flexible Query Answering Systems 7th International Conference*, FQAS 2006, Milan, Italy, June 7-10, 2006, LNCS, 4027, Springer-Verlag, 2006, 121–132.

Robertson, S. E., Walker, S. and Beaulieu, M. (1997). Laboratory Experiments with Okapi: Participation in the TREC Programme. *Journal of Documentation*, **53**(1), 20–34.

Salton, G. (1972). A New Comparison Between Conventional indexing (MEDLARS) and Automatic Text Processing (SMART). *Journal of the American Society for Information Science*, (March–April), 75–84.

Salton, G. (1981). The Smart Environment for Retrieval System Evaluation: Advantages and Problem Areas. In: Spärck Jones, K. (ed.) *Information Retrieval Experiments*. London: Butterworths, 316–329.

Schamber, L. (1994). Relevance and Information Behavior. In: Williams, M. E. (Ed.) *Annual Review of Information Science and Technology (ARIST)*. Medford, NJ: Learned Information, INC., **29**, 3–48.

Spärck Jones, K. (1971). *Automatic Keyword Classification for Information Retrieval*. London: Butterworths.

Spärck Jones, K. (1981a). (ed.) *Information Retrieval Experiments*. London: Butterworths.

Spärck Jones, K. (1981b). Retrieval System Tests 1958–1978. In: Spärck Jones, K. (ed.) *Information Retrieval Experiments*. London: Butterworths, 213–255.

Spink, A., Greisdorf, H. and Bateman, J. (1998). From Highly Relevant to Not Relevant: Examining Different Regions of Relevance. *Information Processing & Management*, **34**(5), 599–621.

Su, L. T. (1992). Evaluation Measure for Interactive Information Retrieval. *Information Processing & Management*, **28**(4), 503–516.

Tague-Sutcliffe, J. (1992). The pragmatics of information retrieval experimentation, revisited. *Information Processing and Management*, **28**(4), 467–490.

Thorne, R. G. (1955). The efficiency of subject catalogues and the cost of information searches. *Journal of Documentation*, **11**(3), 130–148.

Voorhees, E. M. and Harman, D. K. (Eds.) (2005a). *TREC:Experiment and Evaluation in Information Retrieval*. Cambridge, Massachusetts: The MIT Press.

Voorhees, E. M. and Harman, D. K. (2005b). The text Retrieval Conference. In: Voorhees, E. M. and Harman, D. K. (eds) (2005a). *TREC:Experiment and Evaluation in Information Retrieval*. Cambridge, Massachusetts: The MIT Press. 3–19.

Walker, S. and De Vere, R. (1990). *Improving subject retrieval in online catalogues: 2. Relevance feedback and query expansion)*. London: British Library. (British Library Research Paper 72).

Walker, S. (1989). The Okapi online catalogue research projects. In: *The Online Catalogue: Developments and Directions*. London: The Library Association, 84–106.

Wang, P. (2001). Methodologies and methods for user behavioral research. In: Williams, M. E., ed., *Annual Review of Information Science and Technology*, **34**, 1999, 53–9.

3

Multimedia Resource Discovery

Stefan Rüger

3.1 Introduction

Resource discovery is more than just search: it is browsing, searching, selecting, assessing and evaluating, i.e. ultimately accessing information. Giving users access to collections is one of the defining tasks of a library. For thousands of years the traditional methods of resource discovery have been facilitated by librarians: they create reference cards with metadata that are put into catalogues (nowadays, databases); they also place the objects in physical locations that follow certain classification schemes and they answer questions at the reference desk.

The advent of digital documents has radically changed the organisation principles; now it is possible to *automatically* index and search document collections as big as the world wide web *à la* Google and browse collections utilising author-inserted links. It is almost as if automated processing has turned the traditional library access paradigm upside down. Instead of searching metadata catalogues in order to retrieve the document, web search engines search the full content of documents and retrieve their metadata, i.e. the location where documents can be found. Not all manual intervention has been abandoned, though. For example, the Yahoo directory is an edited classification scheme of submitted web sites that are put into a browsable directory structure akin to library classification schemes.

Undoubtedly, however, it is the automated approaches that have made all the difference to the way the vast world wide web can be used. While the automated indexing of text documents has been successfully applied to collections as large as the world wide web for over a decade now, multimedia indexing by content involves different, still less mature and less scalable technologies. Figure 3.1 shows a matrix of different search engines depending on the query (columns) and document repository (rows). Entry a) in this matrix corresponds to a traditional text search engine with a completely different technology than entry b), a system that allows you to express musical queries by humming a tune and that then plays the corresponding song; the three c) entries in Figure 3.1 correspond to a multimodal video search engine, allowing search by motion with example images and text queries, e.g. *find me video shots of the Houses of Parliament in London zooming into its tower's clock face using the following example still image*; in contrast to this d) could be a search engine with a query text box that returns BBC Radio 4 discussions. Section 3.2 summarises different basic technologies involved in multimedia search.

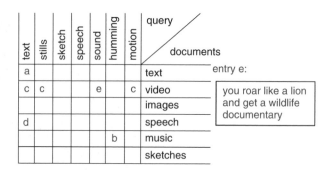

Figure 3.1   New search engine types

Multimedia collections pose their very own challenges; for example, images and videos don't often come with dedicated reference cards or metadata, and when they do, as in museum collections, their creation will have been expensive and time-consuming. Section 3.3 explores the difficulties and limitations of automatically indexing, labelling and annotating image and video content. It briefly discusses the inherent challenges of the semantic gap, polysemy, fusion and responsiveness.

Even if all these challenges were solved, indexing sheer mass is no guarantee of a successful annotation either. While most of today's interlibrary loan systems allow access to virtually any book publication in the world – at least around 98 million book entries in OCLC's Worldcat database (OCLC 2008) and 3 million entries from Bowker's Books In Print – students and researchers alike seem to be reluctant to actually make use of this facility. On the other hand, the much smaller catalogue offered by Amazon appears to be very popular, presumably owing to added services such as subject categories; fault-tolerant search tools; personalised services telling the customer what's new in a subject area or what other people with a similar profile bought; pictures of book covers; media and customer reviews; access to the table of contents, to selections of the text and to the full-text index of popular books; and the perception of fast delivery. In the multimedia context Section 3.4 argues that automated added services such as visual queries, relevance feedback and summaries can prove useful for resource discovery in multimedia digital libraries. Sections 3.4.1 is about summarising techniques for videos, Section 3.4.2 exemplifies visualisation of search results, while Section 3.4.3 discusses content-based visual search modes such as query-by-example and relevance feedback.

Finally, Section 3.5 promotes browsing as resource discovery mode and looks at underlying techniques to automatically structure the document collection to support browsing.

3.2   Basic Multimedia Search Technologies

The current best practice to index multimedia collections is via the generation of a library card, i.e. a dedicated database entry of metadata such as author, title, publication year and keywords. Depending on the concrete implementation, these can be found with SQL queries, text-search engines or XML query language, but all these search modes are based on text descriptions of some form and are agnostic to the structure of the actual objects they refer to, be it books, CDs, videos, newspaper articles, paintings, sculptures, web pages, consumer products, etc.

The left column of the matrix of Figure 3.1 is underpinned by text search technology and requires the textual representation of the multimedia objects, an approach that I like to call *piggy-back text retrieval*. Other approaches are based on an automatic classification of multimedia objects and on assigning words from a fixed vocabulary. This can be a certain camera motion that can be detected in a video (zoom, pan, tilt, roll, dolly in and out, truck left and right, pedestal up and down, crane boom, swing boom etc.); a genre for music pieces such as jazz or classics; a generic scene description in images such as inside/outside, people, vegetation, landscape, grass, city-view etc. or specific object detection such as faces and cars. These approaches are known as *feature classification* or *automated annotation*.

The type of search that is most commonly associated with multimedia is *content-based*: The basic idea is that still images, music extracts, video clips themselves can be used as queries and that the retrieval system is expected to return 'similar' database entries. This technology differs most radically from the thousands-year-old library card paradigm in that there is no necessity for metadata at all. In certain searches there is the desire to match not only the general type of scene or music that the the query represents, but instead one and only one exact multimedia object. For example, you take a picture of a painting in a gallery and submit this as a query to the gallery's catalogue in the hope of receiving the whole database record about this particular painting, and not a variant or otherwise similar exhibit. This is sometimes called *fingerprinting* or *known-item search*.

The rest of this section outlines these four basic multimedia search technologies.

3.2.1 Piggy-back text retrieval

Amongst all media types, TV video streams arguably have the biggest scope for automatically extracting text strings in a number of ways: directly from closed-captions, teletext or subtitles; automated speech recognition on the audio and optical character recognition for text embedded in the frames of a video. Full-text search of these strings is the way in which most video retrieval systems operate, including Google's TV search engine http://video.google.com or Blinkx-TV http://www.blinkx.tv. In contrast to television, for which legislation normally requires subtitles to assist the hearing impaired, videos stored on DVD don't usually have textual subtitles. They have *subpicture* channels for different languages instead, which are overlayed on the video stream. This requires the extra step of optical character recognition, which can be done with a relatively low error rate, owing to good quality fonts and clear background/foreground separation in the subpictures. In general, teletext has a much lower word error rate than automated speech recognition. In practice, it turns out that this does not matter too much as query words often occur repeatedly in the audio – the retrieval performance degrades gracefully with increased word error rates.

Web pages afford some context information that can be used for indexing multimedia objects. For example, words in the anchor text of a link to an image, a video clip or a music track, the file name of the object itself, metadata stored within the files and other context information such as captions. A subset of these sources for text snippets are normally used in web image search engines.

Some symbolic music representations allow the conversion of music into text, such as MIDI files which contain a music representation in terms of pitch, onset times and duration of notes. By representing differences of successive pitches as characters one can, for example, map monophonic music to one-dimensional strings. A large range of different text matching techniques can be deployed, for example the edit distance of database strings with a string representation of a query. The edit distance between two strings computes the smallest number of deletions, insertions or character replacements that is necessary to transform one string into the other. In the case of query-by-humming, where a pitch tracker can convert the hummed query into a MIDI-sequence (Birmingham *et al.* 2006), the edit distance is also able to deal gracefully with humming errors. Other techniques create fixed-length strings, so called *n*-grams, with windows that glide over the sequence of notes. The resulting strings can be indexed with a normal text search engine. This approach can also be extended to polyphonic music, where more than one note can be sounded at any one time (Doraisamy and Rüger 2003).

3.2.2 Automated annotation

Automatically annotating images with text strings is less straightforward. Methods attempting this task include dedicated machine vision models for particular words (such as 'people' or 'aeroplane'). However, the most popular and successful *generic* approaches are based on classification techniques. This normally requires a large training set of images that have annotations and from these one can extract features, or characteristics (see Section 3.2.3), and somehow correlate these with the existing annotations of the training set. For example, images with tigers will have orange-black stripes and often green patches from surrounding vegetation, and their existence in an unseen image can in turn bring about the annotation 'tiger'.

In particular, the following techniques have been studied: machine translation methods that link image regions (blobs) and words in the same way as corresponding words in two text documents written in different languages, but otherwise of same contents (Duygulu *et al.* 2002); co-occurrence models of low-level image features of tiled image regions and words (Mori *et al.* 1999); cross-lingual information retrieval models (Jeon *et al.* 2003; Lavrenko *et al.* 2003); inference networks that connect image segments with words (Metzler and Manmatha 2004); probabilistic modelling with latent Dirlichet allocation (Blei and Jordan 2003), Bernoulli distributions (Feng *et al.* 2004) or non-parametric density estimation with EMD kernels (Yavlinsky *et al.* 2005); support vector machine classification and relevance feedback (Izquierdo and Djordjevic 2005); information-theoretic semantic indexing (Magalhães and Rüger 2007); and simple scene-level statistics (Torralba and Oliva 2003). All these methods have in common that a controlled vocabulary of limited size (of the order of 500 more or less general terms) is used to annotate images based on a large training set.

Both those machine vision methods that model one particular item and the more generic machine learning methods that create individual models for every term of a limited vocabulary show a varying degree of success and, in general, relatively large error rates. As such these automated annotation methods are more suitable in the context of browsing or in conjunction with other search methods. If you want to 'find shots of the front of the White House in the daytime with the fountain running' (TRECVid 2003, topic 124), then a query-by-example search in a large database may be solved quicker and better by emphasising those shots that were classified as 'vegetation', 'outside', 'building', etc., even though the individual classification may be wrong in a significant proportion of cases.

Machine learning methods are not limited to images at all. For example one can extract motion vectors from MPEG-encoded videos and use these to classify a video shot independently into categories such as 'object motion from left to right', 'zoom in', etc. In contrast to the above classification tasks the extracted motion vector features are much more closely correlated to the ensuing motion label than image features are to text labels, and the corresponding learning task is much simpler.

Musical genre classification can be carried out on extracted audio-features that represent a performance by its statistics of pitch content, rhythmic structure and timbre texture (Tzanetakis and Cook 2002): timbre texture features are normally computed using short-time Fourier transform and Mel-frequency cepstral coefficients that also play a vital role in speech recognition; the rhythmic structure of music can be explored using discrete wavelet transforms that have a different time resolution for different frequencies; pitch detection, especially in polyphonic music, is more intricate and requires more elaborate algorithms. For details, see the work of Tolonen and Karjalainen (2000). Tzanetakis and Cook (2002) report correct classification rates of between 40% (rock) and 75% (jazz) in their experiments with 10 different genres.

3.2.3   Content-based retrieval

The query-by-example paradigm extracts features from the query, which can be anything from a still image, a video clip, humming, etc., and compares these with corresponding features in the database. The idea is to return to the user those items in the multimedia database that are most similar to the query, i.e. whose features are most similar to the features of the query. Figure 3.2 shows a typical architecture of such a system.

The most common way of indexing the visual content of images is by extracting low-level features, which represent colour usage, texture composition, shape and structure, localisation or motion. These representations are often real-valued vectors containing summary statistics, e.g. in the form of histograms. Their respective distances act as indicators whether or not two images are similar with respect to this particular feature. Design and usage of these features can be critical, and there is a wide variety of them, as published by participants in the TRECVID conference (TRECVid 2003).

Once created, those features will allow the comparison and ranking of images in the database with respect to images submitted as a query, the *query-by-example* paradigm. Summary statistics are very

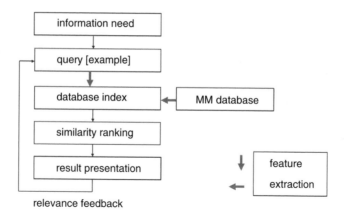

Figure 3.2 Content-based multimedia retrieval: generic architecture

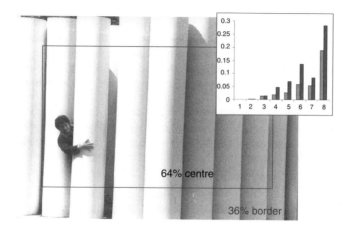

Figure 3.3 Centre–border intensity histogram of an image

crude indicators of similarity. For example an eight-bin histogram of intensity values of an image is a simple approximation of its brightness distribution, and many images do indeed share the same histogram. The image in Figure 3.3 shows a woman in the middle of bright column sculptures, but an image of a skier in snow is likely to have the same intensity histogram. The other disadvantage of global histograms computed over the whole image is that they lose any locality information. This can be alleviated by computing separate histograms for different areas of an image such as the centre, the border region, or in a number of tiles. Figure 3.3 is an example of a centre/border intensity histogram, where two histograms are computed for two different areas.

Normally, each multimedia document m gives rise to a number of low-level features $f_1(m), f_2(m), \ldots, f_k(m)$, each of which would typically be a vector or numbers representing aspects like colour distribution, texture usage, shape encodings, musical timbre, pitch envelopes etc. Most systems accumulate the distances of these features to the corresponding features of the query q in order to define an overall distance:

$$D_w(m, q) = \sum_{i=1}^{k} w_i d_i (f_i(m), f_i(q)) \qquad (3.1)$$

between multimedia documents m and a query q. Here $d_i(\cdot,\cdot)$ is a specific distance function between the vectors from the feature i, and $w_i \in \mathbb{R}$ is a weight for the importance of this feature. Note that the overall distance $D_w(m, q)$ is the number that is used to rank the multimedia documents in the database, and that the ones with the smallest distance to the query are shown to the user as query result, see Figure 3.2. Note also that the overall distance and hence the returned results crucially depend on the weight vector $w = (w_1, \ldots, w_k)$. In most interfaces the user can either set the weights explicitly (as in the interface shown in Figure 3.8) or the system can change the weights implicitly if the user has given feedback on how well the returned documents fit their needs.

The architecture presented here is a typical, albeit basic one; there are many variations and some radically different approaches that have been published in the past. A whole research field has gathered around the area of video and image retrieval, as exemplified by the annual International ACM Conferences on Video and Image Retrieval (CIVR), Multimedia (ACM MM) and Multimedia Information Retrieval (MIR) and the TREC video evaluation workshop TRECVid; there is another research field around music retrieval, see the annual International Conference on Music Information Retrieval (ISMIR).

3.2.4   Fingerprinting

Multimedia fingerprints are unique indices in a multimedia database. They are computed from the contents of the multimedia objects, are small, allow the fast, reliable and *unique* location of the database record and are robust against degradation or deliberate change of the multimedia document that do not alter their human perception. Audio fingerprints of music tracks are expected to distinguish even between different performances of the same song by the same artist at perhaps different concerts or studios.

Interesting applications include services that allow broadcast monitoring companies to identify what was played, so that royalties are fairly distributed or programmes and advertisements verified. Other applications uncover copyright violation or, for example, provide a service that allows you to locate the metadata such as title, artist and date of performance from snippets recorded on a (noisy) mobile phone.

Cano *et al.* (2002) review some audio fingerprinting methods and Seo *et al.* (2004) propose an image fingerprinting technique.

3.3   Challenges of Automated Visual Indexing

There are a number of open issues with the content-based retrieval approach in multimedia. On a perceptual level, those low-level features do not necessarily correlate with any high-level meaning the images might have. This problem is known as the *semantic gap*: imagine a scene in which Bobby Moore, the captain of the English National Football team in 1966, receives the World Cup trophy from Queen Elizabeth II; there is no obvious correlation between low-level colour, shape and texture descriptors and the high-level meaning of victory and triumph (or defeat and misery if you happened to support the West German team). Some of the computer vision methods go towards the bridging of the semantic gap, for example the ability to assign simple concrete labels to image parts such as 'grass', 'sky', 'people', 'plates'. A consequent use of an ontology could explain the presence of higher-level concepts such as 'barbecue' in terms of the simpler labels.

Even if the semantic gap could be bridged, there is still another challenge, namely *polysemy*: images usually convey a multitude of meanings so that the query-by-example approach is bound to under-specify the real information need. Users who submit an image such as the one in Figure 3.3 could have a dozen different information needs in mind: 'find other images with the same person', 'find images of the same art scene', 'find other bright art sculptures', 'find images with gradual shadow transitions', ... It is these different interpretations that make further user feedback so important.

User feedback can change the weights in Equation (3.1), which represent the plasticity of the retrieval system. Hence, putting the user in the loop and designing a human–computer interaction that utilises the user's feedback has been one of the main approaches to tackle these perceptual issues. Amongst other methods there are those that seek to reformulate the query (Ishikawa *et al.* 1998) or those that weight the various features differently, depending on the user's feedback. Weight adaptation methods include cluster analysis of the images (Wood *et al.* 1998); transposed files for feature selection (Squire *et al.* 2000); Bayesian network learning (Cox *et al.* 2000); statistical analysis of the feature distributions of relevant images and variance analysis (Rui *et al.* 1998); and analytic global optimisation (Heesch and Rüger 2003). Some approaches give the presentation and placement of images on screen much consideration to indicate similarity of images amongst themselves (Santini and Jain 2000; Rodden *et al.* 1999) or with respect to a visual query (Heesch and Rüger 2003).

On a practical level, the multitude of features assigned to images poses a *fusion problem*; how to combine possibly conflicting evidence of two images' similarity? There are many approaches to carry out fusion, some based on labelled training data and some based on user feedback for the current query (Aslam and Montague 2001; Bartell *et al.* 1994; Shaw and Fox 1994; Yavlinsky *et al.* 2004).

There is a *responsiveness problem*, too, in that the naïve comparison of query feature vectors to the database feature vectors requires a linear scan through the database. Although the scan is eminently scalable, the practicalities of doing this operation can mean an undesirable response time, of the order of seconds rather than the 100 milliseconds that can be achieved by text search engines. The problem is that high-dimensional tree structures tend to collapse to linear scans above a certain dimensionality (Weber *et al.* 1998). As a consequence, some approaches for fast nearest-neighbour search use compression techniques to speed up the disk access of linear scan as in Weber *et al.* (1998) using VA-files; or they approximate the search (Nene and Nayar 1997; Beis and Lowe 1997); decompose the features componentwise (de Vries *et al.* 2002; Aggarwal and Yu 2000) saving access to unnecessary components; or deploy a combination of these (Müller and Henrich 2004; Howarth and Rüger 2005).

3.4   Added Services

3.4.1   Video summaries

Even if the challenges of the previous section were all solved and if the automated methods of Section 3.2 enabled a retrieval process with high precision (proportion of the retrieved items that are relevant) and high recall (proportion of the relevant items that are retrieved) it would still be vital to present the retrieval results in a way so that the users can quickly decide to which degree those items are relevant to them.

Images are most naturally displayed as thumbnails, and their relevance can quickly be judged by users. Presenting and summarising videos is a bit more involved. The main metaphor used for this is that of a *storyboard* that contains *keyframes* with some text about the video. Several systems exist that summarise news stories in this way, most notably Informedia (Christel *et al.* 1999) and Físchlár (Smeaton *et al.* 2004). The Informedia system devotes much effort to added services such as face recognition and speaker voice identification, allowing retrieval of the appearance of known people. Informedia also provides alternative modes of presentation, e.g. through film skims or by assembling 'collages' of images, text and other information (e.g. maps) sourced via references from the text (Christel and Warmack 2001). Físchlár's added value lies in the ability to personalise the content (with the user expressing like or dislike of stories) and in assembling lists of related stories and recommendations.

Our very own TV news search engine ANSES (Pickering *et al.* 2003, Pickering 2004) records the main BBC evening news along with the subtitles, indexes them, breaks the video stream into shots (defined as those video sequences that are generated during a continuous operation of the camera), extracts one keyframe per shot, automatically glues shots together to form news stories based on an overlap in vocabulary in the subtitles of adjacent shots (using lexical chains), and assembles a storyboard for each story. Stories can be browsed or retrieved via text searches. Figure 3.4 shows the interface of ANSES. We use the natural language toolset GATE (Cunningham 2002) for automated

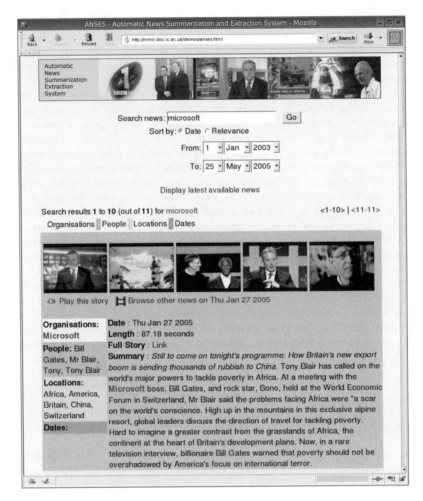

Figure 3.4 News search engine interface   (Reproduced courtesy of © Imperial College London. Screenshot adapted from British Broadcasting Corporation (BBC), www.bbc.co.uk)

discovery of organisations, people, places and dates; displaying these prominently as part of a story-board as in Figure 3.4 provides an instant indication of what the news story is about. ANSES also displays a short automated textual extraction summary, again using lexical chains to identify the most salient sentences. These summaries are never as informative as hand-made ones, but users of the system have found them crucial for judging whether or not they are interested in a particular returned search result.

Dissecting the video stream into shots and associating one keyframe along with text from subtitles to each shot has another advantage: a video collection can essentially be treated as an image collection, where each, possibly annotated image acts as entry point into the video.

3.4.2 New paradigms in information visualisation

The last 15 years have witnessed an explosion in interest in the field of information visualisation, (Hemmje *et al.* 1994; Ankerst *et al.* 1996; Card 1996; Shneiderman *et al.* 2000, Börner 2000). Here we present three new visualisation paradigms, based on our earlier design studies (Au *et al.* 2000; Carey *et al.* 2003). These techniques all revolve around a representation of documents in the form

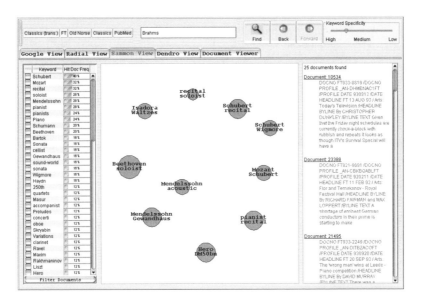

Figure 3.5   Sammon map for cluster-guided search

of bag-of-words vectors, which can be clustered to form groups. We use a variant of the buckshot clustering algorithm for this. Basically, the top, say, 100 documents that were returned from a query are clustered via hierarchical clustering to initialise document centroids for k-means clustering that puts all documents returned by a query into groups. Another common element of our visualisations is the notion of *keywords* that are specific to the returned set of documents. The keywords are computed using a simple statistic; for details see (Carey *et al.* 2003). The new methods are:

Sammon cluster view. This paradigm uses a Sammon map to generate a two-dimensional screen location from a many-dimensional vector representing a cluster centroid. This map is computed using an iterative gradient search (Sammon 1969) while attempting to preserve the pairwise distances between the cluster centres. Clusters are thus arranged so that their mutual distances are indicative of their relationship. The idea is to create a visual landscape for navigation. Figure 3.5 shows an example of such an interface. The display has three panels, a scrolling table panel to the left, a graphic panel in the middle and a scrolling text panel to the right that contains the traditional list of returned documents as hotlinks and snippets. In the graphic panel each cluster is represented by a circle and is labelled with its two most frequent keywords. The radius of the circle represents the cluster size. The distance between any two circles in the graphic panel is an indication of the similarity of their respective clusters – the nearer the clusters, the more likely the documents contained within will be similar. When the mouse passes over the cluster circle a tool-tip box in the form of a pop-up menu appears that allows the user to select clusters and *drill down*, i.e. re-cluster and re-display only the documents in the selected clusters. The back button undoes this process and climbs up the hierarchy (*drill up*). The table of keywords includes box fields that can be selected. At the bottom of the table is a filter button that makes the scrolling text window display only the hotlinks and snippets from documents that contain the selected keywords.

Dendro map visualisation. The Dendro map visualisation represents documents as leaf nodes of a binary tree that is output by the buckshot clustering algorithm. With its plane-spanning property and progressive shortening of branches towards the periphery, the Dendro map mimics the result of a non-Euclidean transformation of the plane as used in hyperbolic maps without suffering from their computational load. Owing to spatial constraints, the visualisation depth is confined to five levels of the hierarchy with nodes of the lowest level representing either documents or subclusters. Different colours

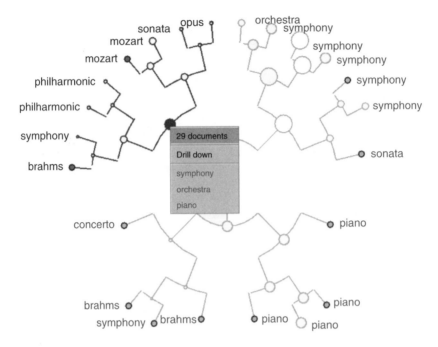

Figure 3.6   Dendro map – A plane-spanning binary tree (query 'Beethoven')

facilitate visual discrimination between individual documents and clusters. Each lowest-level node is labelled with the most frequent keyword of the subcluster or document. This forms a key component of the Dendro Map as it gives the user the cues needed for navigating through the tree. As the user moves the mouse pointer over an internal node, the internal nodes and branches of the associated subcluster change colour from light blue to dark blue while the leaf nodes, i.e. document representations, turn bright red. As in the Sammon map, a tool-tip window provides additional information about the cluster and can be used to display a table with a list of keywords associated with the cluster. The user may drill down on any internal node. The selected node will as a result replace the current root node at the center and the entire display is reorganised around the new root. The multi-level approach of the Dendro map allows the user to gain a quick overview over the document collection and to identify promising subsets.

Radial interactive visualisation. Radial (Figure 3.7) is similar to VIBE (Korfhage 1991), to Radviz (Hoffman *et al.* 1999) and to Lyberworld (Hemmje *et al.* 1994). It places the keyword nodes round a circle, and the position of the document dots in the middle depend on the force of invisible springs connecting them to keyword nodes: the more relevant a keyword for a particular document, the stronger its spring pulls on the document. Hence, we make direct use of the bag-of-words representation without explicit clustering. Initially, the twelve highest ranking keywords are displayed in a circle. The interface lets the user move the keywords, and the corresponding documents follow this movement. This allows the user to manually cluster the documents based on the keywords they are interested in. As the mouse passes over the documents, a bubble displays a descriptive piece of text. The location of document dots is not unique, owing to dimensionality reduction, and there may be many reasons for a document to have a particular position. To mitigate this ambiguity in Radial the user can click on a document dot, and the keywords that affect the location of document are highlighted. A choice of keywords used in the display can be exercised by clicking on two visible lists of words. Zoom buttons allow the degree of projection to be increased or reduced so as to distinguish between documents around the edges of the display or at the centre. The Radial visualisation appears to be a good interactive tool to structure the document set according to one's own preferences by shifting keywords around in the display.

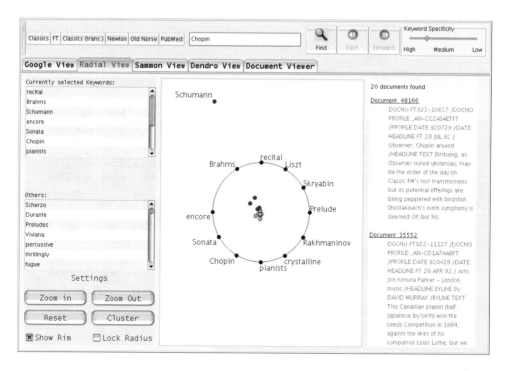

Figure 3.7 Radial visualisation

Unified approach. The integration of the paradigms into one application offers the possibility of browsing the same result set in several different ways simultaneously. The cluster-based visualisations give a broader overall picture of the result, while the Radial visualisation allows the user to focus on subsets of keywords. Also, as the clusters are approximations that highlight particular keywords, it may be useful to return to the Radial visualisation and examine the effect of these keywords upon the whole document set. The Radial visualisation will perhaps be more fruitful if the initial keywords match the user's area of interest. The Sammon map will let the user dissect search sets and re-cluster subsets, gradually homing in on target sets. This interface was developed within the joint NSF-EC project CHLT (http://www.chlt.org); it was evaluated from a human – computer interaction point of view with encouraging results (Chawda *et al.* 2005) and has proven useful in real-world multi-lingual scholarly collections (Rydberg-Cox *et al.* 2004).

3.4.3 *Visual search and relevance feedback*

The visual query-by-example paradigm discussed in Section 3.3 gives rise to relatively straightforward interfaces; an image is dragged into a query box, or, e.g. specified via a URL, and the best matching images are displayed in a ranked list to be inspected by the user, see Figure 3.8(a). A natural extension of such an interface is to offer the selection of relevant results as new query elements. This type of relevance feedback, a.k.a. *query point moving*, is shown in Figure 3.8(b).

One other main type of relevance feedback, *weight space movement*, assumes that the relative weight of the multitude of features that one can assign to images (e.g. structured metadata fields such as author, creation date and location; low-level visual features such as colour, shape, structure and texture; free-form text) can be learned from user feedback. Of the methods mentioned in Section 3.3 our group chose analytic weight updating as this has a very small execution time. The idea is that users can specify the degree to which a returned image is relevant to their information needs. This is done by having a visual representation; the returned images are listed in a spiral, and the distance of an image

(a)

(b)

Figure 3.8 Visual search for images of dark doors starting with a bright-door example. (a) Query by example (left panel) with initial results in the right panel; (Reproduced courtesy of © Imperial College London. Images reproduced from Corel Gallery 380,000 CD, © Corel Corporation. All Rights Reserved.) (b) a new query made of three images from (a) results in many more dark-door images (Reproduced courtesy of © Imperial College London. Images in screenshot reproduced from Corel Gallery 380,000 CD, © Corel Corporation. All Rights Reserved)

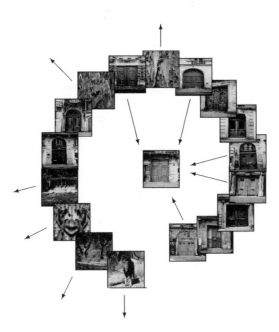

Figure 3.9 A relevance feedback model   (Reproduced courtesy of © Imperial College London. Images in screenshot reproduced from Corel Gallery 380,000 CD, © Corel Corporation. All Rights Reserved)

to the centre of the screen is a measure of the relevance that the search engine assigns to a specific image. Users can now move the images around with the mouse or place them in the centre with a left mouse click and far away with a right click. Figure 3.9 shows this relevance feedback model. We evaluated the effectiveness of negative feedback, positive feedback and query point moving, and found that combining the latter two yields the biggest improvement in terms of mean average precision (Heesch and Rüger 2003).

A new and relatively unexplored area of relevance feedback is the exploitation of social context information. By looking not only at the behaviour and attributes of the user, but also his past interactions and also the interactions of people he has some form of social connection with could yield useful information when determining whether search results are relevant or not. Browsing systems could recommend data items based on the actions of a social network instead of just a single user, using more data to yield better results.

The use of such social information is also becoming important for multimedia metadata generation, particular in the area of folksonomies where the feedback of users actively produces the terms and taxonomies used to describe the media in the system instead of using a predetermined, prescribed dictionary (Voss 2007). This can be seen being effectively used in online multimedia systems such as Flickr (http://www.flickr.com) and del.icio.us (http://del.icio.us).

3.5 Browsing: Lateral and Geotemporal

The idea of representing text documents in a nearest-neighbour network was first presented by Croft and Parenty (1985), albeit, as an internal representation of the relationships between documents and terms, not for browsing. Document networks for interactive browsing were identified by Cox (1992; 1995). Attempts to introduce the idea of browsing into content-based image retrieval include Campbell's work (2000); his ostensive model retains the basic mode of query based retrieval, but in addition allows browsing through a dynamically created local tree structure. Santini and Jain's *El niño* system (2000) is another attempt to combine query-based search with browsing. The system tries to display configurations of images in feature space such that the mutual distances between images are preserved as well as possible. Feedback is given in the same spirit as in Figure 3.9 by manually forming clusters of images that appear similar to the user. This in turn results in an altered configuration with potentially new images being displayed.

Other network structures that have increasingly been used for information visualisation and browsing are Pathfinder networks (Dearholt and Schvaneveldt 1990). They are constructed by removing redundant edges from a potentially much more complex network. Fowler *et al.* (1992) use Pathfinder networks to structure the relationships between terms from document abstracts, between document terms and between entire documents. The user interface supports access to the browsing structure through prominently marked high-connectivity nodes.

Our group (Heesch and Rüger 2004) determines the nearest neighbour for the image under consideration (which we call the *focal* image) for *every* combination of features. This results in a set of what we call *lateral neighbours*. By calculating the lateral neighbours of all database images, we generate a network that lends itself to browsing. Lateral neighbours share some properties of the focal image, but not necessarily all. For example, a lateral neighbour may share text annotations with the focal image, but no visual similarity with it at all, or it may have a very similar colour distribution, but no structural similarity, or it may be similar in all features except shape, etc. As a consequence, lateral neighbours are deemed to expose the polysemy of the focal image. Hence, when they are presented, the user may then follow one of them by making it the focal image and explore its lateral neighbours in turn. The user interaction is immediate, since the underlying network was computed offline.

We provide the user with entry points into the database by computing a representative set of images from the collection. We cluster high-connectivity nodes and their neighbours up to a certain depth using the Markov chain clustering algorithm (van Dongen 2000), which has robust convergence properties and allows one to specify the granularity of the clustering. The clustering result can be seen as a image database summary that shows highly connected nodes with far-reaching connections. The right panel of Figure 3.10(a) is such a summary for our TRECVid (2003) database. The user may select any of these images as an entry point into the network. Clicking on an image moves it into the centre around which the lateral neighbours are displayed, see the nearest-neighbour panel on the left side of figure 3.10(a). If the size of the lateral neighbour set is above a certain threshold the actual number of images displayed is reduced to the most salient ones.

If a user wanted to find 'video shots from behind the pitcher in a baseball game as he throws a ball that the batter swings at' (TRECVid 2003, topic 102) then they might explore the database in Figure 3.10 by clicking on the falcon image. The hope is that the colour of a baseball field is not far off from the green colour of that image. The resulting lateral neighbours, displayed in the left panel of Figure 3.10(a), do not contain the desired scene. However, there is an image of a sports field. Making that the focal image, as seen in the left part of Figure 3.10(b), reveals it has the desired scene as a lateral neighbour. Clicking that will unearth a lot more images from baseball fields, see the right-hand side of Figure 3.10(b). The network structure, a bit of lateral thinking and three mouse clicks have brought the desired result.

In the same way, and again with only three clicks, one could have started from the football image in the database overview to find 'video shots from behind the pitcher in a baseball game as he throws a ball that the batter swings at' (Figure 3.11). Heesch (2005) has shown that this is no coincidence; lateral neighbour networks computed in this way have the so-called *small world property* (Watts and Strogatz 1998) with only 3–4 degrees of separation even for the large TRECVid (2003) database that contains keyframes from 32 000 video shots. Lateral browsing has proven eminently successful for similar queries (Heesch *et al.* 2003).

Geotemporal browsing takes the idea of timelines and automatically generated maps, e.g. as offered in the Perseus Digital Library (Crane 2005), a step further. It integrates the idea of browsing in time and space with a selection of events through a text search box. In this way, a large newspaper or TV news collection can be made available through browsing, based on what happened where and when, as opposed to by keyword only.

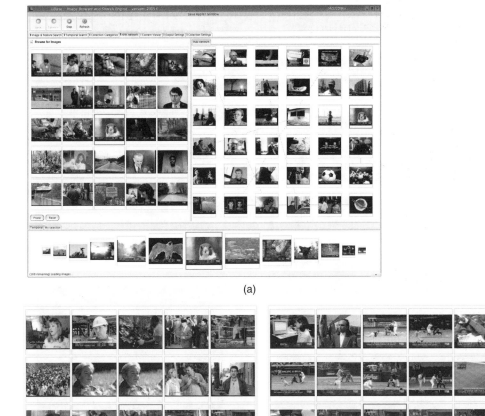

(a)

(b)

Figure 3.10 Lateral browsing for an image 'from behind the pitcher in a baseball game...' (a) Initial visual summary of the database (right panel) from which the user chooses the falcon, its nearest lateral neighbours are then displayed in the left panel; (Reproduced courtesy of © Imperial College London. Images in screenshot adapted from TREC Video Retrieval Evaluation 2003 (TRECVID), http://www-nlpir.nist.gov/projects) (b) clicking on any image will make it the centre of the nearest neighbours panel and display is associated lateral neighbours around it (Reproduced courtesy of © Imperial College London. Images in screenshot adapted from TREC Video Retrieval Evaluation 2003 (TRECVID), http://www-nlpir.nist.gov/projects)

Figure 3.11 Alternative ways to browse for images 'from behind the pitcher ...' Starting with the football image (upper left) from the database overview, one of its lateral neighbours is an image of a lawn with a sprinkler; when this is made the focal image (upper right) there are already images from baseball scenes. Clicking on one of them (lower left) reveals that there are more of this kind; they can be enlarged and the corresponding video played in the 'viewer tab' (lower right) (Reproduced courtesy of © Imperial College London. Images in screenshot adapted from TREC Video Retrieval Evaluation 2003 (TRECVID), http://www-nlpir.nist.gov/projects)

The interface in Figure 3.12 is a design study in our group that allows navigation within a large news event dataset along three dimensions: time, location and text subsets. The search term presents a text filter. The temporal distribution can be seen in lower part. The overview window establishes a frame of reference for the user's region of interest. In principle, this interface could implement new zooming techniques, e.g. speed-dependent automatic zooming (Cockburn and Savage 2003), and link to a server holding a large quantity of maps such as National Geographic's MapMachine (http://plasma.nationalgeographic.com/mapmachine/ as of May 2005) with street-level maps and aerial photos.

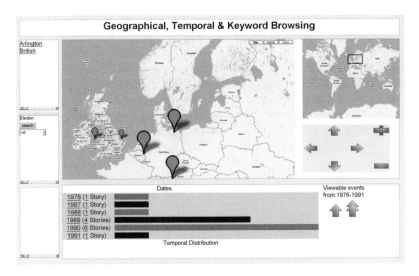

Figure 3.12   Geotemporal browsing in action

3.6   Summary

This chapter has introduced basic concepts of multimedia resource discovery technologies for a number of different query and document types; these were the piggy-back text search, automated annotation, content-based retrieval and fingerprinting. The paradigms we have discussed include summarising complex multimedia objects such as TV news, information visualisation techniques for document clusters, visual search by example, relevance feedback and methods to create browsable structures within the collection. These exploration modes share three common features: they are automatically generated, depend on visual senses and interact with the user of the multimedia collections.

Multimedia resource discovery has its very own challenges in the semantic gap, in polysemy inherently present in under-specified query-by-example scenarios, in the question how to combine possibly conflicting evidence and the responsiveness of the multimedia searches. In the last part of the chapter we have given some examples of user-centred methods that support resource discovery in multimedia digital libraries. Each of these methods can be seen as an alternative mode to the traditional digital library management tools of metadata and classification. The new visual modes aim at generating a multi-faceted approach to present digital content: *video summaries* as succinct versions of media that otherwise would require a high bandwidth to display and considerable time by the user to assess; *information visualisation* techniques help the user to understand a large set of documents that match a query; *visual search* and *relevance feedback* afford the user novel ways to express their information need without taking recourse to verbal descriptions that are bound to be language-specific; alternative resource discovery modes such as *lateral browsing* and *geotemporal browsing* will allow users to explore collections using lateral associations and geographic or temporal filters rather than following strict classification schemes that seem more suitable for trained librarians than the occasional user of multimedia collections. The cost for these novel approaches will be low, as they are automated rather than human-generated. It remains to be seen how best to integrate these services into traditional digital library designs and how much added value these services will bring about (Bainbridge *et al.* 2005).

The following four books provide further reading for content-based multimedia retrieval; they complement the material of this chapter:

A del Bimbo (1999). *Visual Information Retrieval*. Morgan Kaufmann.

M Lew (ed.) (2001). *Principles of Visual Information Retrieval*. Springer.

Y-J Zhang (ed.) (2006). *Semantic-based Visual Information Retrieval*. Idea Group Inc.

S Rüger (to appear 2009): *Multimedia Information Retrieval*. Morgan & Claypool.

Acknowledgements.   The paradigms outlined in this chapter and their implementations would not have been possible without the ingenuity, imagination and hard work of all the people I am fortunate to work with or to have worked with: Paul Browne, Matthew Carey, Daniel Heesch, Peter Howarth, Partha Lal, João Magalhães, Alexander May, Simon Overell, Marcus Pickering, Adam Rae, Jonas Wolf, Lawrence Wong and Alexei Yavlinsky.

Exercises

3.1 Search types

Look at the matrix of search engine types in Figure 3.1. Give a search and retrieval scenario for each element of this query retrieval matrix. What would be the most appropriate search technology (piggy-back text search, feature classification, content-based, fingerprint) for each scenario?

3.2 Colour histograms

Colour is perceived as a point (r, g, b) in a *three-dimensional* space. Each colour component (red, green and blue) is usually encoded as an integer between 0 and 255 (1 byte); there are two principal methods to create colour histograms: you can compute three 1-D histograms of each of the r, g and b components independently or you can compute one 3-D colour histogram.

(a) If you divide each of the red, green and blue axes into n_r, n_g, n_b equal intervals, respectively, then the 3-D colour cube $[0, 255]^3$ is subdivided into $n_r n_g n_b$ cuboids or bins. Show that by mapping (r, g, b) to

$$\left\lfloor \frac{n_r r}{256} \right\rfloor n_g n_b + \left\lfloor \frac{n_g g}{256} \right\rfloor n_b + \left\lfloor \frac{n_b b}{256} \right\rfloor$$

you get an enumeration scheme $0, \ldots, n_r n_g n_b - 1$ for the cuboids. A 3-D colour histogram of an image is a *list of the numbers of the pixels in an image that fall into each of the different cuboids* in colour space. In other words, you look at each pixel in an image, compute the index from its colour (r, g, b) and increment the corresponding variable, which records the number of pixels that fall into this colour cuboid.

(b) Compute both types of colour histograms for the image below (the colours of the stripes are given by the table next to the image). Use $64 = 4^3$ bin cubes for the 3-D-bin-histogram and 22 bins for each of the three colour-component histograms (yielding 66 bins altogether), so that both histogram types have a similar number of bins.

R	G	B	
0	0	0	black
255	0	0	red
0	255	0	green
0	0	255	blue
0	255	255	cyan
255	0	255	magenta
255	255	0	yellow
255	255	255	white

(c) Which of the two colour histogram methods has retained more information about the colour distribution in the original picture? How came it about that one of the two methods lost vital information despite having roughly the same number of bins (64 vs 66)?

(d) Why are usually *normalised* histograms computed and stored for content-based retrieval? Normalised histograms store the proportions of pixels in the respective bins not the absolute number of pixels.

3.3 Image search

Sketch the block diagram of a colour-and-texture-based image search engine for curtain fabrics. Explain the general workings of a content-based search engine and contrast it with the workings of a text search engine in terms of retrieval and indexing technology.

References

Aggarwal, C. and Yu, P. (2000). The IGrid index: reversing the dimensionality curse for similarity indexing in high dimensional space. In *ACM International Conference on Knowledge Discovery and Data Mining*, pp. 119–129.

Ankerst, M., Keim, D. and Kriegel, H. (1996). Circle segments: A technique for visually exploring large multidimensional data sets. In *IEEE Visualization*.

Aslam, J. and Montague, M. (2001). Models for metasearch. In *ACM International Conference on Research and Development in Information Retrieval*, pp. 276–284.

Au, P., Carey, M., Sewraz, S., Guo, Y. and Rüger, S. (2000). New paradigms in information visualisation. In *ACM International Conference on Research and Development in Information Retrieval*, pp. 307–309.

Bainbridge, D., Browne, P., Cairns, P., Rüger, S. and Xu, L-Q. (2005). Managing the growth of multimedia digital content. *ERCIM News: special theme on Multimedia Informatics 62*.

Bartell, B., Cottrell, G. and Belew, R. (1994). Automatic combination of multiple ranked retrieval systems. In *ACM International Conference on Research and Development in Information Retrieval*, pp. 173–181.

Beis, J. and Lowe, D. (1997). Shape indexing using approximate nearest-neighbour search in high-dimensional spaces. In *International Conference on Computer Vision and Pattern Recognition*, p 1000.

Birmingham, W., Dannenberg, R. and Pardo, B. (2006). Query by humming with the vocalsearch system. *Commun. ACM* **49**(8), 49–52.

Blei, D. and Jordan, M. (2003). Modeling annotated data. In *ACM International Conference on Research and Development in Information Retrieval*, pp. 127–134.

Börner, K. (2000). Visible threads: A smart VR interface to digital libraries. In *International Symposium on Electronic Imaging 2000: Visual Data Exploration and Analysis*, pp. 228–237.

Campbell, I. (2000). *The ostensive model of developing information-needs*. PhD Thesis, University of Glasgow.

Cano, P., Batlle, E., Kalker, T. and Haitsma, J. (2002). A review of algorithms for audio fingerprinting. In *International Workshop on Multimedia Signal Processing*, pp. 169–173.

Card, S. (1996). Visualizing retrieved information: A survey. *IEEE Computer Graphics and Applications* **16**(2), 63–67.

Carey, M., Heesch, D. and Rüger, S. (2003). Info navigator: a visualization interface for document searching and browsing. In *International Conference on Distributed Multimedia Systems*, pp. 23–28.

Chawda, B., Craft, B., Cairns, P., Rüger, S. and Heesch, D. (2005). Do 'attractive things work better'? An exploration of search tool visualisations. In *BCS Human–Computer Interaction Conference*, vol. 2, pp. 46–51.

Christel, M. and Warmack, A. (2001). The effect of text in storyboards for video navigation. In *IEEE International Conference on Acoustics, Speech, and Signal Processing*, pp. 1409–1412.

Christel, M., Warmack, A., Hauptmann, A. and Crosby, S. (1999). Adjustable filmstrips and skims as abstractions for a digital video library. In *IEEE Forum on Research and Technology Advances in Digital Libraries*, p 98.

Cockburn, A. and Savage, J. (2003). Comparing speed-dependent automatic zooming with traditional scroll, pan and zoom methods. In *BCS Human-Computer Interaction Conference*, pp. 87–102.

Cox, I., Miller, M., Minka, T., Papathomas, T. and Yianilos, P. (2000). The Bayesian image retrieval system, PicHunter. *IEEE Trans on Image Processing* **9**(1), 20–38.

Cox, K. (1992). Information retrieval by browsing. In *International Conference on New Information Technology*, pp. 69–80.

Cox, K. (1995). *Searching through browsing*. PhD Thesis, University of Canberra.

Crane, G. (ed.) (2005). *Perseus Digital Library Project*. Tufts University, 30 May 2005, http://www.perseus.tufts.edu.

Croft, B. and Parenty, T. (1985). Comparison of a network structure and a database system used for document retrieval. *Information Systems* **10**, 377–390.

Cunningham, H. (2002). GATE, a general architecture for text engineering. *Computers and the Humanities* **36**, 223–254.

Dearholt, D. and Schvaneveldt, R. (1990). Properties of Pathfinder networks. In Schvaneveldt, R (ed.), *Pathfinder associative networks: Studies in knowledge organization*. Norwood, pp. 1–30.

van Dongen, S. (2000). A cluster algorithm for graphs. Technical Report INS-R0010, National Research Institute for Mathematics and Computer Science in the Netherlands.

Doraisamy, S. and Rüger, S. (2003). Robust polyphonic music retrieval with n-grams. *Journal of Intelligent Information Systems* **21**(1), 53–70.

Duygulu, P., Barnard, K., de Freitas, N. and Forsyth, D. (2002). Object recognition as machine translation: Learning a lexicon for a fixed image vocabulary. In *European Conference on Computer Vision*, pp. 97–112.

Feng, S., Manmatha, R. and Lavrenko, V. (2004). Multiple Bernoulli relevance models for image and video annotation. In *International Conference on Computer Vision and Pattern Recognition*, pp. 1002–1009.

Fowler, R., Wilson, B. and Fowler, W. (1992). Information navigator: An information system using associative networks for display and retrieval. Technical Report NAG9-551, 92-1, Dept of Computer Science, University of Texas.

Heesch, D. (2005). *The NN^k technique for image searching and browsing*. PhD Thesis, Imperial College London.

Heesch, D., Pickering, M., Rüger, S. and Yavlinsky, A. (2003). Video retrieval using search and browsing with key frames. In *TREC Video Retrieval Evaluation*.

Heesch, D. and Rüger, S. (2003). Performance boosting with three mouse clicks – relevance feedback for CBIR. In *European Conference on Information Retrieval*, pp. 363–376.

Heesch, D. and Rüger, S. (2004). NN^k networks for content based image retrieval. In *European Conference on Information Retrieval*, pp. 253–266.

Hemmje, M., Kunkel, C. and Willet, A. (1994). Lyberworld – a visualization user interface supporting fulltext retrieval. In *ACM International Conference on Research and Development in Information Retrieval*.

Hoffman, P., Grinstein, G. and Pinkney, D. (1999). Dimensional anchors: a graphic primitive for multi-dimensional multivariate information visualizations. In *New Paradigms in Information Visualisation and Manipulation in Conjunction with ACM CIKM*, pp. 9–16.

Howarth, P. and Rüger, S. (2005). Trading precision for speed: localised similarity functions. In *International Conference on Image and Video Retrieval*, pp. 415–424.

Ishikawa, Y., Subramanya, R. and Faloutsos, C. (1998). MindReader: Querying databases through multiple examples. In *International Conference on Very Large Databases*, pp. 218–227.

Izquierdo, E. and Djordjevic, D. (2005). Using relevance feedback to bridge the semantic gap. In *International Workshop on Adaptive Multimedia Retrieval*, pp. 19–34.

Jeon, J., Lavrenko, V. and Manmatha, R. (2003). Automatic image annotation and retrieval using cross-media relevance models. In *ACM International Conference on Research and Development in Information Retrieval*, pp. 119–126.

Korfhage, R. (1991). To see or not to see – is that the query? In *ACM International Conference on Research and Development in Information Retrieval*, pp. 134–141.

Lavrenko, V., Manmatha, R. and Jeon, J. (2003). A model for learning the semantics of pictures. In *Neural Information Processing Systems*, pp. 553–560.

Magalhães, J. and Rüger, S. (2007). Information-theoretic semantic multimedia indexing. In *International ACM Conference on Image and Video Retrieval*, pp. 619–626.

Metzler, D. and Manmatha, R. (2004). An inference network approach to image retrieval. In *International Conference on Image and Video Retrieval*, pp. 42–50.

Mori, Y., Takahashi, H. and Oka, R. (1999). Image-to-word transformation based on dividing and vector quantizing images with words. In *International Workshop on Multimedia Intelligent Storage and Retrieval Management*.

Müller, W. and Henrich, A. (2004). Faster exact histogram intersection on large data collections using inverted VA-files. In *International Conference on Image and Video Retrieval*, pp. 455–463.

Nene, S. and Nayar, S. (1997). A simple algorithm for nearest neighbor search in high dimensions. *IEEE Trans Pattern Anal Mach Intell* **19**(9), 989–1003.

OCLC (2008). Online computer library center, worldcat: About: Worldcat facts and statistics: World-cat statistics by format as of March 2008. http://www.oclc.org/worldcat/statistics (accessed March 2008).

Pickering, M. (2004). *Video Retrieval and Summarisation*. PhD Thesis, Imperial College London.

Pickering, M., Wong, L. and Rüger, S. (2003). ANSES: Summarisation of news video. In *International Conference on Information and Knowledge Management*, pp. 481–486.

Rodden, K., Basalaj, W., Sinclair, D. and Wood, K. (1999). Evaluating a visualization of image similarity. In *ACM International Conference on Research and Development in Information Retrieval*, pp. 36–43.

Rui, Y., Huang, T. and Mehrotra, S. (1998). Relevance feedback techniques in interactive content-based image retrieval. In *Storage and Retrieval for Image and Video Databases*, pp. 25–36.

Rydberg-Cox, J., Vetter, L., Rüger, S. and Heesch, D. (2004). Approaching the problem of multi-lingual information retrieval and visualization in Greek and Latin and Old Norse texts. In *European Conference on Digital Libraries*, pp. 168–178.

Sammon, J. (1969). A nonlinear mapping for data structure analysis. *IEEE Transactions on Computers* **C-18**(5), 401–409.

Santini, S. and Jain, R. (2000). Integrated browsing and querying for image databases. *IEEE Multimedia* **7**(3), 26–39.

Seo, J. S., Haitsma, J., Kalker, T. and Yoo, C. D. (2004). A robust image fingerprinting system using the radon transform. *Signal Processing: Image Communication* **19**, 325–339.

Shaw, J. and Fox, E. (1994). Combination of multiple searches. In *Text Retrieval Conference*, pp. 243–252.

Shneiderman, B., Feldman, D., Rose, A. and Ferré Grau, X. (2000). Visualizing digital library search results with categorical and hierarchical axes. In *ACM Digital Libraries*, pp. 57–66.

Smeaton, A., Gurrin, C., Lee, H., McDonald, K., Murphy, N., O'Connor, N., O'Sullivan, D., Smyth, B. and Wilson, D. (2004). The Físchlár-news-stories system: Personalised access to an archive of TV news. In *RIAO Conference on Coupling Approaches, Coupling Media and Coupling Languages for Information Retrieval*, pp. 3–17.

Squire, D., Müller, W., Müller, H. and Pun, T. (2000). Content-based query of image databases: inspirations from text retrieval. *Pattern Recognition Letters* **21**(13–14), 1193–1198.

Tolonen, T. and Karjalainen, M. (2000). A computationally efficient multi-pitch analysis model. *IEEE Transactions on Speech and Audio Processing* **8**, 708–716.

Torralba, A. and Oliva, A. (2003). Statistics of natural image categories. *Network: Computation in Neural Systems* **14**, 391–412.

TRECVid (2003). Trec video retrieval evaluation. http://www-nlpir.nist.gov/projects/tv2003/ last accessed February 2006.

Tzanetakis, G. and Cook, P. (2002). Musical genre classification of audio signals. *IEEE Transactions on Speech and Audio Processing* **10**(5), 293–302.

Voss, J. (2007). Tagging, folksonomy and co – renaissance of manual indexing?

de Vries, A., Mamoulis, N., Nes, N. and Kersten, M. (2002). Efficient *k-nn* search on vertically decomposed data. In *ACM International Conference on Management of Data*, pp. 322–333.

Watts, D. and Strogatz, S. (1998). Collective dynamics of 'small-world' networks. *Nature* **393**, 440–442.

Weber, R., Stock, H-J. and Blott, S. (1998). A quantitative analysis and performance study for similarity search methods in high-dimensional space. In *International Conference on Very Large Databases*, pp. 194–205.

Wood, M., Campbell, N. and Thomas, B. (1998). Iterative refinement by relevance feedback in content-based digital image retrieval. In *ACM Multimedia*, pp. 13–20.

Yavlinsky, A., Pickering, M., Heesch, D. and Rüger, S. (2004). A comparative study of evidence combination strategies. In *IEEE International Conference on Acoustics, Speech, and Signal Processing*, pp. 1040–1043.

Yavlinsky, A., Schofield, E. and Rüger, S. (2005). Automated image annotation using global features and robust nonparametric density estimation. In *International Conference on Image and Video Retrieval*, pp. 507–517.

4

Image Users' Needs
and Searching Behaviour

Stina Westman

4.1 Introduction

Images are sought after in many contexts, on the Web and in closed collections, both at work and during leisure time. The amount of visual information available has grown dramatically in the past few years, brought about by the proliferation of digital cameras, both in professional and consumer use, as well as the sharing of images and video online by their producers. The nature of visual information creates some challenges for retrieval in both image indexing and searching. The aim of this chapter is to create an understanding of image searching behaviour in work and leisure contexts by discussing the image attributes which are useful to searchers and how users interact with image collections and retrieval systems. A fuller understanding of image searching behaviour could lead to improved image indexing and support for retrieval in different contexts.

This chapter reviews studies of the needs and searching behaviour of actual image users with a focus on their searching activities rather than the retrieval techniques applied in systems. There is a discussion about the attributes of images which may be employed in searches, the types of queries currently posed to image collections and the motives for searching for images. We focus further on the phases of the image search process, the contexts in which image searching occurs and the types of interactions users have with image retrieval systems. Lastly, we discuss some recent developments in image retrieval, aimed at providing enhanced access to images.

Currently, users are mostly limited to image retrieval systems that use text retrieval methods (methods that employ text data, such as image keywords, to perform searches). Although different content-based image retrieval applications (systems that use visual characteristics of images to perform queries) have been implemented, information about their actual usage is very limited. Thus the studies reviewed in this chapter will focus on text-based still image searching behaviour for which much is known. The studies are also mostly limited to photographs, although some also deal with art images, illustrations or moving imagery.

4.2   Image Attributes and Users' Needs

This section reviews existing approaches to how image content may be described, and provides empirical results on the kinds of image content descriptions users employ while querying and describing images. User needs in image retrieval may be understood via categorisations of searched content, search tasks and intended image use, among others. For this purpose we have compiled typologies of the various attributes of images which may be important to searchers, the types of searches conducted and the uses made of the found images.

4.2.1   Image attributes

Images have various properties which may be important to users and there exist several conceptual frameworks on image content to illustrate this. Some theories structure images themselves (Burford *et al.* 2003; Eakins *et al.* 2004; Shatford 1986) while others classify descriptions of images (Hollink *et al.* 2004b; Jörgensen 1998), and some are meant to be used in the indexing of images (Jaimes and Chang 2000). Nevertheless, they all classify perceived image content.

Panofsky (1955) first described three basic levels of meaning in art images:

1. Pre-iconographical – description of the generic image elements and objects
2. Iconographical – analysis of the subject matter depicted in the image
3. Iconological – interpretation of the meaning or theme represented by the subject matter

Shatford (1986) extended Panofsky's theory in order to provide a general classification scheme for images. Shatford categorised the subjects of images as generic of, specific of and about. Here *ofness* refers to the factual content of the image, *aboutness* to its expressional content. Shatford added four facets (who/what/where/when) to each level creating an analytical matrix structure. The Panofsky/Shatford facet matrix has become a widespread model for describing image content.

Jaimes and Chang (2000) developed a conceptual model for describing visual content. The model relies in its classification on the amount of knowledge required to identify and index attributes on each level. The first four levels are so-called perceptual levels, on which no world knowledge is needed: Type/technique, Global distribution, Local structure, Global composition. The six remaining levels are conceptual levels: Generic object, Generic scene, Specific object, Specific scene, Abstract object, Abstract scene. General, specific or abstract world knowledge is required to formulate image descriptions on these levels. In addition to these, Jaimes and Chang also present three classes of non-visual image information: Biographical information, Associated information and Physical attributes.

Hollink *et al.* (2004b) presented a three-level framework for the classification of image descriptions including three description levels: non-visual, perceptual and conceptual. For a complete description of a specific image, these levels may all be used at once. Hollink *et al.*, like Shatford (1986) and Jaimes and Chang (2000) also divide the conceptual level, dealing with image semantics, into three sublevels: general, specific and abstract.

Burford *et al.* (2003) and Eakins *et al.* (2004) proposed a taxonomy of image content based on a survey of literature on computer science, art history and psychology. They list altogether 10 categories of information associated with an image: Perceptual primitives, Geometric primitives, Visual relationships, Visual extension, Semantic units, Contextual abstraction, Cultural abstraction, Technical abstraction, Emotional abstraction and Metadata. The first nine categories are thought to be roughly hierarchical and to reflect the way meaning is constructed in images.

Jörgensen (1998) analysed user behaviour in different image description and sorting tasks. She presented 12 classes of image attributes used by the subjects: Object, People, Colour, Visual elements, Location, Description, People-related attributes, Art historical information, Content/story, Abstract concepts, External relationships and Viewer response. Jörgensen further classified these attributes classes into perceptual, interpretational and reactive.

Table 4.1 illustrates these frameworks to show differences and commonalities regarding the types of attributes identified in images. The far left column provides a summarizing classification of image

Table 4.1 Image attributes and content levels in various frameworks

Attribute type	Panofsky (1955); Shatford (1986)	Jaimes and Chang (2000)	Hollink et al. (2004b)	Burford et al. (2003); Eakins et al. (2004)	Jörgensen (1998)
Non-visual					
Bibliographical		Bibliographical information (Non-visual)	Date, material, style/period, culture,	Metadata	Artist, format, medium, time reference, style, type, technique, representation (Art historical information)
Physical		Physical attributes (Non-visual) Type/technique	ID number, title, creator, rights (Non-visual) Type/technique (Perceptual)		
Contextual		Associated information (Non-visual)	Relation (Non-visual)		Comparison, similarity, reference (External relation)
Syntactic					
Global		Global distribution	Colour, texture, shape (Perceptual)	Perceptual primitives	Colour, colour value, texture, shape (Visual elements)
Local		Local structure		Geometric primitives	
Compositional		Global composition	Composition, position, spatial relation of elements (Perceptual)	Visual relationships, Visual extension	Composition, motion, orientation, perspective, focal point, visual component (Visual elements)
					Location-general, location-specific (Location)

(continued overleaf)

Table 4.1 (continued)

Attribute type	Panofsky (1955); Shatford (1986)	Jaimes and Chang (2000)	Hollink et al. (2004b)	Burford et al. (2003); Eakins et al. (2004)	Jörgensen (1998)
Semantic					
Generic	Pre-iconography/ generic 'of'	Generic object	General conceptual level: object, scene, event, place, time	Semantic units	Object, body part, clothing, text (Objects)
		Generic scene			People (People)
	Iconography/ specific 'of'	Specific object	Specific conceptual level: object, scene, event, place, time		Activity, time aspect, event, setting, category (Content/story)
		Specific scene			Number, description (Description)
Abstract	Iconology/ 'about'	Abstract object	Abstract conceptual level: object, scene, event, place, time	Contextual, cultural and technical abstraction	Abstract, state, symbolic aspect, theme (Abstract concepts)
		Abstract scene			Relationship, social status (People-related attributes)
				Emotional abstraction	Emotion (People-related attributes)
					Atmosphere (Abstract concepts)

attribute classes. Images may be described, indexed and queried with attributes on three main levels by using non-visual image information, syntactic image information or semantic image information. As one goes down the table, each listed syntactic and semantic attribute type represents a higher level of abstraction than its predecessor.

Non-visual image information refers to information not present in the image itself, but rather associated with the image and its production or presentation. Non-visual information may contain *biographical* attributes (e.g. creator, date, title), *physical* attributes (e.g. type, technique, location) or *contextual* attributes (e.g. caption, reference). These may not be derived from low-level image content as such but often take the form of metadata attached to the image.

Syntactic image information (image syntax) is information present in the visual characteristics of images, in the low-level visual properties of the image. It is also called perceptual image information, based on the idea that it is conveyed directly through perception. Syntactic attributes may address three levels of image syntax. *Global distribution* refers to the image-wide distribution of low-level visual content such as colour, texture and sharpness. *Local structure* entails individual non-representational image components such as geometric shapes. *Image composition* refers to the spatial layout of the components in two or three dimensions.

Semantic image information (image semantics) refers to conceptual image content. In contrast to syntactic information, the interpretation of semantic information requires some previous personal or cultural knowledge. Semantic attributes may pertain to different levels: generic, specific or abstract. *Generic* semantic attributes describe types of objects or scenes while *specific* attributes refer to identified and named objects or scenes. *Abstract* attributes refer to what the image represents, such as its symbolic aspects or theme. They may also refer to the emotions or mood interpreted from the image or elicited in the viewer. Semantic attributes on any level may refer to various issues, such as people, places, events, settings, time, etc. Furthermore, image semantics may be analysed both at the level of discrete *objects* appearing in the image or holistic *scenes*, that is, complete images.

The number and types of individual attributes needed to convey the content and meaning of an image is still under debate. The semantic meaning of an image may even be seen to emerge from a user's interaction with the image collection, making the appropriate description levels dependent on both the collection and its users' needs (Santini *et al.* 2001). In any case, the image attributes which are considered useful in a particular retrieval context should be offered as access points to searchers. This may be achieved by textual annotations or analysis of image features.

4.2.2 Image attributes in queries

In order to develop techniques and systems for image retrieval, it is necessary to understand how users would interact with the systems and what types of queries users might construct. Several researchers have addressed this question, though to date there is no clear consensus on user needs. A common research strategy is to analyse the queries or requests made to particular collections as a means of categorising the uses made of these collections. Some studies have also employed questionnaires describing searches; and user studies have included interviews and observations of image search sessions. Research has also been conducted on other types of image-related tasks, such as image description.

Several studies have categorised the content of image queries or requests within specific domains or professions. Queries may be either posed to digitised image collections directly by the searcher, or take the form of written or verbal image requests to intermediaries such as librarians, archivists or curators. The users of these collections are for example book/magazine/newspaper publishers (Enser 1993; Markkula and Sormunen 2000; Westman and Oittinen 2006), advertising and design companies (Jörgensen and Jörgensen 2005), television and audiovisual companies (Enser 1993), health professionals (Keister 1994), and various other professional and academic users (Armitage and Enser 1997). Less work has gone into studying the types of image searches conducted on Web search engines (Cunningham and Masoodian 2006; Goodrum and Spink 2001).

Image needs from these users may concern all the image attributes shown in Table 4.1. Across these studies conducted in different user populations and image collections, some tendencies may be noticed in the use of the different attributes. Various researchers have also suggested typologies of image searches based on empirical studies or literature reviews. Two main types of search typologies can be distinguished – those based on different types of image attributes used in the search (Eakins 1996; 1998; Jörgensen 1999) and those based on levels of complexity of the search task (Conniss et al. 2000; Jörgensen 1999; Hung 2005).

4.2.2.1   Semantic image information

Images are most often queried or requested based on image semantics. These searches may be based on many different types of semantic attributes, and the resulting searches may be characterised for example as searches for topics, events, concepts or emotions (Jörgensen 1999). Needs may refer to scenes (e.g. sunbathing) or standalone objects (e.g. dog) within the image. Semantic image needs may further include specifications of events, places and times (e.g. World Cup Match, Argentina–Cameroon, Milan 1990). The semantic query terms naturally vary due to the need and collection used. The terms vary widely even among different queries put into the same collection. The most frequently used individual query terms are present only in a few percent of the queries on the Web and in closed collections (Goodrum and Spink 2001; Jörgensen and Jörgensen 2005).

It is however possible to differentiate between searches on the three semantic levels: general, specific and abstract (Jörgensen 1999). General searches consist of finding images of a general topical or subject category: a kind of person, group, thing, event, location or action. Specific searches consist of finding images of an individually named person, group, thing, event, location or action, that is, a specific instance of a general category. Queries may be designed to find types of objects (e.g. computer) or specific objects (e.g. Big Ben). An abstract search involves finding images that communicate certain emotional or abstract concepts (e.g. warmth). Image queries and selections may also be based on the moods and emotions which could be associated with the image or elicited in the viewer (e.g. sadness).

Image queries or requests for (specific or general) objects are most common, entailing over half of all queries or requests (Jörgensen and Jörgensen 2005; Westman and Oittinen 2006). The specificity of queries seems to depend heavily on the domain. In some collections such as news imagery specific queries dominate (Enser 1993; Markkula and Sormunen 2000; Westman and Oittinen 2006) while in other areas such as stock imagery general needs are clearly the most frequent (Jörgensen and Jörgensen 2005; Hollink et al. 2004b). Also general web image searches seem to be mostly specific (Cunningham and Masoodian 2006).

Queries referring to the complete scene portrayed in the image (as opposed to individual objects within the image) appear to account for roughly one quarter of image queries (Jörgensen and Jörgensen 2005; Westman and Oittinen 2006). Activities represented by verbs or terms referring to action comprise nearly one tenth of image query terms (Jörgensen and Jörgensen 2005; Westman and Oittinen 2006). Abstract concepts and reactive terms seem to compose 5% of all query terms (Hollink et al. 2004b; Westman and Oittinen 2006). The share of affective image needs seems to be slightly higher in casual web image searching than in professional image searches (Cunningham and Masoodian 2006).

Facet analysis of image requests following the Panofsky/Shatford framework suggests that out of all facets present in requests, a little over 50% are specific, a little less than 50% generic and less than 5% are abstract (Armitage and Enser 1997; Westman and Oittinen 2006). Especially the specific who (who is portrayed?) and where (what location is portrayed?) facets seem to be pronounced as is the generic who (what kind of a being is portrayed?) facet.

Nearly half of all image queries are modified with refiners concerning for example desired or undesired time, location, actions, events or technical attributes (Enser 1993; Markkula and Sormunen 2000; Westman and Oittinen 2006). Refiners may refer to the whole image or the objects within the image. The refiners may serve to refine general terms (e.g. girl) into more specific visual requests (e.g. blonde girl) or even into more abstract requests (e.g. beautiful girl) (Goodrum and Spink 2001).

4.2.2.2 Syntactic image information

Images may also be retrieved based on their syntax. These queries concern the local or global levels of syntactic features or the spatial location of image elements (Eakins 1996). Image needs may concern the presence of a certain colour or texture (e.g. red tarmac). They may also be concerned with particular shapes within the image (e.g. stars and stripes) or with the composition of the image (e.g. small circles in the foreground).

Syntactic image information is employed less in queries and requests than semantic content. Although Keister (1994) found that one-third to one-half of the requests analysed were based on visual constructs, later studies have shown that perceptual and compositional terms only make up a small percentage of all image queries (Jörgensen and Jörgensen 2005, Hollink *et al*. 2004b). Syntactic image attributes are sometimes used as refiners in otherwise conceptual queries, adding specifications of for example colour or shooting distance (Markkula and Sormunen 2000). Sharpness has been evaluated to be important in image queries and at the selection stage (Eakins *et al*. 2004; Markkula and Sormunen 2000; Westman and Oittinen 2006).

4.2.2.3 Non-visual image information

Non-visual attributes of images may serve as access points to both syntactic and semantic image content. They are also crucial in image searches which are concerned with the context of production of the image (Jörgensen 1999). These needs could be satisfied with the aid of non-visual image information such as metadata (e.g. known photographer, format, specific image ID). Classifications often ignore image queries that are based on retrieval by associated metadata, since these are considered a text retrieval issue. However, currently the most widely available image retrieval systems use text search only and retrieval is based on image annotations or captions that attempt to cover the image content on multiple levels. When these textual annotations may offer the only access point to images, it seems wise to include them in image retrieval considerations.

Furthermore, image needs may refer to a specific item (Conniss *et al*. 2000; Jörgensen 1999). These needs may only be satisfied with a single, known item (e.g. Le Baiser de l'Hôtel de Ville, Paris, by Robert Doisneau), while others may be fulfilled with many different kinds of images (e.g. a couple kissing). Searches for previously known images may account for up to one tenth of all searches in a particular collection (Markkula and Sormunen 2000; Westman and Oittinen 2006).

4.2.3   Attributes beyond queries

When attempting to understand image needs via the analysis of requests and queries one must remember that searchers express their information need in a way they think the retrieval system can handle. In essence, users modify their original image need in order to be able to submit a query or otherwise search in a retrieval system. Several user studies have shown differences between free image descriptions and queries and annotations.

Describing images for retrieval, i.e. querying, seems to result in more semantic terms and less syntactic attributes than free description of image content (Hollink *et al*. 2004b; Jörgensen 1998). This may be due to users adjusting their image needs to the perceived capabilities of an image search engine. In addition to differences brought on by different goals of the description, also constraining the description by for example limiting the number of terms to be used modifies the descriptions users give. Greisdorf and O'Connor (2002) found that users employed perceptual terms when selecting query terms from a list, but did not generate terms in this category themselves. Laine-Hernandez and Westman (2006) found that constrained image annotation resulted in more terms depicting the story, setting and theme of the image than a free description task, which led test subjects to enumerate individual objects and describe their locations within the image. Imposing either a retrieval system or a format of description (preselected terms, limited number of terms) on users may change the way they describe images.

4.2.4   Image needs

4.2.4.1   Facets of users' image needs

The discussion in Section 4.2.2 reflects the fact that image users may have very specific and concrete needs or they may be more interested in material that conveys particular moods or abstract concepts. Another dimension of image needs is how well-defined they are. Image searches may be vague, where those searching are unsure of what they are looking for exactly, or where they have only vague or fuzzy criteria (Conniss *et al.* 2000). Some users may not want a specific image at all; instead they want to browse a collection for inspiration or ideas. The vagueness of the need may also make the need difficult to verbalise and transform into a query.

We have summarized both the content and form dimensions of image needs described above in Figure 4.1.

4.2.4.2   Motives for searching for images

Research into image searching has emphasised the expressed need, as interpreted by the formulation of the query. However, it does not tell us much about what the *actual need* is nor what *use* will be made of the retrieved images. Conniss *et al.* (2000) presented seven classes of image use based on their study of image searching behaviour:

- *Illustration:* Images are used in conjunction with some accompanying media, as a means of representing what is being referred to, e.g. magazine illustrations where images are shown with accompanying text. The relation between images and other media may be various, e.g. images and text may be related in many ways (Marsh and White 2004).
- *Information processing:* The use of the data contained within the image is of primary importance, e.g. diagnosis by radiologists.
- *Information dissemination:* The image is a stand-alone piece of information transmitted to someone else, e.g. the dissemination of a mugshot to police officers.
- *Learning:* Gaining knowledge from the image content, e.g. illustrations in a textbook.

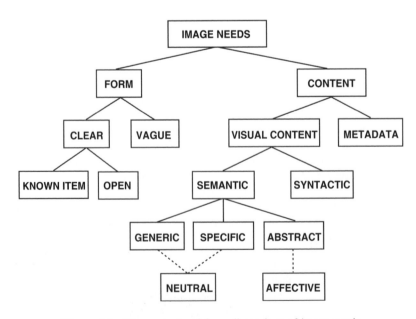

Figure 4.1   The content and form dimensions of image needs

- *Generation of ideas:* Images are used to provide inspiration or provoke thought patterns, such as in design.
- *Aesthetic value:* Images are simply required for decoration, e.g. posters on a wall.
- *Emotive/persuasive purposes:* Images are used to stimulate emotions or convey a particular message as in advertising and the media.

These uses do not necessarily occur exclusively since on any given occasion users may seek images that cover more than one class of use. For example, images may be used for illustration of a topic in a publication, but also constitute a process of learning about the topic.

Another way to typify image uses is suggested by Fidel (1997) who maps image retrieval tasks onto a continuum between the data pole and the objects pole. At the data pole, images are used as information sources and are retrieved based on their containing certain data. An example of this kind of use would be verifying from a photograph of the White House how many pillars there are at the rear facade. At the objects pole images are needed as objects of their own merit. For example, another user may intend to use the same image to illustrate a talk on US politics. When comparing the two categorisations presented here, information processing, information dissemination and learning would be closer to the data pole while illustration, aesthetic value and emotive purposes sit closer to the objects pole. Generation of ideas could reflect the use of images as either data (an information element in the image provokes thoughts) or objects (the image provides inspiration as a whole).

Interestingly, these uses do not tell us what is actually done with the images once they are found. Access to images is not just about finding the right image. Users may want to manipulate the images in some way: images may be cropped, re-sized, synthesised or enhanced – so as to produce another image. In all sorts of contexts, image re-use or re-packaging is becoming common.

This section has summarised the image needs that users may have which function as the basis and motivation for their image searching behaviour. Classifications of query types may serve in illustrating the strengths and limitations of different retrieval techniques, such as those presented in Chapter 3.

4.3   Image Searching Behaviour

Users' interaction with an image retrieval system encompasses more than the initial formulation of the image need in the form of a query. The image search process includes several steps which relate to choosing the image resources to be used and the search strategies to be employed, interactions with the search functionalities and finally selecting the image(s) from the retrieved set. All these steps are affected by several personal and contextual influences (Conniss *et al.* 2000). These contextual factors shown in Figure 4.2 may have a mediating effect on understanding how and why people search for and use images.

4.3.1   Search process

Conniss *et al.* (2000) have developed a generic model of image search processes in work environments, consisting of six phases: starting, scoping, applying, selecting, iterating and ending. In the starting phase the searcher identifies an image need or receives an image request from someone. The user then defines the image need in order to develop criteria for assessing the suitability of the images found. At this point searchers seem to form a mental model of the target image (Frost 2001). This model may be based on the intended use of the image, previous information on the topic or what the searcher can expect to find (Westman and Oittinen 2006). Defining the selection criteria may require discussions with colleagues for sharing of expertise or ideas if the subject matter is technical, unusual or outside the user's expertise. Specifying the criteria is particularly important in vague and abstract searches and it may be performed iteratively throughout the search process. Searchers gain new knowledge of their retrieval task and the image collection through the search process and the image sets retrieved (Greisdorf and O'Connor 2002). This means that their perception of both the search task and the image selection criteria is dynamic.

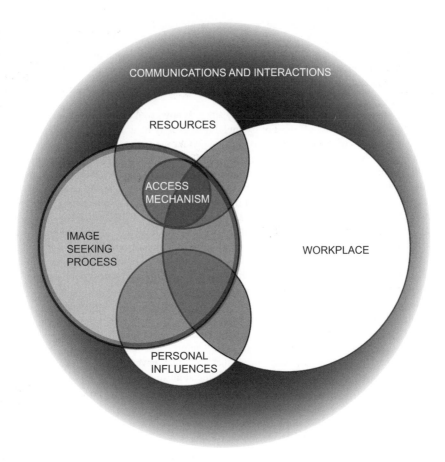

Figure 4.2   Overview of the interaction between factors in the contextual framework and the image searching process (Conniss *et al.* 2000)   (Reproduced by permission of © 2000 Northumbria University, Newcastle, UK)

The scoping phase of the image search process consists of the searcher deciding upon the breadth of the search approach based on the resources available and the nature of the task. Most searchers in the study carried out by Conniss *et al.* preferred executing broad searches by using simple text search, searching across categories, setting parameters to widest ranges or using wild cards, resulting in maximal inclusion of relevant material. If users preferred to see only a small number of definitely relevant images, they opted for a narrow approach instead. Scoping may be performed iteratively during a search session by searchers narrowing or expanding their approaches based on factors such as the number of images returned.

The applying phase includes deciding which image resources and search interfaces to use in carrying out the search. The availability of image collections, constraints such as time and money, and the need for involving a search intermediary all affect the decisions of where and how to search. Image searching behaviour may be tailored towards different resources, contents and access mechanisms. Different search tasks may require gathering more knowledge. The users' expertise and knowledge also affect the choices on which resources to use and also how they are able to use them. Regardless of the functionalities offered by a particular interface, users may select different search strategies for their image search sessions. These are discussed in more detail in Section 4.3.2.

Selecting images may occur as a single action or in stages by first selecting a subset of relevant images and then selecting the image(s) to be used. The selection may be based on the visual properties of the images or the textual information detailing their content. This links the selection to the display features of the interface. When users look at the retrieved results, they assess them in relation to the selection criteria established in the starting phase. Depending on image availability the user may need to be flexible with the criteria. Instead of finding the absolute best images, users may make just acceptable selections (Markkula and Sormunen 2000). Sometimes, an image is preferred because it would suit many possible uses (Westman and Oittinen 2006). The selection or relevance criteria applied by image searchers is discussed in Section 4.3.3. Usually information retrieval is thought to include a selection phase where the document is downloaded or accessed. In some cases of image retrieval, browsing alone can satisfy information needs and the goal may be the enjoyment of the search rather than the retrieval of a result image (Cunningham and Masoodian 2006).

Image searches most often include some iteration to the first search instance. These modifications done during the iterating phase are discussed in Section 4.3.2.4. Searchers also keep track of their own search which is discussed in Section 4.3.2.5.

The ending phase is related to how the search is finalised and when the user stops searching. The search may be *exhaustive* where user has looked through all resources and search permutations and found suitable images depending on their existence. The search may also have been *non-exhaustive*, where the user has searched through only a proportion of the image collection, possible query permutations or images returned by a system and already found suitable images or decided to stop searching. In addition to a best match or no match the result of the search may also have been several matches to choose from. Feedback from others may also affect the choice of whether to keep searching, stop searching, or indeed, recommence searching.

4.3.2   Search strategies

Information search strategies may be analysed in detail by taking a look at the tactics searchers employ in attempt to further their search. These search tactics may relate to constructing queries, navigating the collection or keeping track of search activities. In image retrieval, tactics relate to which search methods (e.g. textual query, content-based query, category search, browsing) to use, whether the constructed queries are simple or complex and whether the queries are modified in the course of a search. Search strategies, consisting of combinations of tactics, also include relevance assessments. Strategies may be analysed at different levels: session (transactions identified by e.g. a unique session ID), search (related queries addressing the same topic), query (one or more search terms or attributes in a single interaction) or search term (single terms appearing in query) (Goodrum and Spink 2001; Jörgensen and Jörgensen 2005).

4.3.2.1   Choice of search tactics

The choice of which search methods to use depends on both the image need and the functionalities available (Conniss *et al.* 2000). Generally speaking, searchers may opt for simplistic search techniques (generic terms, single keywords) or more complex ones (combined terms, Boolean modifiers, wild card, truncation, spelling/syntax alternatives, filters). Broad search approaches or lack of knowledge tend to lead to simplistic searches while a narrow approach and searching experience lead to complex searches. Deciding upon the breadth of the approach depends mostly on the time available, but also on the expected amount of images to be found and the users' background. The experience users have with image retrieval systems affects their searches in at least two ways (Eakins and Graham 1999). An expert user is able to use advanced search features in ways that are suited to the particular need but also, as the user gains knowledge on how image retrieval systems work, the query terms themselves are affected.

Different types of search topics produce different kinds of search tactics both in professional and casual searches. General search topics easily lead to multiple queries and heavy browsing while specific needs are more likely to result in just one or two queries and shorter browsing sessions (Cunningham and Masoodian 2006; Hung 2005; Markkula and Sormunen 2000). Conceptual and abstract image needs seem to lend themselves more naturally to browsing rather than querying (Conniss *et al.* 2000), most likely because they may be difficult to translate into query terms. Different query input methods seem to be preferred for different types of image searches. Eakins *et al.* (2004) found that when using text-based search, users prefer to type in semantic search terms rather than select terms from a hierarchical list. However, users expressed a preference for selecting content-based search elements such as shape or texture from a menu, rather than entering them freehand.

Several factors influence whether users browse through the returned image set or modify the query: the time available, the number of irrelevant images retrieved, and the difficulty of finding suitable images (Conniss *et al.* 2000). The methods used to modify queries are discussed in Section 4.3.2.4. Markkula and Sormunen (2000) found that willingness to browse also depends on the motivation to find a particularly good image. Regular, professional searchers can browse hundreds of images if necessary and possible (Conniss *et al.* 2000; Markkula and Sormunen 2000). There are also situations where users purposefully look through all retrieved images to either make sure they selected the best image(s) or to gather knowledge about the collection (Conniss *et al.* 2000).

Different domains typically have different types of search tasks which tie tactic selection into the domain: for example, users from advertising seem to search more on the level of generic objects or scenes while specific needs dominate in journalistic searches for particular people. The need type also influences wider searching behaviour as specific searches sometimes require that the user gathers more subject knowledge (Conniss *et al.* 2000). The choice of search tactic is naturally also dependent on personal preferences and capabilities. When using an intermediary, some searchers make a very narrow request because they want the intermediary to return a single image, whereas others formulate a very broad request because they want to select the image themselves (Ørnager 1995).

4.3.2.2   Length of queries, searches and sessions

Textual queries for images tend to be short. Typical image queries consist of either a single term or a single phrase. Image queries on the web contain on average a little over two terms (Cunninghan and Masoodian 2006; Goodrum and Spink 2001; Goodrum *et al.* 2003), whereas specialist queries in closed collections include a little less than two terms (Jörgensen and Jörgensen 2005; Westman and Oittinen 2006). The number of different facets in image requests can also shed light on how many aspects expressed image needs include. According to studies that have used facet analysis, image requests include an average of 1.5 facets specifying the who, what, where or when of the photograph (Armitage and Enser 1997; Westman and Oittinen 2006). This apparent shortness of image queries is not in contradiction with the queries possibly being complex. Image needs are sometimes vague or fuzzy and can not be fully explicated. These types of complex image needs may be boiled down to a few query terms by for example naming a critical object that should appear in the image (Westman and Oittinen 2006).

There are less empirical results on image searches or search sessions than on queries due to problems in identification. Determining session length is difficult, especially when the search is conducted on a web application. Also, deciding when a change in query terms represents a change of topic and thus a new search is not trivial. Image search sessions have been shown to last roughly 20 minutes and to include on average 3 queries (Goodrum and Spink 2001; Jörgensen and Jörgensen 2005).

4.3.2.3   Role of browsing

Browsing is a central search tactic in image retrieval. Through browsing the user either gains enough knowledge to begin query interactions or already finds a suitable image (Conniss *et al.* 2000). Browsing differs from querying in that it allows users to recognise what interests them within the collection rather than needing to formulate a precise query in advance (Frost *et al.* 2000). This makes browsing an attractive search strategy, especially for users with little prior knowledge about the domain or collection. When searching for images on the web, non-professional users may also navigate directly to a website they feel is likely to contain desired images or use a search engine to identify a suitable website to browse (Cunningham and Masoodian 2006).

Image searchers spend a large portion of their time browsing image collections and retrieved image sets. Browsing activities seem to account for nearly four times as much searching activity as querying (Goodrum *et al.* 2003; Hung 2005). Image searchers browse in almost all search sessions after the initial textual query, and professional searchers seem to browse on average 100 images per session (Jörgensen and Jörgensen 2005; Markkula and Sormunen 2000).

Browsing is often used in combination with broad and simple textual queries. Markkula and Sormunen (2000) found that searchers browsed because they wanted to avoid excluding potentially relevant images from the result set by formulating too narrow queries. The searchers also claimed that browsing often required less effort and time than formulating a well-thought-out query. Users have expressed preference towards a multimode search – combining both browse and query, since it provides alternate modes of access (Frost *et al.* 2000). An image always includes and infers more details and meanings than may be gathered in a textual description of the image. Some image characteristics such as colours and colour impressions are difficult to verbalise, but easy to verify from an image (Heidorn 1999). This means that it is often necessary for the searcher to verify some image characteristics by looking at the images, that is, by browsing them. Browsing has been observed to be more common when searching for abstract or subjective topics, suggesting that users adopt a browsing strategy when they are unable to formulate a verbal query. Browsing thus compensates the difficulties of forming a query and is at its most useful in highly visual image needs or when the user has ill-defined goals for the task (McDonald and Tait 2003).

4.3.2.4 Modifications to queries

Searching usually involves modifications to the first query. Also in image searches most queries are modified after the initial query (Goodrum *et al.* 2003; Jörgensen and Jörgensen 2005). Image queries seem to be modified more often than text queries on the web (Goodrum and Spink 2001). Various types of query modifications exist at the query level. Query terms may be added, deleted, substituted, narrowed, broadened; related terms may be added or substituted; users may revert to a previous query, or use retrieved information as query terms (Jörgensen and Jörgensen 2005). An example from the study of Jörgensen and Jörgensen (2005) shows the use of several techniques in twelve subsequent queries, such as replacing a noun with an adjective, using a visual construct, using synonyms, and replacing terms with related terms:

brat

spoiled

kid sticking tongue out

brat child

spoiled child

mean child

small kid big kid

bossy child

bossy

bossy brothers

mean brothers

brat sisters

Jörgensen and Jörgensen note that users are willing to attempt relatively complex search strategies and modify queries, but do not know how to go about the process, leading to a large number of query modifications. According to their results, users are over ten times more likely to move to a synonym than to a broader or narrower query term.

There are some particular situations where modifications generally take place (Conniss *et al.* 2000). One apparent situation is when either too many or too few results are returned or when clearly too many irrelevant images are returned. The modifications are then aimed at either removing non-relevant images from the result set or including more relevant images (Jörgensen and Jörgensen 2005). These modifications are based on inspecting the result sets. Another situation that requires query modification is when no images or no suitable images are returned. The user may also wish to progressively explore search permutations in order to ensure the best possible results.

Query modifications may also be due to changes in the search task or new retrieval ideas created during the search. The test subjects of Hung (2005) changed query terms to match the content or textual description of the images that resulted from the initial search. Jörgensen and Jörgensen (2005) found that over 10% of query modifications were due to users employing terms that appeared in the captions of images browsed during the session. These strategies can be seen as a change in search strategy towards a textual query-by-example. On a related note, Conniss et al. (2000) describe image searchers using the categorisation of images to learn more about the subject area.

4.3.2.5   Monitoring the search

Search sessions may include multiple searches where users test different search ideas, viewpoints or strategies. Each search then becomes an attempt to find images using a particular viewpoint on a trial-and-error basis (Markkula and Sormunen 2000). Users may also systematically vary their search parameters. These differing search approaches and complex search topics require the user to keep track of their search activities, that is, to monitor the search. This may happen, for example by taking notes (either mental or written) on queries made or search features used (Conniss et al. 2000).

When searching for an image, the user is actually searching for a representation that will match her internal model of the image needed (Heidorn 1999). The existence of such a mental image does not constraint the search, however. Frost (2001) found that one-fifth of image searchers were satisfied with an image that did not match their initial mental image. In these cases the selected image corresponded to a new idea discovered in the process of the search.

A search session may involve several queries and the retrieved image sets may be large, making it impractical to remember which images are relevant. To facilitate relevance assessments, users frequently retrieve a subset of images to act as a buffer from which they can choose later (Conniss et al. 2000). Images may be printed out at this stage or simply kept in mind and retrieved later (Markkula and Sormunen 2000). The final selection out of this set might occur much later, depending on when the images are needed (Markkula and Sormunen 2000). These later selections and comparisons are based on more detailed examinations of the visual properties or textual descriptions of the images (Conniss et al. 2000) and comparison of the candidate images (Westman and Oittinen 2006). This gradual selection strategy helps to structure the selection task into several subtasks, making it easier to handle.

4.3.3   Relevance criteria

The relevance of an image refers to its match to the image need the user has. Image relevance may be judged on multiple levels. Topicality seems to often function as the first relevance criterion and it may be the most important criterion (Markkula and Sormunen 2000; Choi and Rasmussen 2002). Beyond topicality, image selection is based on both compositional and informational criteria. Conniss et al. (2000) found that, depending on the context of use, images may be selected on the basis of content, salient features, quality, visual appearance, level of accuracy, semantic value, image tone, physical size, layout, recency, and anonymity. Consideration of context of use also covers such factors as image overuse, format, download time, audience, copyright restrictions, and importance.

The textual descriptions associated with images are an important source of information when judging the (topical) relevance of images (Choi and Rasmussen 2002; Hung 2005; Markkula and Sormunen 2000). Westman and Oittinen (2006) found that searchers tended to alternate between viewing the textual description and the image proper during the selection process. It may be impossible to interpret the image or judge its relevance without seeing a textual description. Also, initial judgments may change if the description becomes available later on (Choi and Rasmussen 2002). It is often simply not enough that the image 'looks relevant': the searchers need to verify what the image means from its textual description. The text is especially important when the search topic is previously unknown to the searcher (Westman and Oittinen 2006).

As mentioned earlier, the intended use of the image provides a context for the search and information about the intended use is employed in creating task-related relevance criteria. For example, the relevance criteria and judgments of journalists depend on their work task and related contextual factors such as the article to be illustrated, the page layout and the illustration styles of the particular newspaper (Markkula and Sormunen 2000; Westman and Oittinen 2006). In a specific retrieval context there may also be implicit relevance criteria, e.g. in newspaper photo illustrations it is desirable to give an impression of actuality, and thus select archive photos of the current season (Markkula and Sormunen 2000). The employment of these criteria requires extensive domain knowledge.

The final criteria for image selection include subjective, abstract and affective criteria such as visual aesthetics and emotional reactions to images (Choi and Rasmussen 2002; Conniss *et al.* 2000, Markkula and Sormunen 2000). The final selection criteria can also be clearly dynamic, born of the proceeding search process and result sets viewed. The searchers may focus on features that separate the candidate images from each other or especially catch their attention and, in the final stage of the selection process, apply these as relevance criteria (Westman and Oittinen 2006).

4.4   New Directions for Image Access

The development of image retrieval systems and interfaces requires knowledge about users and their interaction with visual information. Users have complex and dynamic image needs embedded in real situations and contexts so new approaches for supporting the resulting adaptive and multiple search strategies are needed. Many new paradigms have come to light in the image retrieval research community in the last few years. These developments concern all parts of the process of image retrieval, ranging from acquiring annotations to the images to supporting the search process by an end-user. Some of these new directions for image access are discussed next.

4.4.1   Social tagging

Having expert archivists manually annotate images is time-consuming and costly. Alternative annotation approaches to expert annotators have been experimented with both in research and practice. Folksonomies are used in various Web applications including image applications such as Flickr (http://www.flickr.com/) in which content is collaboratively tagged by its creators and consumers. Some applications such as Google Labeler have packaged the image annotation task as a game (http://images.google.com/imagelabeler/) and others like Phetch intend to gather complete descriptions of images instead of just a few keywords (http://www.peekaboom.org/phetch/).

Tagging has been called consensus-based indexing, following the naming convention of concept-based annotation of semantic image attributes and content-based analysis of syntactic image features (Jörgensen 2007). Tagging-based applications may improve image accessibility on the web (von Ahn *et al.* 2006) and seem to create motivation for users to provide descriptions to images. Tagging has also been used in connection with content-based features extracted automatically from the images to create more complete annotations by combining the strengths of both methods (Aurnhammer *et al.* 2006). Aurnhammer *et al.* also note that tags reflect the way users navigate through an archive and could be used to aid navigation of image collections.

4.4.2   Images in context

The different attributes of images discussed in Section 4.2.1 are used to describe the content and meaning of images for indexing, and subsequently, retrieval. The concept of emergent semantics however states that images do not have any intrinsic meaning of their own, but rather their meaning emerges from the interaction with a user and from the context of the image (Santini *et al.* 2001). This means that the meaning of an image is contextual (dependent upon the user and her situation) and differential (born of differences and similarities with other images). Thus, image context, such as an image collection or the result set from a query, plays an equally important role as image content in determining the meaning of an image.

Cool and Spink (2002) suggest focusing also on context at query level: the changes in the meanings that terms and images acquire in different contexts. Due to the context of the image need, very different intended targets or uses may lie behind similar queries. Currently this context information is lost at the query stage though it may be present in image requests (Westman and Oittinen 2006).

The construction of meaning in images has been considered too complex an issue to be thoroughly indexed, but emergent semantics can be seen to turn this restriction into an enabler, by allowing the creation of meaning in interaction with images and capturing this meaning for others to access (Jörgensen 2007). The basic idea of emergent image semantics may be seen reflected in searches, as users guide their queries towards a certain group of images in a previous result set, or judge image relevance based on features which differentiate images.

4.4.3   Visualisations

Browsing is currently the main tactic for image retrieval beyond the first short text query. Browsing will be central also in the future due to the visual nature of the information sought after and the need to see the images in order to inspect their relevance. Users are also quite willing to browse large image sets. The question then becomes how to support this visual and cognitive activity of browsing, even for large image sets. A key issue to browsing is that the clusters must make sense to the user (Barnard and Forsyth 2001). This is why providing a meaningful clustering and entry point to the set is important.

Several researchers have suggested the use of content-based image retrieval methods for result set visualisations in visual retrieval (Markkula and Sormunen 2000; Hollink *et al.* 2004a). The content-based methods could be applied to structuring the set of images retrieved by textual methods. The retrieved set could be arranged spatially with multivariate methods so that visually similar images would be closer to each other and dissimilar images wider apart. This would provide the user with a visualisation of different image subsets within the retrieved image set.

4.4.4   Workspaces

Users like browsing, partly due to the control it provides them over the search. A prevalent theme among suggestions for image systems relating to this sense of control is allowing users to compare images in sets which they can create themselves (Frost *et al.* 2000). The ability to track searches is also important to users (Conniss *et al.* 2000). Some implementations of these workspaces for image retrieval exist. They provide users with functionalities regarding image selection, comparison, grouping and query-by-example. The user evaluation of one such workspace (Urban and Jose 2007) found that users created their own semantic image categories with the tools provided and were able to diversify their searches, exploring the search task from multiple perspectives in one search session.

These workspaces could also facilitate collaborative image retrieval. In illustration processes image selections are sometimes made collaboratively and the final selections may be done by a different person than the preceding search (Markkula and Sormunen 2000). A new combined image search and organisation tool could assist in communicating the search history and selection criteria to all collaborators.

4.5   Summary

This chapter discussed image needs and searching behaviour based on reviewing empirical studies of image retrieval. We provided basic typologies of image content and image queries made by users based on several theories and studies. We discussed the process and phases of image searching, the choices of different search strategies including the length of queries, searches and sessions, the role of browsing as well as the contextual aspects of image searching. We also presented research results on how image searches are modified and monitored by the searcher. Finally, we discussed new directions in image annotations and search interface functionalities aimed at providing enhanced image access, including multimode access, social tagging, image context and emergent semantics, result set visualisations, and workspaces. These new approaches to image description and retrieval may be employed in conjunction with the classical concept-based image annotation and text query methods.

Different types of users with different types of image needs require different access points to images, thus creating a need for multifaceted image annotation or content analysis. People mostly use high-level semantic image descriptors so these need to be offered as access points. Syntactic image content has been evaluated to be of lesser use for image retrieval, but some visual attributes such as sharpness have been mentioned as important selection criteria (Choi and Rasmussen 2002; Conniss *et al.* 2000; Markkula and Sormunen 2000; Westman and Oittinen 2006). This suggests that the possibilities of content-based image analysis have not been exhausted, especially when used in combination with concept-based methods.

Interface components are needed to satisfy image querying and browsing as well as other types of supporting behaviour in image searching. The knowledge on users' image searching behaviour presented in this chapter may be used to create basic evaluation criteria for visual retrieval systems. The systems may be assessed based on whether or not they support user needs regarding:

- Querying by non-visual attributes (bibliographical, physical and contextual metadata);
- Querying by visual features on different syntactic levels (global and local syntax, composition);
- Querying by concepts on different semantic levels (generic, specific, abstract, affective);
- Querying by visual and textual examples;
- Specifying interaction-based image attributes created through social tagging and emergent semantics;
- Specifying image quality criteria;
- Browsing images and associated textual information;
- Modifying queries and exploring the search from several viewpoints;
- Visualising retrieved image sets by content-based methods;
- Making image selections and comparing images.

For further reading on image searching behaviour, please see the work by Conniss *et al.* (2000) and Jörgensen (2003).

Acknowledgements. The author would like to sincerely thank Margaret E. Graham who initially put forth the idea of writing this chapter. She also provided her time and input in the form of valuable comments and suggestions for improvement of the manuscript.

Exercise

4.1 Analyse a sample of image tags from a website which uses collaborative tagging, for example Flickr. You may take any selection of tags e.g. your own tags, popular tags or tags in the latest images uploaded. Which attributes of images content from Table 4.1 do the tags describe? Do they provide a useful structure for browsing and searching for images? You should be able to identify tags on differing levels of image content and usefulness for retrieval.

References

Armitage, L. and Enser, P. G. B. (1997) Analysis of user need in image archives. *Journal of Information Science*, **23**(4), 287–99.

Aurnhammer, M., Hanappe, P. and Steels, L. (2006) Integrating collaborative tagging and emergent semantics for image retrieval. *Proceedings of WWW2006 Collaborative Web Tagging Workshop*, May 22, Edinburgh, Scotland.

Barnard, K. and Forsyth, D. (2001) Learning the Semantics of Words and Pictures. *Proceedings of the 8th International Conference on Computer Vision (ICCV2001)*, July 7–14, Vancouver, BC, Canada, pp. 408–415.

Burford, B., Briggs, P. and Eakins, J. (2003) A taxonomy of the image: On the classification of content for image retrieval. *Visual Communications*, **2**(2), 123–61.

Choi, Y. and Rasmussen, E. (2002) Users' relevance criteria in image retrieval in American history. *Information Processing and Management*, **38**(5), 695–726.

Conniss, L. R., Ashford, A. J. and Graham, M. E. (2000) Information Seeking Behaviour in Image Retrieval: Visor I Final Report. *Library and Information Commission Research Report 95*. Institute for Image Data Research, Newcastle upon Tyne.

Cool, C. and Spink, A. (2002) Issues of context in information retrieval (IR): an introduction to the special issue. *Information Processing and Management*, **38**(5), 605–11.

Cunningham, S. and Masoodian, M. (2006) Looking for a picture: an analysis of everyday image information searching. *Proceedings of the 6th ACM/IEEE-CS Joint Conference on Digital Libraries*, June 11–15, Chapel Hill, NC, USA. ACM Press, New York, NY, pp. 198–199.

Eakins, J. P. (1996) Automatic image content retrieval – are we getting anywhere? *Proceedings of the 3rd International Conference on Electronic Library and Visual Information Research (ELVIRA3)*, April 30–May 2, Milton Keynes, UK, pp. 123–135.

Eakins, J. P. (1998) Techniques for image retrieval. *Library and Information Briefings, 85*. South Bank University, Library Information Technology Centre, London, pp. 1–15.

Eakins, J. P., Briggs, P. and Burford, B. (2004) Image Retrieval Interfaces: A User Perspective. *Lecture Notes in Computer Science*, **3115**, 628–37.

Eakins, J. P. and Graham, M. E. (1999) Content-based Image Retrieval. *JISC Technology Applications Programme Report 39.*

Enser, P. G. B. (1993) Query analysis in a visual information retrieval context. *Journal of Document and Text Management*, **1**(1), 25–52.

Fidel, R. (1997) The image retrieval task: implications for the design and evaluation of image databases. *The New Review of Hypermedia and Multimedia*, **3**, 181–199.

Frost, C. O. (2001) The Role of Mental Models in a Multi-Modal Image Search. In Information in a Networked World: Harnessing the Flow. *Proceedings of the 64th Annual Meeting of the American Society for Information Science and Technology*, November 3–8, Washington, D.C., USA. Information Today. Medford, NJ, pp. 52–57.

Frost, C. O., Taylor, B., Noakes, A. *et al.* (2000) Browse and Search Patterns in a Digital Image Database. *Information Retrieval*, **1**(4), 287–313.

Goodrum, A. and Spink, A. (2001) Image searching on the Excite Web search engine. *Information Processing and Management*, **37**(2), 295–311.

Goodrum, A. A., Bejune, M. M. and Siochi, A. C. (2003) A State Transition Analysis of Image Search Patterns on the Web. *Proceedings of CIVR 2003*, July 24–25, Urbana, IL, USA. *Lecture Notes in Computer Science*, **2728**, 281–90. Springer.

Greisdorf, H. and O'Connor, B. (2002) Modelling what users see when they look at images: a cognitive viewpoint. *Journal of Documentation*, **58**(1), 6–29.

Heidorn, P. B. (1999) Image retrieval as linguistic and nonlinguistic visual model matching. *Library Trends*, **48**(2), 303–25.

Hollink, L., Nguyen, G. P., Koelma, D. *et al.* (2004a). User Strategies in Video Retrieval: A Case Study. *Proceedings of CIVR 2004*, July 21–23, Dublin, Ireland. *Lecture Notes in Computer Science* **3115**, 6–14.

Hollink, L., Schreiber, A. Th., Wielinga, B. J. and Worring, M. (2004b) Classification of User Image Descriptions. *International Journal of Human – Computer Studies*, **61**(5), 601–26.

Hung, T-Y. (2005) Search Moves and Tactics for Image Retrieval in the Field of Journalism: A Pilot Study. *Journal of Educational Media and Library Sciences*, **42**(3), 329–46.

Jaimes, A. and Chang, S-F. (2000) A Conceptual Framework for Indexing Visual Information at Multiple Levels. *Proceedings of IS&T/SPIE Internet Imaging* I, January 28–30, San Jose, CA, USA. *SPIE vol. 3964.* pp. 2–15.

Jörgensen, C. (1998) Attributes of images in describing tasks. *Information Processing and Management*, **34**(2/3), 161–74.

Jörgensen, C. (1999) Access to Pictorial Material: A Review of Current Research and Future Prospects. *Computers and the Humanities*, **33**(4), 293–318.

Jörgensen, C. (2003) *Image Retrieval: Theory and Research*, Scarecrow Press, Lanham, MD.

Jörgensen, C. (2007) Image Access, the Semantic Gap, and Social Tagging as a Paradigm Shift. *Proceedings of the 18th Workshop of the American Society for Information Science and Technology (ASIS&T) Special Interest Group in Classification Research*, October 20, Milwaukee, Wisconsin, USA. ASIS&T, Medford, NJ.

Jörgensen, C. and Jörgensen, P. (2005) Image querying by image professionals. *Journal of the American Society of Information Science and Technology*, **56**(12), 1346–59.

Keister, L. H. (1994) User types and queries: impact on image access systems. *Proceedings of the 57th Annual Meeting of the American Society for Information Science (ASIS)*, October 17–20 Alexandria, VA, USA. Learned Information, NJ, pp. 7–22.

Laine-Hernandez, M. and Westman, S. (2006) Image Semantics in the Description and Categorization of Journalistic Photographs. *Proceedings of the 69th Annual Meeting of the American Society for Information Science and Technology (ASIS&T)*, November 3–8, Austin, Texas, USA. ASIS&T, Medford, NJ.

Markkula, M. and Sormunen, E. (2000) End-User searching challenges indexing practices in the digital photograph archive. *Information Retrieval*, **1**(4), 259–85.

Marsh, E. E. and White, M. D. (2004) A taxonomy of relationships between images and text. *Journal of Documentation*, **59**(6), 647–72.

McDonald, S. and Tait, J. (2003) Search Strategies in Content-Based Image Retrieval. *Proceedings of the 26th ACM SIGIR Conference on Research and Development in Information Retrieval (SIGIR)*, July 28–August 1, Toronto, Canada. ACM Press, pp. 80–87.

Ørnager, S. (1995) The Newspaper Image Database: Empirical Supported Analysis of Users' Typology and Word Association Clusters. *Proceedings of the 18th Annual International ACM SIGIR Conference on Research and Development in Information Retrieval*, July 9–13, Seattle, Washington, USA. ACM Press, pp. 212–218.

Panofsky, E. (1955) *Meaning in the Visual Arts. Papers in and on Art History*, Doubleday Anchor, New York, NY.

Santini, S., Gupta, A. and Jain, R. (2001) Emergent Semantics through Interaction in Image Databases. *IEEE Transactions on Knowledge and Data Engineering*, **13**(3), 337–51.

Shatford, S. (1986) Analyzing the Subject of a Picture: A Theoretical Approach. *Cataloging and Classification Quarterly*, **6**(3), 39–62.

Urban, J. and Jose, J. (2007) Evaluating a workspace's usefulness for image retrieval. *Multimedia Systems*, **12**(4/5), 355–73.

von Ahn, L., Ginosar, S., Kedia, M. *et al.* (2006) Improving accessibility of the web with a computer game. *Proceedings of the SIGCHI Conference on Human Factors in Computing Systems*, April 22 – 27, 2006, Montréal, Québec, Canada. ACM Press, pp. 79–82.

Westman, S. and Oittinen, P. (2006) Image Retrieval by End-Users and Intermediaries in a Journalistic Work Context. *Proceedings of the 1st IIiX Symposium on Information Interaction in Context*. Royal School of Library and Information Science, October 18–20, Copenhagen, Denmark. ACM Press, pp. 171–187.

5

Web Information Retrieval

Nick Craswell and David Hawking

5.1 Introduction

This chapter outlines some distinctive characteristics of web information retrieval, starting with a broad description of web data and the needs of web searching users, then working through ranking and design issues that arise. It is intended as an overview of what makes web information retrieval different, and provides an introduction to some fundamental literature in the area.

5.2 Distinctive Characteristics of the Web

The characteristics that make web search different from some other types of search stem mainly from characteristics of the data and from distinctive web user behaviour.

5.2.1 Web data

Web documents are known as 'pages', each of which can be addressed by an identifier called a uniform resource locator (URL). For example: `http://www.acm.org/pubs/contents/proceedings/series/sigir/`. Web pages are usually grouped into 'sites', sets of pages published together. For example: `http://www.acm.org`.

Pages are remarkably versatile. A web page can play the same roles as a news article in a conventional IR corpus, such as providing information for a report or helping to answer a question – indeed these days most news articles are also published as web pages. However, the web also contains pages of many other types.

Many web pages exist purely to help users navigate to other pages in the site, such as the 'entry page' or 'home page' of a site. Some pages provide an interactive service such as a search form. Some pages provide a commercial service, allowing users to shop for products online. Some pages are generated dynamically and intended to be used once, such as a search engine results list page. Other pages are front-ends to databases, so each page represents part of an underlying relational or XML database. That database might be generally useful (`http://imdb.com`) or very large and specialised (`http://www.ebi.ac.uk/flybase/` – a database of the *Drosophila* genome) to the extent that it is not clear that a general-purpose engine should index it. Some pages are error pages, indicating

Information Retrieval: Searching in the 21st Century   edited by A. Göker & J. Davies
© 2009 John Wiley & Sons, Ltd

that the user has reached a bad URL, and others are intended to direct the user to other pages ('you are now leaving NASA' or 'doorway page to my site'). Some pages exist to contain a media file or an interactive game. Web pages are used in these ways and many more, and we often observe a single page that can be used in several of these ways. The notion of types of page, and types of user interaction, will be revisited throughout this chapter.

Web search engines discover pages by 'crawling' the Web, discovering new pages by following hyperlinks. Access to particular web pages may be restricted in various ways. For example, pages on a corporate intranet may be visible only to those who can authenticate themselves as company employees. Web publishers may also impose additional restrictions on access by search engines, using the `robots.txt` Robots Exclusion Protocol (Koster 1994). They may do this to prevent load on their servers, to help search engines avoid wasting resources crawling ephemeral or low-value content, or for commercial reasons. Finally, search engines can, in general, only crawl pages which are linked to from pages already in the crawl. A page with no incoming links is extremely unlikely to be indexed by a general web search engine.

The set of web pages which can not be included in search engine indexes is often called 'the hidden web', 'the deep web', or 'web dark matter' (Bailey *et al.* 2000).

5.2.2 Web structure

Web pages are connected by two major structures, the link graph and the URL hierarchy. We describe basic characteristics of the structure here, and then the techniques that arise from the structure in later sections.

The link graph is formed when web pages hyper-link to each other. The source page contains a reference to the URL of the link's target page. In many link graphs, there is a power-law degree distribution, which means that the probability of a link having degree i is proportional to $1/i^x$. One large-scale study (Broder *et al.* 2000) found $x > 2.1$. The power-law distribution means that most pages have very few incoming links, but a few have a very large number.

An outgoing hyperlink is represented by an 'anchor' in a web page. Usually, when the page is displayed in a web browser, the anchor is highlighted. The person may click on it in order to follow the link. If the anchor is textual we refer to it as the link's 'anchor text'. Anchor text often provides a useful short description of the link target, The anchor text of all the incoming links to a particular page can be aggregated for retrieval purposes into a useful surrogate for or adjunct to the actual document text. Anchors may be interpreted as labels on arcs in the general link graph, permitting the creation of topic-specific sub-graphs.

Incoming anchor text can permit retrieval of useful pages (e.g. images, or pages in another language) whose textual content doesn't match the query. For example, the incoming anchor text for the page `http://www.robotstxt.org` includes many different descriptive phrases, such as 'robots.txt' (most common), 'http://www.robotstxt.org/' (very common), 'robots.txt system', 'RobotsTXT.org', 'Robots Exclusion Protocol', 'Robots Text','robots', 'here', 'robots.txt (Robots Exclusion Standard)', 'robots.txt file', 'robots.txt protocol', 'Martijn Koster's site about the Robot Exclusion Protocol.', 'Le protocole robots.txt', 'Crawler', 'den sedvänja som gäller på internet', ...

The URL hierarchy is a separate structure from the hyperlink graph. Taking as an example, the URL `http://blah.dog.com/dir1/dir2/file.html`, we can say that any two pages that are both part of `.com` are loosely related. Pages that are part of `dog.com` are more strongly related and two pages from the same host `blah.dog.com` even more so. Then those on the same host that share some part of their directory path, such as `/dir1/` are further related. Site entry points, which are often the desired result of navigational queries, tend to be shallow in a host's URL hierarchy. Huberman and Adamic (1999) found that the number of pages on each host also follows a power-law distribution. Using sampling techniques, Lawrence and Giles (1999) estimated that the mean number of pages per server was 289.

The link graph and URL hierarchy are related. For example, Bharat *et al.* (2001) found that 76% of links in each of three separate crawls are within-host.

5.2.3   User behaviour

Much of the work that makes web information retrieval distinctive has arisen from understanding user needs. Users can research topics of interest, which is a traditional information retrieval activity. However, web pages link to each other; the user may wish to reach a page so they can browse further. Web pages are interactive; the user may wish to visit the page so they can shop, book travel or search a database.

The seminal contribution in describing web user needs was the taxonomy due to Broder (2002), first presented during the web plenary session at TREC-2000 and later published in SIGIR Forum. Three types of information need were defined: *navigational* where the immediate intent is to reach a particular page or site; *informational* where the user seeks information relevant to their topic of interest; *transactional* where the intent is to perform some web-mediated activity. In a followup study by Rose and Levinson (2004), roughly 60% the queries were classified as informational, 25% as resource/transactional and 15% navigational.

Study of navigational search proved to be the most immediately fruitful, because it can be evaluated using a single answer. Even before the taxonomy was presented, a number of studies were underway with navigational evaluation (Singhal and Kaszkiel 2001; Craswell *et al.* 2001; Bharat and Mihaila 2001). Since then there have been further experiments in the context of the TREC Web Track (Hawking and Craswell, 2005). Ranking methods that were developed to answer navigational searches are described in the next section.

5.2.4   User interaction data

Another important way in which web search differs from traditional informational retrieval is in the truly massive amount of user interaction data collected by the major search engines. The most popular web search engines are believed to log of the order of a billion interaction records each day. (A gigabyte of query text is accumulated approximately every 60 million queries.) These interaction logs include the queries people typed and the result links they clicked on. From them, search engines can assign general popularity ratings to pages (analogous to counting incoming links; Culliss 1999). They can also use clicks to associate queries with pages and then use the associated queries in ranking (analogously to anchor text; Xue *et al.* 2004; Hawking *et al.* 2006). More detailed interaction data collected by the user's own computer can be used to improve search rankings (Agichtein *et al.* 2006).

Click patterns can be used to deduce relationships between pairs of queries and/or pairs of documents (Jones *et al.* 2006; Craswell and Szummer 2007). Search engines may use click data as low-cost relevance judgments for evaluating and tuning their systems (Joachims 2002; Joachims *et al.* 2005). Interaction sequences can also be used to suggest spelling corrections or related queries.

Unfortunately, academic researchers have little access to search engine logs because of privacy concerns. Interaction sequences in logs may reveal a great deal of private information, *even if the data contains no usernames or IP addresses*. A good-faith attempt to make anonymised logs available to researchers in 2006, led to the unfortunate consequences described by Barbaro and Zeller (2006).

5.3   Three Ranking Problems

To be retrieved and presented to a user by a web search engine, a page may have to pass three 'hurdles'. First the page has to rank sufficiently well in crawler prioritisation, otherwise it will never make it into the index. Second, if the system prunes its searches and uses a global ordering of pages, which is one technique for efficient query processing on very large indexes, the page must be high enough in the index to avoid being pruned. Finally, having been crawled and not pruned, the page must rank highly enough in the results list that the user sees it.

This gives three ranking problems. Effective ranking given a query has been studied for decades in the field of information retrieval (see Chapter 1), but in the web case specific techniques are needed to satisfy certain types of search. Here we focus on the navigational search type. The other two ranking problems, in the crawl and in the index, are query-independent and have not been studied as rigorously. This section describes all three ranking problems.

5.3.1 Retrieval

Traditional relevance ranking technologies attempt to retrieve pages that contain relevant text. So for the query 'australian government' a good result is any page that talks about important aspects of the topic. However, another plausible scenario for the same query is that the user is performing a navigational search, where there is one correct answer http://www.australia.gov.au. In that case, different ranking techniques are required. Regardless of whether the user's need is informational, transactional or navigational, that site is likely to be a very good answer.

First consider the evidence available[1]. The goal is to get that page to the top of the ranking in a set of over 100 million matching pages (estimates are taken from Google). Using traditional ranking based on term frequencies (TF), assuming no stemming, the page has TF(australian)=2 and TF(government)=8. This is not strong enough to guarantee a top ranking for the page, as other pages may contain more (and a greater density of) occurrences of the query terms. New evidence is needed.

The two most successful sources of evidence for improving navigational ranking come from the link structure and URL hierarchy. The www.australia.gov.au page has almost 6000 incoming links which in the power-law distribution of the web means that this page is unusually heavily linked. The URL signal that helps this page is the fact that it is a short URL and that it is the root page for a hostname.

Many incoming links for this page are likely to contain the words 'australian' and/or 'government' in their anchor text.

To be useful, query-independent evidence such as URL type, URL length, PageRank (Page et al. 1998), inlink count, click popularity, etc. must be combined with query-dependent relevance scores derived from document text and with scores derived from external text such as URL words and anchor text. The best-known web search engines are believed to use ranking functions which combine hundreds of features.

Chapter 1 presents a number of IR models which are capable of incorporating multiple sources of evidence. The reader is referred to Kraaij et al. (2002), Craswell et al. (2005), and Ogilvie and Callan (2003) for web-oriented examples of combining evidence.

5.3.2 Selective crawling

Crawling can be thought of in terms of a queue, containing the URLs that the crawler has not yet requested. In the beginning, the queue contains a set of seed URLs. Then the crawler takes an URL from the queue, fetches that page (if available) and extracts its links. If the URLs have not been seen before, the crawler adds them to the queue. This can lead to the discovery of a massive set of pages, if unrestrained. Today, large engines have crawls with billions of pages. In practice the crawl may not be strictly behaving as in queueing because of mechanisms used to achieve scalability, including distribution of the queue across multiple crawling machines and processes (Heydon and Najork 1999), but it is still useful to think of it in terms of queueing. Recent large-scale crawls for research purposes are described by Baeza-Yates et al. (2005) and Fetterly et al. (2003).

The scope of the crawl may be limited in a number of ways. The crawler may have a set of rules based on URLs which restrict the pages it covers, for example, only crawling pages from a given list of hosts and domains. It may also have depth limits, only adding pages that are a certain number of link 'hops' from a seed page. Crawls usually also have an overall time limit. In theory it would be possible to continue the crawl until the queue becomes empty, getting 'all' the pages in a given scope. In practice, the time limit is useful because at some stage the crawl will become unproductive. The queue will contain only pages that are duplicates of existing pages, or perhaps pages from a very

---

[1] These observations were made around May 2005, and may have changed since. However, they still illustrate the point.

large site that is of low value. If there are crawler traps that generate an infinite number of interlinked URLs without useful new content, the crawl queue may never become empty.

Given that the crawl will stop early, with a non-empty queue, crawl priority becomes an interesting question. When the crawl terminates, a set of pages will have been crawled and a set of pages will have been missed. Which pages make it into the crawled set is determined by crawl priority.

The simplest crawl priority is given by a breadth-first traversal, which is equivalent to a first-in first-out (fifo) crawl queue. This can also be thought of as a 'generational' crawler, where the first generation is the seed set, the second generation is those pages of link distance 1 from the seed set and so on. Given that the seeds are important, it is not difficult to imagine that pages a shorter distance from the seeds will tend to be more important, under a basic locality assumption. Also, informal observations have shown that SPAM network pages and pages generated by an infinite crawler trap will be rare early in the crawl (assuming a good seed set) and will dominate later generations of the crawl. Najork and Wiener (2001) observed that crawling breadth-first also yields pages with high PageRank, which is further evidence that breadth-first traversal is desirable, in applications where crawling high-PageRank pages is desirable.

Another approach is to use a priority queue, rather than a fifo queue (Baeza-Yates et al. 2005). In that case, each URL in the queue is assigned a priority score, and the top-scoring page is dequeued. An important paper (Cho et al. 1998) studied such prioritisation in the context of the Stanford web, using indegree, PageRank and match with a driving query. Chakrabarti et al. (1999) also considered crawling focused to a particular query as did a number of other studies (De Bra et al. 1994; Diligenti et al. 2000; Aggarwal et al. 2001). The general conclusion is that it is possible to effectively prioritise uncrawled pages, for example to selectively fetch pages that match a certain topic, based on the content of linking pages that have already been crawled. This relies on anchor text that describes the target page and topic locality, that two pages on the same topic are more likely to be interlinked than two pages that are not (Davison 2000).

The crawl queue prioritisation discussed so far was in terms of a newly created crawl. In the case of crawl maintenance it is possible to consider crawl priority and crawl updating as two separate concerns (Cho and Garcia-Molina, 2000). For crawl priority, mechanisms such as PageRank prioritisation or match with a driving query can still be used. The crawler must prioritise the crawled pages against the known uncrawled pages. Using numbers from Richardson et al. (2006), if we have a 5 billion page crawl, we can identify a further 20 billion URLs that are not yet crawled. So the selection problem can be seen as picking the best 5 billion from a set of 25 billion known URLs. The update problem is separate, involving revisiting the 5 billion known URLs to see if they have changed or been removed, and updating the index accordingly (Edwards et al. 2001).

In a general crawl, rather than a topic-focused one, PageRank seems to be the most promising method. Good navigational answers and high PageRank are associated, as described in the previous subsection, so prioritising high-PageRank pages will tend to ensure that the navigational answers are selected first. If there are only one or two good navigational answers per site, and each site has on average 100 pages, then it's not surprising that PageRank-ordered crawling can get these top $1-2\%$ of navigational answers. However, with a more sophisticated notion of which pages are desirable to crawl (many past studies have both prioritised using partial-crawl PageRank and evaluated using full-crawl PageRank (Cho et al. 1998; Baeza-Yates et al. 2005), it is possible that a much more sophisticated and effective crawl priority system could be developed. This system could take PageRank as an input, but also other aspects of the link structure, URL hierarchy and even page content.

5.3.3   Index organisation

There are many efficiency and engineering aspects of indexing and query processing which are critical to real web search engines, but which are not covered here. The reader is referred to Chapter 12 for a discussion of parallel IR and to Zobel and Moffat (2006) for a survey of state-of-the-art indexing techniques.

Recent papers (Long and Suel 2003; Michel et al. 2005) have described a method for index organisation and pruning which works well for large-scale retrieval, but requires a global ordering of pages. The method is as follows. Rank all crawled pages according to some query-independent metric such as PageRank (Long and Suel 2003), assigning document identifiers (docids) in increasing order. Then arrange each postings list in docid order. This is a global order because it is used to order all the postings lists, as opposed to other schemes such as impact ordering (Anh and Moffat, 2002). Use an AND query and skip lists (Pugh 1990) so that for a multi-term query, the page-at-a-time processing can skip to pages containing all terms, avoiding processing of pages that only contain some terms. Broder *et al.* (2003) show how the AND requirement can be weakened while maintaining efficiency.

Finally, given that PageRank is used in the results ranking as well as the index ordering, the pages that are processed first will be those with the highest potential score (due to their high PageRank component). This means that processing can be pruned (stopped) when it is decided that it is unlikely that any further pages will be found that outscore the pages already processed.

5.4   Other Web IR Issues

5.4.1   Stemming

Despite the consistent observation in TREC that moderately aggressive stemming, (e.g. Porter 1980), tends to increase effectiveness, applying the TREC stemming approach in web search is problematic for a number of reasons.

First, the chance of stemming errors is higher on the web because the web is multilingual and its vocabulary size is orders of magnitude larger than that of TREC, boosted by huge numbers of business names, product names, domain names, acronyms, placenames and names of people. It is feasible to recognise the language of web documents with reasonable accuracy and thus to apply a language-specific stemmer during indexing. However, recognising the language of a one- or two-word query is far less reliable, particularly when spelling errors are common, and where searchers often lack the ability to produce accented letters. Similarly, although one could selectively avoid stemming based on the case of a word, searchers are notorious for failing to supply reliable case information in their queries.

Second, stemming errors are more likely to change the overall meaning of a query when queries are only one or two words. One of the present authors is particularly sensitive on this subject, his name having been confused by at least one search engine with a football club (the Hawks).

Third, stemming errors resulting in non-matching documents at the top of a search engine ranked list are more visible than in TREC *ad hoc*. In the latter, one may need to read the whole document to be sure it is irrelevant. Furthermore, a completely off-topic document returned a the top of the ranking has only a tiny depressing effect on average precision, particularly when averaged over 50 topics. In contrast, the same misfortune in a web search is made obvious through display to the searcher of the result's title, URL and highlighted query words. Anecdotally, searchers view such failures with amusement, frustration, or bewilderment.

In practice, Web search engines typically apply a very light form of stemming in which the benefit strongly outweighs the risk of error.

5.4.2   Treatment of near-duplicate content

Established IR collections include instances of duplicate or overlapping content (such as different versions of the same news report) but on the web the problem is magnified by the presence of systematic causes of duplication and by the resource implications of unnecessary crawling and indexing.

Certain websites set up to distribute information or software may be subject to excessive load at peak times. Their content may be replicated on webservers in other countries to distribute the load, or for purposes of reliability or preservation. In some cases this is handled imperceptibly behind the scenes, the content always published from a single web address. In other cases, the replicated information is published on a different site or *mirror*, meaning that identical documents appear on the web with different URLs. e.g. `mysoft.com/downloads/` may appear also as `mirror.xyz.ac.uk/mysoft/downloads/`. The reader is referred to Bharat and Broder (1999) for a discussion of techniques for detecting mirroring.

Aliasing of hostnames and of pathnames within a host is frequently practised. For example, the webhost `jupiter.xyz.ac.uk` may also be referencable as `www.xyz.ac.uk`, meaning that all its pages are accessible via two different URLs. Similarly, some (but not all) webservers treat URLs case-insensitively, meaning that e.g. `www.xyz.ac.uk/Sociology/Intro.HTM` can equivalently be accessed and linked to by millions of case variants. Often the content of pages fetched from different but equivalent URLs is exactly identical and the equivalence can be detected by simple checksum matching. However, when documents are dynamically generated, equivalent pages may be published with slight additions or alterations. For example, a page may include its own URL, the date at which it was accessed or a visitor counter. In these cases, more sophisticated techniques must be used to reliably detect the equivalence. See Broder (1997) for a discussion of *shingling*.

Dynamic generation allows a fixed unit of content to be presented in a profusion of different forms, potentially with different colours and fonts, different images and with different sets of links. This may happen when an organisation operates a number of separate businesses, each with their own branding. A more extreme example is provided by the online book retailers who generate websites to sell books on behalf of the bookselling giant Amazon, receiving a small referral commission. They all extract the basic content from the same Amazon database, but compete for traffic by packaging and presenting it differently.

Dynamic URLs often include one set of parameters which identify the basic content, another set which control how it is presented, and a session-id parameter which may affect what is presented or which may merely enable tracking of an individual's browsing activity, e.g. `x.y.z/gen.asp?docid=123&stylesheet=look3&sid=11293129`. Search engines typically attempt to simplify URLs to avoid unnecessary fetching of the same basic content. They may strip off session-ids and presentation parameters. This can dramatically reduce wasted crawling effort, but may sometimes result in missed content.

Correct treatment of near-duplicate content affects search engine operation in important ways. Undetected duplication wastes valuable crawling time, increases network traffic costs, consumes extra disk space, extends indexing time and potentially clutters up results lists with repetitions. Not only that but duplicates add phantom nodes to the web graph with adverse consequences to graph-based ranking measures such as PageRank.

5.4.3   Spelling suggestions

Web search engines receive a significant number of misspelled queries. Some provide a very helpful, 'did you mean X?' service. Due to the previously noted multilingual, neologism-prone characteristics of web publishing, it is not at all feasible to make spelling suggestions by approximate searching within a normal dictionary word list. Nor is it useful to perform simple-minded approximate matching against the full vocabulary list of the web, as web authors, like web searchers, are highly prone to spelling errors. Very few misspellings would be detected by this method and suggestions made could easily be foreign words or spelling errors.

Details of commercial search engine spelling suggestion algorithms have not been published to our knowledge, but it is very likely that they are based on analysis of query logs. Cucerzan and Brill (2004) describe and evaluate methods for spelling suggestions based on logs.

5.4.4   Spam rejection

To quote Broder once more, web search engines operate in an adversarial information retrieval environment. Many people and organisations who make money via the web have a strong financial incentive to try to ensure that their sites appear prominently in web search results. A search engine optimisation industry has grown up to service this need, providing tools and services and even running conferences on how to improve website visibility within search engine rankings.

At the benign end of the optimisation spectrum, web publishers may be encouraged to use good publishing practices – ensuring that titles, anchor text and content match the queries for which the page is a good answer and creating simple site and link structures to encourage linking and facilitate crawling. At the other end of the spectrum, spammers pursue their clients' interests regardless of harm to the interests of searchers, search engines and other publishers. They try to have their target pages highly ranked against queries which bear no relation to the subject of the page. They deliver different content to crawlers than to site visitors (*cloaking*) and they set up forests of interlinked artificial sites to increase PageRank and anchortext scores.

If search engines published the techniques they use to reject or down-weight SPAM pages, spammers would immediately be at work on better methods to defeat them. However, it seems clear that search engines use a combination of automatic SPAM classifiers and manual blacklists to cut down on unnecessary crawling and to keep unwanted pages out of search results. A SPAM score may be assigned to a page based on the presence of unusual features in the text, on its participation in unusual link structures, and on its link distance from known black sites (trustrank). This score could be used in setting crawl priorities and in assigning global document numbers (see Section 5.3.3.).

5.4.5 Adult content filtering – genre classification

Web search engines typically provide family-friendly filtering of search results in an attempt to prevent the display of sexually explicit material to minors. Much of this material can be detected on the basis of the presence of certain words and phrases, but a more sophisticated classifier might take into account words in anchor text, words in URLs, links from other sexually oriented sites, and even the relative prevalence of flesh tones in images.

Sexually explicit pages are one example of a web page genre. It is possible to build classifiers for other genres (e.g. Glover 2001; see also Chapter 3) and potentially to use them either in normal ranking or in personalised or contextualised result rankings.

5.4.6 Query-targeted advertisement generation

Web search engine companies rely on advertising revenue to fund their operations and to generate profits. In the early days of the web, advertisements were indiscriminately published on search result pages for a small per-view charge to the advertiser, but conversion rates were low and revenues declined. The introduction of targeted advertising by Overture and later Google showed that, when advertisements are related to the search query, conversion rates are much higher and advertisers are willing to bid substantial sums of money for query triggers related to their business. At the same time searchers are happier, because fewer advertisements are displayed and those which are displayed are more likely to be useful.

Search engines must operate a complex infrastructure to support the auctioning of query words and the accurate billing of advertisers. There is also a science in optimising which advertisements are displayed in order to maximise revenue to the search company. These issues were discussed by Ribeiro-Neto *et al.* (2005).

5.4.7 Snippet generation

Results presented by early web search engines provided a hint as to the content of each page in the form of a fixed snippet from the head of the text. Such snippets can be generated by means of a simple lookup. More recent search engines now provide one or two short excerpts from the document, selected on the basis of localised relevance to the query. Tombros and Sanderson (1998) describe a basic method for query-biased summarisation and evaluate its effectiveness, but do not address the issue of efficient generation of such summaries. Turpin *et al.* (2007) have investigated the use of static compression schemes, caching and sentence reordering in improving the efficiency of snippet generation.

For the major search engines, snippet generation poses significant computational load. Since ten results are typically generated per query, the processing of a hundred million queries requires the generation of a billion query-biased summaries.

5.4.8   Context and web information retrieval

Even if a web search engine returns search results which maximise utility across all the users who submit that query, a minority may be poorly served. For example, those searching for information on a Tony Blair other than the former UK Prime Minister.

Contextualisation of search results, as described in Chapter 7, can be applied by web search engines to address this type of case. At a simple level, major search engines now routinely detect the country from which each search originates (based on IP address) and use this to:

- Set the language of the interface;
- Present advertising relevant in the local market;
- Offer spelling suggestions appropriate to the language; and
- Bias search results toward sites operating in that country.

For example, in response to the query 'bank', a major web search engine currently (May 2008) returns Bank of America, Migros Bank, Barclays Bank, or Commonwealth Bank in first rank, depending upon whether the search is conducted in the US, Swiss, UK or Australian contexts. In each case, the search interface assumes that the searcher speaks the dominant language of that country and is most interested in advertisements targeting that market.

This is called *localisation*.

Personalisation of search attempts to provide search results tailored for an individual rather than a nation.

A search engine can keep track of a user's interactions (e.g. the queries they issued and the results they clicked on) using cookies and search logs and can potentially use this data to disambiguate or expand queries based on inferred interests or preferences.

However, much of the research on web search personalisation assumes that the personalisation context is deduced at the client side and is communicated to the search engine in the form of an expanded or transformed query. Personalisation at the client side means that all of a person's local electronic information (e.g. office documents, emails, and web pages downloaded) as well as a full record of their Web interactions can be mined for personalisation context. Teevan *et al.* (2005) and Chirita *et al.* (2007) claim significant improvements in the quality of search results as a result of client-side personalisation.

5.4.8.1   Risks of personalisation and localisation

Personalisation and localisation can make things worse if incorrect assumptions are made about the searcher's skills, preferences or tasks. A holiday in Brazil doesn't necessarily make an English woman proficient in Portuguese! Furthermore, despite relaxing in Rio, she may still want to perform searches oriented to her job or to her life in the UK

A good search engine will allow searchers to override automated personalisation and localisations.

Personalisation of search results has potential privacy implications, as illustrated in the following example. A woman uses her husband's laptop to search the web. For some strange reason, her innocent, but ambiguous query produces results which all relate to an unfamiliar sexual practice ...!

5.5   Evaluation of Web Search Effectiveness

Progress of academic research in Web IR in the 1990s was impeded by the use of an established but inappropriate evaluation methodology. The task, judgments and measures used in the first TREC Web

Track (Hawking *et al.* 1999) were inherited from TREC *ad hoc*. Inherent in this methodology are assumptions which can be seen to be violated by the majority of Web searches:

1. The purpose of all search is to find documents whose text contains information relevant to the topic of the query.
2. The searcher's goal is to find as many relevant documents as possible (implicit in the use of average precision as the key measure in TREC-8).
3. It is appropriate to regard as relevant documents which contribute very little information and those which repeat information contained in other relevant documents.
4. All relevant documents are of equal value.

These assumptions are most inappropriate when search is navigational in intent, but even when this is not the case, Web searchers typically prefer site homepages and pages which provide access to a service. It seems safe to assume that the majority of people who type the query 'passport' to government or whole-of-web search facilities, want to obtain or renew a passport, rather than write an essay about passports! Webpages describing the history of passports or incidentally mentioning the use or loss of passports are of almost no value in this context.

The typical web searcher clicks on only one or two results. A study of search logs for the Australian Stock Exchange website showed that only one search in 35 000 resulted in clicks on all of the first ten results.

The evaluation procedures initially adopted in the TREC Web Track were unable to reveal any benefit for the link-based methods on which successful commercial engines rely.

Hawking and Craswell (2005) document the evolution of the Web Track toward greater web realism through: engineering a more representative small corpus, making use of queries from search engine logs, and studying a series of deliberately web-oriented tasks.

5.5.1   TREC-9 Web Track: Realistic queries, rich link structure, traditional IR task

The TREC-9 'Main Web' task (run in 2000) went a long way toward web realism in everything but the task. It used a corpus (WT10g) which had been specifically engineered to achieve a high degree of interlinking and to mimic other properties of the Web (Bailey *et al.* 2003). It also made use of queries selected from a public web search engine log and augmented the normal TREC *ad hoc* relevance scale with a 'highly relevant' category. The selected queries were slightly longer (3.4 words) than the reported web average of 2.3 words.

Queries were selected by NIST assessors who imagined an information need behind the query and documented it in the description and narrative fields of the TREC topic statement.

For example, topic 488: 'newport beach california' was interpreted as

Description. What forms of entertainment are available in Newport Beach, California?
Narrative. Any document which refers to entertainment in Newport Beach is relevant. This would include spectator and participation sports, shows, theatres, tourist attractions, etc.

This interpretation is only one of the possible 'needs behind the query'. Other searchers posing this query might be interested in getting a general overview of the place where their boyfriend goes for family holidays, or in finding: the nearest airport to Newport Beach, photographs of Newport Beach, history of Newport Beach, accommodation in Newport Beach, driving instructions, a map, real estate investment potential, retirement homes, etc.

In TREC-9, 35 documents were judged relevant to 'newport beach california' and one was judged highly relevant. The latter was `www.commpro.com/anaheim/tourism/beaches.html`. It contained information about Newport, Laguna, Huntington, San Clemente and Balboa beaches, but no Newport-specific links. The only specific information about Newport Beach was:

> At Newport, there's also no shortage of sights and activities. The Upper Newport Bay Ecological Reserve is a 700-acre site that serves as home to wildlife and migratory birds. The protected coves of Corona del Mar Marine Life Refuge also offer tide pools alive with sea urchins, starfish and octopus. And at Corona del Mar State Beach, barbecue pits, showers, a snack bar and restrooms offer an ideal place for a family outing. And, early risers can watch as the fishermen of the Dory Fishing Fleet set out their catches of the day for sale at the foot of McFadden's Pier.

This would be a poor result for a search engine to return at rank one, because it fails to address the needs of searchers who are not seeking leisure activities. Even for those who are, there are no links from which accomodation or tours may be booked and none to maps or driving instructions.

In theory at least, successful web search engines would attempt to maximise the utility of their search results page, summed over all searchers.

This might be achieved by including *portal* pages providing search-and-browse access to all or most aspects of the topic, such as site entry pages or hub pages oriented toward the broad topic. It would be appropriate to include results which are useful for more specific searcher needs, but diversified across the range of likely needs. In these more specific cases, too, access to services, pictures and links to other resources may be more useful than paragraphs of informational text.

At the time of writing, the three largest web search engines all return `www.city.newport-beach.ca.us` as the top ranked result. It is published by the Newport Beach city council and provides links to comprehensive and authoritative information about just about every aspect of Newport life and commerce, but no informational text. This page was not included in the WT10G collection, but even if it had been, it is unclear how it would have rated with respect to the task as defined.

We suggest that the reader try this query themselves on their favourite whole-of-Web search engine, noting the characteristics of the results returned and judging them, both against the range of needs which might prompt a person to pose that query, and as if the task were to find paragraphs of relevant text.

5.5.2   Evaluation using web-specific tasks

In the TREC-2001 Web Track, the same corpus was used as in the previous year, but a task was introduced which makes no sense outside a web context – given a name as a query, find the homepage of an entity of that name.

In this task, retrieval methods using anchor text and URL priors were shown to clearly outperform methods based on text content alone.

Subsequent Web Track tasks included named page finding, where web-specific methods were found to bring some, but not as much benefit, and topic distillation, where the task was to find entry pages of authoritative sites relating to a broad topic.

5.5.3   Future directions for web IR evaluations

The web-oriented evaluations conducted in the TREC Web Track are closer to web reality than the earlier attempts to apply TREC *ad hoc* methodology. However, they represent a substantial oversimplification.

A person may enter a query such as 'newport beach california' without knowing whether there is in fact a website devoted to that topic. It may be true that such a person obtains greatest value from home pages of official Newport sites, but they may still gain substantial value from other types of resource, such as recent news reports and individual pages from which Newport holidays or accommodation may be booked.

In reflecting the value of a page of search results to an individual, a scoring system may need to encompass a very wide range of utility values for individual pages, and to sum scores only as far down the ranking as the person is prepared to look, while ignoring or even penalising results which are not useful to that person or which provide information or services which are very similar to those already identified.

In evaluating the performance of a search engine on this query, as opposed to its value to an individual or to a class of similar individuals, it is necessary to sum its individual ratings over all the individuals who might pose the query.

5.5.4   Comparing results lists in context

Most measures employed in web and general IR, rate a results list by summing relevance values (possibly discounted, see Järvelin and Kekäläinen 2000) previously assigned to individual pages in isolation. This approach has the strong advantage (at least in moderate sized collections) that judgments are reusable. However, the score assigned to a result list in this way may substantially overestimate the actual value to a searcher:

- By assigning individual page scores which don't match the searcher's own assessment;
- By double-counting duplicate pages or pages with overlapping content;
- By counting pages which present information the searcher already possesses;
- By presenting information about aspects or interpretations of the topic which don't contribute to satisfaction of the searcher's goal; and/or
- By counting pages further down the list than the searcher is prepared to look.

Thomas and Hawking (2006) propose an alternative method in which pairs of results lists are compared side-by-side by searchers in real-life situations. In one experiment, volunteers substituted the side-by-side search interface for their normal web search engine, whenever they conducted a search in the course of their work, study or leisure activities. Participants were, of course, able to opt out for any particular search. After they entered a query, they were shown two sets of anonymised results, one from search engine A and one from engine B. Assignment of A and B to left and right was randomised. The searcher was given the option to record a preference for the list on the left or the one on the right or to indicate that the two lists were of equal value. A and B were then compared on the basis of the number of searchers who overall favoured one engine versus the other.

Although comparisons between pairs of results lists are not reusable and may not be sensitive enough to detect very small differences, this evaluation methodology has a number of advantages. Judgments are of naturally occurring information needs, conducted by the actual searcher, in the full context of the need, using whatever judging criteria matches their purpose in conducting the search.

5.5.5   Evaluation by commercial web search companies

Web search engine companies have strong commercial incentives to evaluate the quality not only of ranking algorithms, but of result presentation, summary generation, spelling suggestion and, particularly, advertisement placement systems.

Tuning of ranking functions involving hundreds of variables, such as those used by the major web search engine companies, requires sophisticated evaluation over very large sets of queries. It is well known that these companies invest considerable resources in obtaining judgments for both competitive analysis and tuning.

Taylor *et al.* (2006) describe a method for optimising a function of 375 variables using 5-level relevance judgments for training sets for 2048 queries, and speculate that even larger sets are used in commercial practice.

As mentioned earlier, search engine companies are also able to use the vast amounts of user interaction data (click logs) they accumulate as low-cost relevance judgments. Such judgments may be in the form of relevance scores for individual documents or as preference judgments between pairs of documents. As noted by Joachims *et al.* (2005), when a person scanning a list of search results skips over a document D1 and clicks on another D2 at lower rank, this is a strong indication of preference for D2 over D1.

Search engine companies are able to randomly assign a fraction of the user population to a 'flight' which receives search results generated or presented in a non-standard way. Comparison of interaction data across flights can be used to estimate the value of a new feature or a new ranking algorithm.

5.6   Summary

In this chapter we have attempted to convey the key differences between IR on the web and IR in other settings, starting with the challenges of identifying and prioritising candidate documents for inclusion in the index. Next, we noted that web pages are interrelated, both by hierarchical site structure and in a hyperlink graph, and pointed out that measures derived from page interrelationships can be used as prior probabilities to improve search result quality. The effectiveness of web ranking can also be improved through exploitation of textual descriptions (such as anchor text) external to the documents themselves. On the web, such descriptions are prevalent and reliably identifiable.

One of the main reasons why different sources of ranking evidence are highly beneficial on the web, is that web searchers behave differently to those modelled in non-web test collections. It is common for people to use web search engines for navigational and transactional purposes which are not supported in non-web retrieval settings. Major behavioural differences are even observed when the search intent is informational: searchers may value entry pages to authoritative sites on a topic and pages with useful topic links above documents containing chunks of relevant text.

Different user behaviour requires different evaluation methodologies, such as the ones mentioned in this chapter, and encourages the development of new ranking methods. It is now routine for Web search engines to *localise* search results to the country and language deduced from the searcher's IP address. Search results may also be personalised, either at the client side or using interaction histories maintained by a Web search engine.

We have briefly covered a number of topics, such as snippet generation, spelling suggestions, exploitation of user behaviour data, advertisement triggering and so on which are by no means restricted to web environments, but which have been developed further or are currently more heavily used in the web context.

Further reading. In this brief chapter, it has only been possible to provide a cursory overview of a broad topic. Readers are encouraged to read the cited references in areas of particular interest and to seek out other relevant papers in recent proceedings of conferences such as *World Wide Web (WWW)*, *Web Search and Data Mining (WSDM)*, *SIGIR* and *CIKM*. In addition to well-known information retrieval journals such as *ACM Transactions on Information Systems (TOIS)*, *Information Retrieval*, *Information Processing and Management*, two recently established ACM journals, *Transactions on the Web (TWEB)* and *Transactions on Internet Technologies (TOIT)*, publish a proportion of web IR papers. Chapters 19–21 of a recent CUP textbook (Manning *et al.* 2008) provide more detail on some aspects of web search, while Langville and Meyer (2006) cover the PageRank and related algorithms in considerable detail.

Exercises

5.1 Draw an architecture diagram of a breadth-first web crawler. Focus on data structures. How could a depth-first crawler work?

5.2 Try to characterize the behaviour of a major web search engine by submitting queries of different types (e.g. a single stopword such as 'the', queries of different lengths, queries with spelling mistakes, queries for which there is an obvious right answer, ...) and observing the results:

    (a) Does the search engine tailor results to the country it thinks you are in?

    (b) Does the estimated number of results increase or decrease as you add words to a query?

    (c) Can you guess the algorithm used to match advertisements to the queries you submit?

    (d) Does the search engine ever eliminate stopwords from your query?

    (e) Does the search engine appear to make use of stemming?

    (f) Are the top ranked results for a single word query the documents with the greatest density of occurrences of that query word?

    (g) How consistent are search results? If you submit the same query to the search engine in five minutes time, tomorrow or next week, do the answers vary?

    (h) How frequently are websites crawled? (Try the query 'current time glasgow' and look at the result for a current time site such as www.timeanddate.com.)

    (i) How well does the spelling suggestion mechanism work when you type a misspelled query?

5.3 Contrast evaluation using judged queries vs evaluation using click logs. Where/how can mismatch between evaluation and real users arise?

References

Aggarwal, C. C. *et al.* (2001). On the design of a learning crawler for topical resource discovery. *ACM Transactions on Information Systems* **19**(3): 286–309.

Agichtein, E. *et al.* (2006). Improving web search ranking by incorporating user behavior information. In *Proceedings of ACM SIGIR 2006*, pp. 19–26, New York, NY, USA. ACM Press.

Anh, V. N. and Moffat, A. (2002). Impact transformation: effective and efficient web retrieval. In *Proceedings of ACM SIGIR 2002*, pp. 3–10.

Baeza-Yates, R. *et al.* (2005). Crawling a country: better strategies than breadth-first for web page ordering. In *Proceedings of WWW 2005*, pp. 864–872, New York, NY, USA. ACM Press.

Bailey, P. *et al.* (2000). Dark matter on the web. In *WWW9 Poster Proceedings*.

Bailey, P. *et al.* (2003). Engineering a multi-purpose test collection for Web retrieval experiments. *Information Processing and Management* **39**(6): 853–871.

Barbaro, M. and Zeller, T. *Jr.* (2006). A face is exposed for AOL searcher no. 4417749. *The New York Times*, http://www.nytimes.com/2006/08/09/technology/ogaol.html?ex=1312776000.

Bharat, K. and Broder, A. (1999). Mirror, mirror on the web: a study of host pairs with replicated content. In *Proceedings of WWW8*, pp. 1579–1590.

Bharat, K. *et al.* (2001). Who links to whom: mining linkage between web sites. In *Proceedings of IEEE ICDM-01*, pp. 51–58.

Bharat, K. and Mihaila, G. A. (2001). When experts agree: using non-affiliated experts to rank popular topics. In *Proceedings of WWW 2001*, pp. 597–602, New York, NY, USA. ACM Press.

Broder, A. (1997). On the resemblance and containment of documents. In *SEQUENCES '97: Proceedings of the Compression and Complexity of Sequences 1997*, p. 21, Washington, DC, USA. IEEE Computer Society.

Broder, A. (2002). A taxonomy of web search. *ACM SIGIR Forum* **36**(2): 3–10.

Broder, A. *et al.* (2000). Graph structure in the web. In *Proceedings of WWW9*, pp. 426–434. Amsterdam.

Broder, A. Z. *et al.* (2003). Efficient query evaluation using a two-level retrieval process. In *Proceedings of CIKM 2003*, pp. 426–434, New York, NY, USA. ACM Press.

Chakrabarti, S. *et al.* (1999). Focused crawling: a new approach to topic-specific web resource discovery. In *Proceedings of WWW8*, pp. 1623–1640.

Chirita, P. A. *et al.* (2007). Personalized query expansion for the web. In *SIGIR '07: Proceedings of the 30th annual international ACM SIGIR conference on Research and development in information retrieval*, pp. 7–14, New York, NY, USA. ACM.

Cho, J. and Garcia-Molina, H. (2000). The evolution of the web and implications for an incremental crawler. In *Proceedings of VLDB 2000*, pp. 200–209.

Cho, J. *et al.* (1998). Efficient crawling through URL ordering. In *Proceedings of WWW7*, pp. 161–172, Brisbane, Australia.

Craswell, N. *et al.* (2001). Effective site finding using link anchor information. In *Proceedings of ACM SIGIR 2001*, pp. 250–257, New Orleans.

Craswell, N. *et al.* (2005). Relevance weighting for query independent evidence. In *Proceedings of ACM SIGIR 2005*, pp. 416–423, Salvador, Brazil.

Craswell, N. and Szummer, M. (2007). Random walks on the click graph. In *Proceedings of ACM SIGIR 2007*, pp. 239–246.

Cucerzan, S. and Brill, E. (2004). Spelling correction as an iterative process that exploits the collective knowledge of web users. In *Proceedings of Conference on Empirical Methods in Natural Language Processing (EMNLP)*, pp. 293–300, Barcelona, Spain. Association for Computational Linguistics.

Culliss, G. (1999). User popularity ranked search engines. *Presentation at Infonotics Search Engines Meeting*.

Davison, B. (2000). Topical locality in the Web. In *Proceedings of ACM SIGIR 2000*, pp. 272–279, Athens, Greece.

De Bra, P. *et al.* (1994). Information retrieval in distributed hypertexts. In *Proceedings of the 4th RIAO Conference*, pp. 481–491, New York.

Diligenti, M. *et al.* (2000). Focused crawling using context graphs. In *Proceedings of the 26th VLDB Conference*, pp. 527–534, Cairo, Egypt.

Edwards, J. *et al.* (2001). An adaptive model for optimizing performance of an incremental web crawler. In *Proceedings of WWW 2001*, pp. 106–113, New York, NY, USA. ACM Press.

Fetterly, D. *et al.* (2003). A large-scale study of the evolution of web pages. In *Proceedings of WWW 2003*, pp. 669–678, New York, NY, USA. ACM Press.

Glover, E. J. (2001). *Using Extra-Topical User Preferences to Improve Web-Based Metasearch*. PhD Thesis, University of Michigan. http://ericglover.com/papers/glover_thesis.pdf.

Hawking, D. and Craswell, N. (2005). Very large scale retrieval and web search. In Voorhees, E. and Harman, D. (eds), *TREC: Experiment and Evaluation in Information Retrieval*. pp. 199–232, MIT Press.

Hawking, D. *et al.* (2006). Improving rankings in small-scale web search using click-implied descriptions. *Australian Journal of Intelligent Information Processing Systems. ADCS 2006 Special Issue*. **9**(2): 17–24.

Hawking, D. *et al.* (1999). Overview of TREC-8 Web Track. In *Proceedings of TREC-8*, pp. 131–150, Gaithersburg, Maryland USA.

Heydon, A. and Najork, M. (1999). Mercator: a scalable, extensible web crawler. In *Proceedings of WWW2*, pp. 219–229.

Huberman, B. A. and Adamic, L. A. (1999). Growth dynamics of the world-wide web. *Nature* **401**: 131.

Järvelin, K. and Kekäläinen, J. (2000). IR Methods for retrieving highly relevant documents. In *Proceedings of SIGIR'00*, pp. 41–48, Athens, Greece.

Joachims, T. (2002). Optimizing search engines using clickthrough data. In *Proceedings of ACM KDD 2002*, pp. 133–142.

Joachims, T. *et al.* (2005). Accurately interpreting clickthrough data as implicit feedback. In *Proceedings of ACM SIGIR 2005*, pp. 154–161.

Jones, R. *et al.* (2006). Generating query substitutions. In *Proceedings of WWW 2006*, pp. 387–396, Edinburgh, Scotland. ACM Press.

Koster, M. (1994). The web robots pages. http://www.robotstxt.org/, accessed 24 Aug 2008.

Kraaij, W. *et al.* (2002). The importance of prior probabilities for entry page search. In *Proceedings of ACM SIGIR 2002*, pp. 27–34, Tampere, Finland.

Langville, A. N. and Meyer, C. D. (2006). *Google's PageRank and Beyond: The Science of Search Engine Rankings*. Princeton University Press.

Lawrence, S. and Giles, C. L. (1999). Accessibility of information on the web. *Nature* **400**: 107–109.

Long, X. and Suel, T. (2003). Optimized query execution in large search engines with global page ordering. In *Proceedings of VLDB 2003*, pp. 129–140.

Manning, C. D. *et al.* (2008). *Introduction to Information Retrieval*. Cambridge University Press. Full text on-line at www-csli.stanford.edu/~hinrich/information-retrieval-book.html.

Michel, S. *et al.* (2005). KLEE: a framework for distributed top-k query algorithms. In *Proceedings of VLDB 2005*, pp. 637–648. VLDB Endowment.

Najork, M. and Wiener, J. L. (2001). Breadth-first crawling yields high-quality pages. In *WWW 2001: Proceedings of the 10th International Conference on World Wide Web*, pp. 114–118, New York, NY, USA. ACM Press.

Ogilvie, P. and Callan, J. (2003). Combining document representations for known-item search. In *Proceedings of ACM SIGIR 2003*, pp. 143–150, New York, NY, USA. ACM Press.

Page, L. *et al.* (1998). *The PageRank Citation Ranking: Bringing Order to the Web*. Technical Report, Stanford, Santa Barbara, CA 93106. dbpubs.stanford.edu:8090/pub/1999-66.

Porter, M. (1980). An algorithm for suffix stripping. *Program* **14**(3): 130–137. Reprinted in Spärck Jones and Willett. *Readings in Information Retrieval*. Morgan Kaufmann, San Francisco, CA, 1997.

Pugh, W. (1990). Skip lists: a probabilistic alternative to balanced trees. *Communications of the ACM* **33**(6): 668–676.

Ribeiro-Neto, B. *et al.* (2005). Impedance coupling in content-targeted advertising. In *Proceedings of ACM SIGIR 2005*, pp. 496–503, New York, NY, USA. ACM Press.

Richardson, M. *et al.* (2006). Beyond PageRank: machine learning for static ranking. In *Proceedings of WWW 2006*, pp. 707–715, Edinburgh.

Rose, D. E. and Levinson, D. (2004). Understanding user goals in web search. In *Proceedings of WWW 2004*, pp. 13–19, New York, NY, USA. ACM Press.

Singhal, A. and Kaszkiel, M. (2001). A case study in web search using TREC algorithms. In *Proceedings of WWW10*, pp. 708–716, Hong Kong.

Taylor, M. *et al.* (2006). Optimisation methods for ranking functions with multiple parameters. In *CIKM '06: Proceedings of the 15th ACM International Conference on Information and Knowledge Management*, pp. 585–593, New York, NY, USA. ACM.

Teevan, J. *et al.* (2005). Personalizing search via automated analysis of interests and activities. In *Proceedings of ACM SIGIR '05*, pp. 449–456, New York, NY, USA. ACM.

Thomas, P. and Hawking, D. (2006). Evaluation by comparing result sets in context. In *Proceedings of CIKM 2006*, pp. 94–101.

Tombros, A. and Sanderson, M. (1998). Advantages of query biased summaries in information retrieval. In *Proceedings of ACM SIGIR 1998*, pp. 2–10, Melbourne, Australia.

Turpin, A. *et al.* (2007). Fast generation of result snippets in web search. In *Proceedings of ACM SIGIR 2007*, pp. 127–134.

Xue, G.-R. *et al.* (2004). Optimizing web search using web click-through data. In *Proceedings of ACM CIKM 2004*, pp. 118–126.

Zobel, J. and Moffat, A. (2006). Inverted files for text search engines. *ACM Computing Surveys* **38**(2): 1–56.

6

Mobile Search

David Mountain, Hans Myrhaug and Ayşe Göker

6.1 Introduction: Mobile Search – Why Now?

From a historical perspective, the need for tools dedicated to information retrieval arose from organising paper documents in libraries, and became formalised within the discipline of information science. Later, the emergence of computers with electronic storage, and subsequently the Internet, led to the creation of vast (often unstructured) collections of electronic documents, and the need for search tools that could retrieve relevant information from these collections quickly and efficiently (Rosenfeld and Morville 2006). A current area of considerable activity is mobile search. Mobile search is now an exciting research frontier in information retrieval, where technology moves quickly and the distinct information needs of mobile individuals are often poorly understood.

Mobile search is an expanding area, because of the growing diversity and availability of mobile information environments. Successful mobile search systems can satisfy the information needs of people moving around in the physical world, retrieving relevant information via portable – usually handheld – devices. One key distinction of mobile search is that the user's environment is dynamic, in contrast to information searching in more familiar 'static' or 'desktop' scenarios, where a user's environment is less subject to change. A second distinction of mobile search is the ambition to make information relevant within this dynamic environment, suggesting a much stronger link between information and the physical world than has yet been the case for information retrieval.

The objective of this chapter is to illustrate how mobile search is different from PC-based web searching. We will do this by presenting the growing diversity, availability, and use of mobile information services from a worldwide perspective. Next we will present the distinctive nature of mobile search in terms of where information is stored (portable, embedded, or remote) and with a strong emphasis on the need to filter results such as prefiltering, or the automatic filtering of information based on spatial location, time, or personal preferences. Finally, two mobile search applications from two very different mobile environments are shown: Oslo International Airport and The Swiss National Park.

6.1.1 Technological drivers

Two of the key technological trends of the past 15 years have been the increased use of the Internet, and the widespread consumer take-up of mobile phones. Whilst the rise of the Internet has led to the majority of desktop computers being capable of transferring information remotely over networks,

Information Retrieval: Searching in the 21st Century   edited by A. Göker & J. Davies
© 2009 John Wiley & Sons, Ltd

simultaneously the increasing sophistication of mobile phones has seen them evolve into portable computers, capable of transferring, storing and processing a wider variety of information than just voice data. It can be seen that there has been a convergence between computers and mobile telephony into a new distinct discipline: mobile computing. Computers can now be thought of as a continuum with large, static, high-performance servers at one end and small, mobile, constrained devices at the other. Between these two extremes there exists a range of devices with varying degrees of portability and connectivity, including mobile phones, personal digital assistants (PDAs), tablet PCs, and laptops. From a mobile search point of view, this continuum of manufactured computers affects how to provide search to a growing number of mobile users.

The capacity of mobile devices to exchange voice calls and data saw them spread faster into society than almost any other technological innovation. This take-up continues at an astonishing pace, driven by people in developing countries acquiring a mobile device for the first time, and people in the developed world regularly upgrading their mobile for more sophisticated models. The top five mobile device manufacturers (i.e. Nokia, Samsung, Motorola, LG, and Sony Ericsson) together shipped 2.237 billion handsets worldwide in total from 1 January 2006 until 30 June 2008, according to their quarterly financial reports (available via their own websites as they are publicly listed companies). Even back in 2006, 85% of the UK population owned mobile phones (Stynes *et al.* 2006). Mobile devices are becoming increasingly sophisticated with faster processors, increased memory, high-resolution colour screens, and improved battery performance. We are rapidly approaching the point where more people have access to the web via wireless mobile devices than via traditional desktop machines.

A crucial characteristic of wireless mobile devices is their need to know their approximate location within a network for call routing. Cellular networks have always kept track of the base station that each mobile device is using to connect to the network in order to route calls to that mobile; it is not feasible to broadcast all calls via all base stations. Knowing which base station a mobile device is using to connect to the network identifies that device (and by association, its user) as being in a particular network cell, which may vary in size from tens of metres to tens of kilometres. Technological advancements have allowed devices to connect to more diverse networks. The regions served by wireless *local* area networks (such as Wi-Fi) and *personal* area networks (such as Bluetooth) cover much smaller areas and hence can potentially determine the position of a mobile device more precisely. In parallel, mobile devices can now incorporate the Global Positioning System (GPS) and report the location of a mobile device with spatial coordinates, independently of a wireless cellular network. Increasingly, mobile devices are integrating dedicated positioning-determining technology, such as GPS receivers. This ability to know the location of a mobile user offers great opportunity to designers of mobile information systems, since this can be used to try to make the information retrieved by a user more relevant to their location and situation (Reades *et al.* 2007; Göker and Myrhaug 2008).

6.1.2   Predicted demand for mobile search

The rapid take-up and sustained use of the Internet and search engines suggests that people have a compelling need to retrieve relevant information to fill gaps in their knowledge. We generally access such resources whilst sitting at desktop computers, however there is no reason to suggest we do not have questions and information needs whilst on the move. This suggests a latent demand for mobile search that is currently, to a large degree, unmet. Increasingly people are retrieving digital information in a mobile context, for example: accessing the Internet via a Wi-Fi access point in a coffee shop; browsing on a phone or PDA via the mobile telecommunications network whilst walking, on public transport, or as a passenger in a car; or using a mobile guide whilst roaming in an outdoor recreational area. Rather than simply mirroring desktop search engines, mobile search may be required to consider a broader context, and retrieve information that takes this into account (Jiang and Yao 2006).

The demand for mobile phones and subscription is large and rising rapidly. The worldwide mobile subscriber base crossed 3 billion in late 2007, as illustrated in Figure 6.1, and is expected to cross 5.5 billion by the end of 2013 – according to data from several Portio Research sources (2008a; 2008b; 2008c). The number of mobile subscribers has almost doubled from 2004 to the end of 2007.

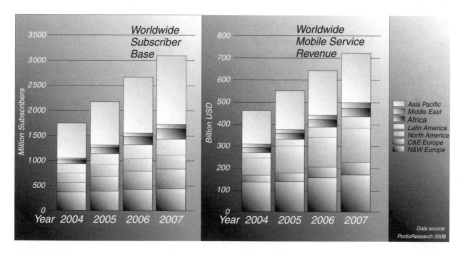

Figure 6.1 Worldwide subscriber base and worldwide mobile service revenue (Reproduced by permission of © AmbieSense™ Ltd)

In 2007, operators worldwide generated USD 803.7 billion in mobile revenues (i.e. USD 717.8 billion in mobile service revenues which excludes sales of mobile handsets, as also shown in Figure 6.1) and this number is expected to reach USD 1094.9 billion by 2013. Although revenues from voice calls still comprise 81% of this worldwide revenue, mobile operators globally now are focusing on mobile data services (non-voice services) as a means to increase their average revenue per user. Non-voice services accounted for the remaining 19% of the worldwide revenues, but this is expected to grow to more than 25.5% by the end of 2012; an estimated worldwide consumer spending exceeding USD 251 billion. Text messaging (SMS) still accounts for approximately 49% of total revenues from mobile data services worldwide, however the percentage contribution in revenue from SMS is expected to diminish in the future, even though SMS traffic volumes will continue to rise. In this light, the percentage contribution of revenue from other mobile data services is expected to increase, from services enabling: e-mail, instant messaging, music, games, video, payment, location-based services (LBS), and maps (Portio Research 2008a; 2008b; 2008c). We anticipate mobile search to be a natural component of a wide range of these mobile Internet services to come.

As previously discussed, the demand for mobile phones themselves is huge, and these devices now function as handheld computers. In order to understand whether mobile search is something that is desired by mobile users, or just something that developers wish to force onto an unwilling market, it is worth considering research that analyses end-user activity trends amongst mobile consumers as a means to forecast the future. Figure 6.2, drawn from data in dotMobi and AKQA Mobile (2008) is in this light interesting because it shows a wide range of activities that mobile subscribers engage in – excluding SMS messaging. Of the 1650 survey respondents, 18% stated they are busy gathering/finding information, and 28% responded they are using maps.[1]

It is generally predicted, by mobile operators, that location-based services (LBS) will be a large percentage contributor to the mobile market. These are tools that tailor retrieved information based upon the location at which a query was made (Brimicombe and Li 2006; Jiang and Yao 2006). Anecdotally, information managers have stated that the main purpose of LBS is to get 'answers to where questions are' (Haller 2004). The scope of LBS is broad and has the potential to include 'where's my nearest' services; searches for local news, weather or sports reports; navigation; adding location-sensitivity to searches of the free-text documents on the Internet; friend-finder services; entertainment; and location-based gaming. Figure 6.3 shows the 70% average growth in user demand for

---

[1] Approximately 45% of respondents stated that they were checking information, but it is possible that the data includes bookmarks, preset menus, or preset device settings which trigger certain home pages to appear.

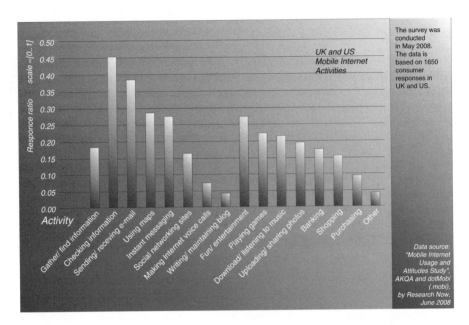

Figure 6.2 Mobile Internet activities in May 2008, 1650 respondents (Reproduced by permission of © AmbieSense™ Ltd)

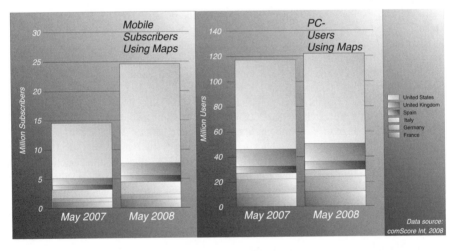

Figure 6.3 Growth in mobile subscribers using maps compared to PC users using maps (Reproduced by permission of © AmbieSense™ Ltd)

using maps on mobile phones (from about 14.4 million to about 24.5 million) from May 2007 to May 2008, for the US and a selection of high population European countries. This is in contrast to the 3.8% average growth in PC users using maps within the same period and countries, i.e. from 116.7 million to 121.2 million users.

We have presented overall statistics that underpin the background and growing demand for mobile search. A large portion of mobile search is currently like a web search, essentially a mirroring of the desktop experience of Internet search on mobiles. There is a significant challenge in trying to bring or adapt the search tools for the current desktop Internet experience on to mobile devices. It is important that we consider the information needs of mobile users in their particular information environments.

Mobile devices are perhaps the most valuable belonging that we have in order to access and interact with diverse information spaces. In this light, mobile search can be seen as offering the opportunity to populate people's hands and pockets with search facilities that satisfy their information needs whilst on the move. In comparison to the amount of information available via our desktop web environments, at present the amount is less for mobiles. However, this is growing. As the content variety and amount available on mobiles continues to grow, so will the demand for mobile search.

6.2   Information for Mobile Search

The first consideration for mobile search must be the nature of information to be used in mobile search systems. The primary issue for local search is establishing associations between information and locations in the physical world. More generic issues for mobile search include where information is stored, and the ownership of information. These issues will be covered in this section.

6.2.1   Linking information to physical space

In the days of paper documents and libraries, there was a strong relationship between documents and physical space: if interested in a particular document you had to travel to a location at which it was stored (for example a library) to view it. The Internet was hailed as the 'death of distance' (Cairncross 1999) since the availability of electronic documents over networks removed this need to travel to information sources, hence the Internet can be seen to have led to a *separation* of information and physical space. Very often however, information refers – either directly or indirectly – to places, and can be thought of as having a *spatial footprint* that links that information to a location or locations in the physical world (Purves and Jones 2006). Examples of this include place names and addresses that appear in documents, and 'points of interest' databases where features are linked to locations described by spatial coordinates, as seen with the Yell™ search engine[2]. For information stored in databases, this spatial information can be stored as fields for each record. For documents found on the Internet, geographic markup languages are increasingly being used to represent the spatial footprint associated with documents, such as the Geography Mark-up Language (GML) – a standard maintained by the Open Geospatial Consortium. Google has also defined the Keyhole Mark-up language (KML) format for representing spatial information that can be indexed and displayed in their web map browser (Google maps) and stand alone mapping application, notably Google Earth.

Seen from this perspective, one of the aims of mobile search is to overlay digital information on the physical world, combining information space and physical space. Spatial information is also a key component in bringing relevant information to specific situations (Myrhaug and Göker 2003; Myrhaug *et al.* 2004). By knowing the location that documents refer to, and the location from which a query arose, local search systems (or location-based services) aim to ensure that information is relevant in terms of the subject, and also relevant to an individual's spatial location (spatial relevance) (Raper 2007a; Mountain and MacFarlane 2007). When this is achieved, information *has its place*, and can be of far greater benefit to users of mobile search systems. Figure 6.4 provides some examples of how digital information can be linked to the physical world. Finding approaches to establishing meaningful associations remains one of the greatest challenges facing an information retrieval researcher.

6.2.2   The storage of information

The architecture for the storage of information is crucial in the design of a mobile information retrieval system, and can influence not only how information is organised, but strategies for ensuring

---

[2] www.yell.com

Information

Aldersgate was a gate in the
London Wall in the City of
London, which has given its
name to a ward and Aldersgate
Street, a road leading north from
the site of the gate, towards
Clerkenwell in the London
Borough of Islington.

Spatial location

(a)

© Wikipedia 2009

1. **The Peasant:**
     240 St John Street, London, EC1V 4PH

2. **The Old Ivy House:**
     166 Goswell Road, London, EC1V 7DT

3. **Jack Beards:**
     32, Hall St, London, EC1V 7NA

(b)

© Heinrich Haller 2008

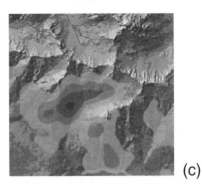

(c)

© Swiss National Park 2007

Figure 6.4   Three examples of how information can be linked to the physical world (a) Place names: a
web page describing historic London includes many place names (top left). These can be identified and
linked to *points* or *areas* on the ground, e.g. the Aldersgate electoral ward (top right).   (Reproduced by
permission of © Crown Copyright/database right 2007. An Ordnance Survey/EDINA supplied service.)
(b) Shops and services: results of a search for bars (middle left) represented as *points* on the ground
(middle right).   (Reproduced by permission of CC-By-SA 2.0 OpenStreetMap.org contributors) (c)
Natural environment: an image of a red deer and the distribution of deer in the park represented as a
density surface (darker represents more deer).   (Reproduced by permission of © 2007 Swiss National
Park)

information is relevant given a user's context. In our view of mobile computing, information may be stored in three distinct *tiers*: (Figure 6.5):

- *Portable* information stored on the mobile device itself;
- *Embedded* information stored on local wireless servers;
- *Remote* information stored on remote servers.

In IR terms this refers to where the document collection is stored, which may influence where one would like to store the IR search index, and additionally it may affect the set of search parameters included for the mobile search. Whilst the user may not be aware of where the information is stored, as Kwon and Kim (2005) also point out, when interacting with a mobile search system, designers must give careful consideration to the storage architecture that they wish to employ, evaluating the benefits and drawbacks of each tier.

6.2.2.1   Portable information

In the portable case, information is stored on the mobile device itself; hence information travels with the user of the system and is available at all times. Most users of mobile phones carry some information with them, for example lists of contacts in 'phone book' applications, text messages, calendar appointments, images and other documents. Whilst presently much of this information is personal and private, increasingly mobile users are carrying publicly available information, such as documents downloaded from the Internet. There is value to users in being able to access this information at all times, regardless of factors such as the availability of a wireless connection. For example, needing to know someone's address to send a postcard, or checking directions and a map when travelling to a new location; the need to complete these tasks does not cease when we roam out of the coverage of our chosen mobile network operator.

Mobile search systems that adopt this portable approach must not only store collections of documents locally, but also design a search engine, which can run on the mobile device, e.g. (Myrhaug *et al.* 2004; Göker *et al.* 2004a; 2004b). There are various strategies that search tools for portable information may employ. A portable search engine may need to create and maintain an index of the information stored locally as files, and use this to satisfy free-text queries. Alternatively, information may be stored using a database on the local device, and design a user friendly interface to this to allow information relevant to a query to be retrieved.

This portable information approach has some drawbacks. First, there is a limit to the volume of information that can be stored on the device. Most mobile devices have a storage capacity typically less than a few gigabytes, at present, although this capacity is increasing through the use of flash memory and potentially from embedded hard disks. Whilst this capacity will be quite sufficient for specific applications, such as a mobile guide to a specific area, the limited size of the cache can never offer a user experience comparable to the diversity of searching the Internet. Second, the need to run a search engine on the device itself can require more sophisticated computational power and memory capacity.

If an ambition of a mobile search service is to retrieve information on the basis of the user's geographic location, there are two requirements. First, it must be possible to establish the location of the user, hence the device must be location-aware. This may be achieved using one of the position-determining technologies described previously, or the user may be prompted to enter their location manually, if they know it. Second, the information they access must be linked in some way to the physical world, and this spatial referencing must be stored explicitly. Knowing the location of the user, and the location associated with information sources, these locations can be compared and filtered to ensure that the information retrieved is relevant to the user's surroundings; this step is described in more detail later.

These requirements when storing information at the portable tier – for sophisticated, location-aware devices – can act as a constraint on the rollout of a mobile search service to a wider audience. Nevertheless, where devices can meet these requirements of fast processors, large storage, and location-aware

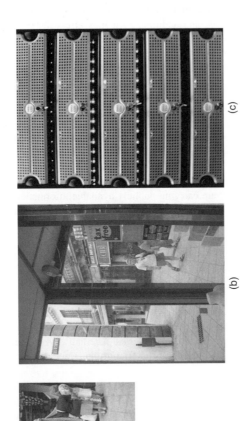

(a)   (b)   (c)

Figure 6.5 Storage of the collection at the portable, embedded and remote tier (a) Portable tier: the information is stored locally on mobile phones (Reproduced by permission of © AmbieSense™ Ltd) (b) Embedded tier: the information is stored on a wireless server embedded in the physical environment. See the top-left corner in the right window. (Reproduced by permission of © AmbieSense™ Ltd) (c) Remote tier: the information is stored on dedicated remote servers. (Reproduced by permission of © David Mountain)

capacity, there are significant advantages. First, the system can operate autonomously, regardless of whether a network connection is available or not. In addition, by storing information on the local device, it can take less time to retrieve since the documents do not need to be transferred over a potentially low-bandwidth network connection, and the user will incur no cost for the transfer of information.

Some sophisticated systems have been developed using this approach – such as the WebPark[3] system (Raper *et al.* 2007c)[4] described later as a case study, however at present, the requirements of the mobile device have acted as a barrier to wider take-up. Some mobile search systems that store information predominantly at the portable tier have adopted a business model where they rent out dedicated devices, with data and the mobile search system preinstalled to visitors in particular areas for short periods of time, which are then returned when the visitors leave the area.

6.2.2.2 Embedded information

An alternative to storing information at the portable tier is to store it on dedicated wireless servers embedded within the physical infrastructure. One major opportunity afforded by this approach is to ensure that the information stored on any particular embedded server takes account of the location – and more broadly, the context – of the people who will use it (Jiang and Yao 2006). The servers may communicate with the mobile devices via a wireless connection of some description, most commonly Wi-Fi or Bluetooth, but possibly RFID or infrared.

The characteristics of this approach have some distinct advantages and drawbacks compared with alternative storage tiers. A major benefit is that larger and more diverse collections of information may be used: the size of the collection is constrained by the storage of the server, rather than the mobile device. Similarly, by delegating much of the search functionality to the embedded server, the system may be compatible with a much wider range of devices, and hence a larger potential audience. Perhaps the major advantage of adopting this embedded approach is that the information stored on each server can be tailored to its particular context. Knowing that the information stored on local servers is likely to be relevant to the surrounding environment, the mobile device itself does not need any positioning-determining technology; it simply needs to connect to those embedded servers that are close by.

This raises questions about how information can be tailored to take account of the surrounding environment and other user contexts (see Chapter 7). Considering examples of how information is distributed on paper: student magazines can be made available at university campuses; bus travel information is signposted at bus stops; financial newspapers may be available in the business lounges of airports. In each particular environment, there is assumed to be a particular group of people with specific information needs. The embedded tier approach is better placed to take account of this context, for example serving information about University social events from a wireless server in the Student Union.

There can nevertheless be challenges to this approach. First is the requirement to embed hardware within the physical infrastructure, which may require permission from place/land owners or local authorities (e.g. Myrhaug *et al.* 2004). Similarly, it can be challenging to scale this embedded approach from a local to a nationwide service due to the need to install dedicated hardware with short-range communications.

6.2.2.3 Remote information

An alternative to storing information at the portable and embedded tiers is to access information stored on remote servers accessible over the Internet. Many mobile devices are shipped with mobile web browsers, which allow them to connect to the Internet and view remotely stored documents. Many web sites reformat their content for mobile web browsers, for example reducing image sizes in

---

[3] http://www.webparkservices.info

[4] Raper *et al.* (2007b) also discusses more widely the potential of such location-based systems.

HTML pages, or offering the same content in Wireless Markup Language (WML) format for Wireless Application Protocol (WAP) browsers. Mobile devices may establish connections to the Internet via the mobile telecommunications network or local Wi-Fi servers. In this environment mobile web browsers can use web search engines to perform Internet searches, and this form of mobile search mirrors the search experience from static, desktop devices. Without additional information about the locations in the physical world that these Web pages refer to, and the location from which the query was made, mobile systems that search 'spaceless' Internet resources are unable to sort or filter information by location to make it more relevant to the user's surroundings.

We saw how for portable information, knowing the position of the user, and spatial location associated with documents, this may be filtered to ensure that only local information is retrieved. This same approach can be adopted for information stored on remote servers where the locations associated with those resources are stored explicitly, and the mobile device has some position-determining technology to establish the user's position. Employing a strategy of accessing information stored at the remote tier has benefits and limitations. The main advantage is that by using remote servers, there is not the same limitation on the size of the information resources as seen when storing information on the portable tier, or even the embedded tier. Much larger document collections, or larger points of interest databases, may be used so the information retrieved by the user may be much richer and more diverse. This remote approach also shares an advantageous characteristic of the embedded tier: the search functionality can be deployed on the server so the mobile device need not be so sophisticated, making the search system accessible to a wider audience. If the need for position-determining technology is removed, the mobile search system may be accessible to anyone with a mobile web browser. This remote approach is more scalable than the embedded approach, since the servers themselves do not have to be located in the physical infrastructure in order to provide a service in a particular area, making it much easier to extend the mobile search service to new regions.

On the downside, the user may incur costs from their mobile operator, or the Wi-Fi service provider when transferring data over the network, and it may take longer to access this information due to limited bandwidth. This approach also has a disadvantage when compared with storing information at the portable tier: the mobile search service will only be available where there is a wireless Internet connection, so there may be no availability in remote areas or when the network is operating at capacity.

6.2.2.4   Storing information at multiple tiers

Given that each of the approaches described above has advantages and drawbacks, systems can be designed to store data at multiple tiers (Figure 6.6). A system may choose to maintain a local cache of the most widely used, or most critical information, so that the mobile search system has some capacity, even when there is no wireless connection. Similarly, the system may allow this portable store of information to grow, adding information retrieved from embedded and remote servers to the portable tier, so users can carry this information with them and access it again, even if they cannot establish a wireless network connection (e.g. the Oslo Airport case study referred to in Section 6.4.1).

The influence of Web 2.0 and user-generated content has increasingly seen information migrating *from* the portable tier (Sui 2008). Social networking sites, discussion boards and web logs (blogs) are common ways of allowing users of the Internet to create and share information. In mobile search systems, users may create some content linked to a particular location, such as a review of a restaurant where they have eaten. In order for other users of the systems to access this content, it must migrate from the portable tier to the embedded or remote tier.

A mobile search system can defer queries from the portable tier to the embedded or remote tier – when they are available – to allow a search of a larger collection of documents. A search system may be designed to first search the cache stored on a mobile client and if this cannot satisfy a particular query, propagate the query to embedded or remote servers. Designers of mobile search systems must consider the limitations of storage at each of the three tiers described above, and either accommodate these constraints, or develop the capacity to cascade queries between tiers. To provide a comprehensive mobile search service, designers may choose to implement search engine

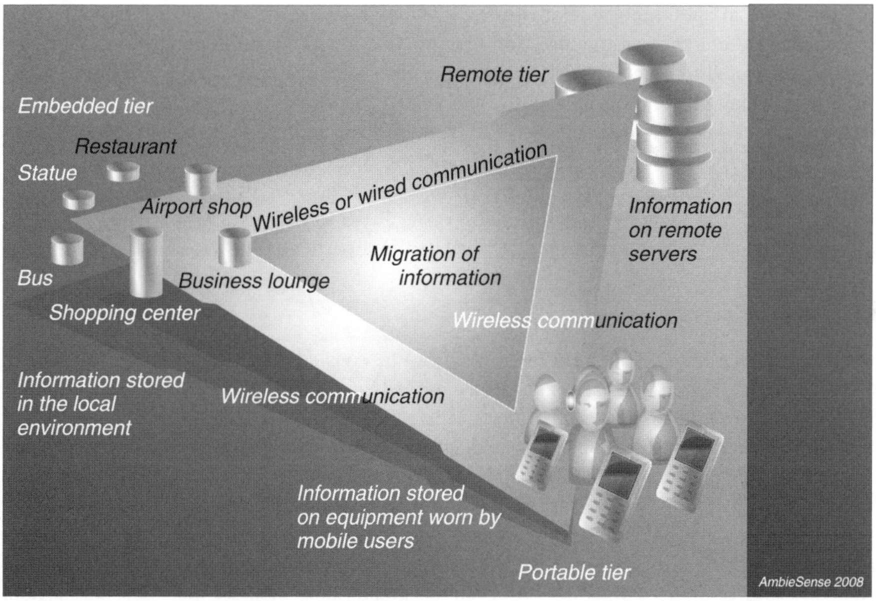
Remote tier
Embedded tier
Restaurant
Statue
Airport shop
Wireless or wired communication
Information on remote servers
Migration of information
Bus
Business lounge
Shopping center
Wireless communication
Information stored in the local environment
Wireless communication
Information stored on equipment worn by mobile users
Portable tier
AmbieSense 2008

Figure 6.6 Storage at, and information migration between, multiple tiers (Reproduced by permission of © AmbieSense™ Ltd)

functionality on all three tiers, requiring the creation and maintenance of document collections and associated indices on all three tiers.

6.2.3 The ownership of information

An increasingly important consideration for mobile and local search, particularly given recent trends towards user generated content and Web 2.0 technologies, is the ownership of the information that will be searched and retrieved (Sui 2008). This has implications for the information that is available for mobile search, whether it is free or must be paid for, and whether it is available to all or restricted to specific groups. Frameworks are required which allow people to access their information at any time via any connected device.

First it is important to distinguish between the unstructured *information* found on web servers and hard drives, and *data* stored in databases, where each record conforms to a specified format. Much of the information made available to mobile users is customised content from commercial point of interest databases: for example, the 'where's my nearest' directory services.

These commercial data repositories offer a useful source of information for mobile search: databases of information that may be queried via a web interface. Sometimes users must pay to access these services – either by subscription or pay per access – sometimes they may be accessed without cost. Each data repository has generally been created and maintained by private organisations, often for some specific purpose, hence the data often has known attributes. A familiar example from the web would be an auction site, with attributes including a product description, a time at which the auction will end, the bids made on an item, and the current highest bid. Another example based upon real-time information useful for mobile search are databases of running information for flight operators, recording for each flight if it is running on time and if delayed, the length of delay. Points-of-interest databases available from private companies may attempt to record, for example, all shops, services and other features for a particular region, providing attributes such as the name of the feature and its location, either as an address, or as spatial coordinates. No single commercial data repository can hope to be as diverse as information found in public information spaces, however they can be useful since each record in the database conforms to a known structure, and may provide the metadata required for mobile search.

Beyond *data* stored in commercial databases, we can make distinctions between the ownership of *information* that may be accessed by mobile devices. The nature of ownership may have implications for assumptions that may be made about datasets: public information sources may be vast, but carry no guarantees about the quality or nature of the content. Shared information spaces tend to represent an online community with a common interest, where the behaviour of other users in the space can result in some information being recommended to you. Personal information spaces on the other hand may contain all the diverse information that is related to a particular individual.

6.2.3.1   Public information space

The documents that comprise the Worldwide Web can be considered a vast public information space, which anyone with a web browser may access either by following hyperlinks or entering URLs directly. Information in this space can be documents, such as HTML or PDF documents, multimedia information such as image files, audio or video clips. Web browsers for handheld devices, such as Opera™, are making this public information space available to mobile individuals, albeit a more limited experience on small screens and slower bandwidths. These public information spaces are distinguished by the fact that they can be accessed by anyone with a web browser: you don't need to pay money or to be part of a particular community to access this information. The public information space can be searched using centralised Internet search engines, and some of these can take into account the location or context information of the user making the query.

This information is useful for mobile search in that it can emulate the Internet as experienced by users of static desktop devices, but as a public information space, with no single owner, there can be no guarantees as to the content of any particular document, the authenticity of the information, nor any additional 'metadata' that may be required. Of particular interest for mobile search, by far the majority of information in this public information space will have no explicit spatial reference that could be used to communicate the location in the physical world to which that information relates, or where it may be relevant. As discussed previously, this information is desirable, and can be communicated using geographic markup languages such as GML or KML, but the majority of creators of information in this public information space at present do not include such information. In future, initiatives to automatically extract spatial information associated with free-text documents may succeed in adding this association between information in the public information space and locations in the physical world (Purves and Jones 2006). This offers the potential to make public information spaces a much more useful source for mobile search.

6.2.3.2   Shared information space

Whilst the public information space is accessible to all, access to shared information spaces may be restricted to members of a particular digital community. A digital community can be seen as a group of people that seeks to create, disseminate and discuss content in an online environment. The members of a digital community are likely to share common interests, and can use the space to give voice to their opinions. It is a microcosm of the public information space, with community members as authors of, and commentators on, the material in that space. Community members may make information available via traditional web documents such as personal web pages, blogs, discussion boards or chat rooms. This trend has exacerbated as the trend towards Web 2.0 has taken off; social networking sites such as Facebook™, MySpace™ and Bebo™ have become widely accepted mechanisms by which digital communities can interact and share content. The content itself may be text, which can be indexed using well established techniques, however increasingly members of digital communities are creating imagery, audio, or video content which is more complex to index and retrieve (see also Chapter 3: Multimedia Resource Discovery, and Chapter 4: Image Users' Needs and Searching Behaviour).

In the mobile domain, some novel applications have been developed which aim to bring a sense of place to the digital community. By adding a spatial location to discussion postings or personal observations, members of a mobile digital community can discover the issues and debates that are occurring around them, and participate in those debates. Potential applications are diverse and implemented examples include the use of *place marks* to report sightings of rare species in National Parks (Krug *et al.* 2003) and spatially referenced discussion threads, allowing members of a local community to debate planned development in their neighbourhood (Lane 2003).

Members may choose to share some or all of the information that they have created with one or more digital communities. If they place no restrictions, the information can be thought of as residing in the public information space. In the most restricted case, a member may choose to make the information space available only to himself or herself: a personal information space.

6.2.3.3   Personal information space

A personal information space is comprised of personal and private information to be accessed by an individual user. The information may reside on any storage tier (remote, embedded, portable) or over multiple tiers, but it should be stored securely and available only to that individual. This may be achieved by restricting access using a password, and encrypting information when it is transferred.

This information may be electronic communications (such as emails) and other private information (such as bank accounts) which we want to be able to access at all times via any device, but do not want others to be able to access. Increasingly this may include images, audio and video information that we do not want to be without. A personal information space might be thought of as the digital equivalent of a shoebox of photographs and old letters under the bed. Just as you may choose to show certain things from this shoebox to certain people, so you may choose to make things in your personal information space available in a shared information space. The personal information space can be a diverse, dynamic collection acting as a diary and personal library, with the opportunity to share some items with family, friends and colleagues, or the wider digital community.

6.3   Designing for Mobile Search

There are many considerations for designers of mobile search systems. The first concern should be the characteristics of mobile usage: particularly what distinguishes this from use of desktop systems. These two cases should not be seen as two non-overlapping communities of people, but as two alternative modes that any user may adopt. A framework for filtering information is described below, which aims to make retrieved information more relevant to mobile users. Further design issues are described, including whether information is filtered manually or automatically, and the use of this framework to make decisions about if and when to push selected information.

6.3.1   Characteristics of mobile usage

Mobile individuals using handheld computers to search for information have some distinct characteristics, which set them apart from users of desktop computers. One of the most significant factors is that mobile usage tends to be characterised by a large number of shorter sessions, compared with desktop usage (Ostrem 2003). Mobile users tend to use their devices to complete a specific task quickly, then put the mobile away (Mountain and MacFarlane 2007). In the context of mobile search, this suggests that they have less time and screen space for browsing long lists of results, or manually expanding or modifying search terms. With this in mind, mobile search systems should consider increasing the precision of results at the expense of recall.

Two other significant characteristics of mobile users can be seen as two sides of the same coin. From a negative perspective, mobile users tend to be operating in dynamic, less predictable, potentially distracting environments. Mobile users can devote less of their attention to the task of searching since the cognitive load from external sources is higher: where desktop users are surrounded by familiar walls, mobile usage tends to take place in less familiar, more dynamic environments, where the context of the search is paramount. A user that is moving adds further complexity, since this context is in a constant state of change. Mobile users are often engaged in a specific task in the real world, relegating mobile search very much to second place in their attention. However this distraction from the outside world can also be seen as an opportunity when considering the specific information needs of mobile users: this engagement with the physical world may itself be the inspiration for an individual's information needs.

'Here' and 'now' are two very important concepts for mobile users: they want information that is not only relevant to the subject of their query, but this information must be *timely* and relevant to their surroundings, or more broadly, their context (Göker *et al.* 2004b; Jiang and Yao 2006). Information retrieval research suggests that locations in the physical world are a major component of web searches: a study by Silva *et al.* (2006) suggests that one in five queries has been found to contain a spatial component, such as a place name. Naturally, the importance of a spatial component can vary from situation to situation, and this figure increases in some cases. For example, a user study considering the information needs of mobile individuals, based on visitors to a National Park (Swiss), found that location was more important still, with 60% of the questions asked by visitors having some spatial component which may be explicitly stated (for example 'How do I get to the town of ...?') or implicitly (for example 'Are ... found *around here*?') (Mountain and MacFarlane 2007). Information needs change as people encounter new surroundings and often it may be the rapid change of situation or environment that prompts a new information need. By adding a spatial component to these queries, or recognising spatial information within documents, it is possible to increase spatial relevance. Similarly, it is possible to reduce frustration that may occur when receiving the out-of-date information or the right information at the wrong time through the use of temporal filter. Simply porting information retrieval systems designed for desktop machines to mobile devices may be insufficient to meet the requirements of mobile individuals. Perhaps the way to get around this challenge is to actually look at the mobile information environment as a different opportunity than the web information environment.

6.3.2   Filters as a framework for mobile search

As discussed above, mobile users tend to devote less time to each individual session: they want to satisfy their information needs quickly, suggesting the need for improved precision in information retrieval at the expense of recall (see Chapter 2: User-centred Evaluation of Information Retrieval Systems for the meaning of this and other evaluation measures). This need for high precision implies the need to *filter* information for mobile users, to reduce the total volume of information, and ensure that only the information most relevant to their information needs is retrieved. This section will consider alternative strategies for filtering information, each of which can be seen as a distinct component of an individual's information need.

Retrieved information may be filtered to return only those documents that are relevant given an individual's particular location. Alternatively, time may be a factor by which information may be filtered. Finally, the particular interests and preferences of individual users of a system may be taken into account when filtering information. These filters may be used at the time of query, or decisions about where to store information may effectively pre-filter information for users in specific scenarios. These three components: space, time, and interest, provide a framework that can be used in the design of mobile search systems. This is shown diagrammatically in Figure 6.7. This diagram represents three overlapping sets of documents, each retrieved from the same parent collection. One is relevant according to a user's current spatial behaviour; one relevant, in some way, to the current time; and one relevant to user preferences. The most relevant information can be found in the intersection of these three sets. Using filters in this way can help increase precision at the expense of recall. The following three sections discuss these filters further.

6.3.2.1   Filtering information by spatial location

Spatial filters can represent the spatial footprint, in the physical world, associated with a mobile user's *query* (Mountain and MacFarlane 2007). Various assumptions can be made about the spatial characteristics of an individual's information needs, and these can be used as a guide for the spatial

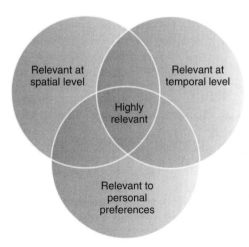

Figure 6.7 Filters as a framework for increasing the relevance of mobile search   (Reproduced by permission of © AmbieSense™ Ltd)

filters that we employ. The first assumption – *nearby* – assumes that the closer the document's spatial footprint is to an individual's location, the more relevant it is (see Figure 6.8a). These filters are familiar from 'where's my nearest' directory services (e.g. the Yell™ search engine) and are perhaps the most intuitive filter for the majority of users of mobile services. A more sophisticated assumption is based upon *accessibility*: where places that can be reached in a shorter period of time are more relevant (see Figure 6.8b). Whilst the nearby filter measures distance in metres, an accessibility filter measures distance in minutes. This can take account of the particular constraints of mobile individuals, considering the locations that are accessible to them given their current location, the amount of time they have free, and their mode of transport (Schwanen and Kwan 2008).

The third assumption – *likely future locations* – is that places that coincide with an individual's likely future location are more relevant than those places they are unlikely to visit (see Figure 6.8c). Algorithms have been developed that can take account of the overall direction that someone is travelling, their speed of movement, and the sinuosity of their path, to create a prediction filter (Mountain and MacFarlane 2007). The fourth assumption – *visibility* – is that places that are visible are more relevant than those which are concealed, reflecting the idea that the external world through which people are moving often acts as a source of inspiration and specific information needs (see Figure 6.8d). Visibility may be restricted by natural features (such as terrain or vegetation), but equally can be dramatically restricted in urban areas by artificial features, predominantly buildings. The visibility filter aims to satisfy queries motivated by the serendipitous observation of some feature in the real world: a query that in day-to-day conversation is more commonly phrased 'What's that?'.

Following a query based on keywords, or a particular subject, these filters can be applied to remove content whose spatial footprint does not coincide with that of the user with the mobile query. The user may choose which of the available spatial filters is most appropriate given their particular context. The relevant filter may then be generated automatically and applied to the retrieved results (Mountain 2005).

6.3.2.2   Filtering information by time

Another approach to filtering information is to consider the *timeliness* of information. Much of the information required by mobile users will have a temporal component. To consider some examples from the human environment, shops and services have opening times each day, and these may vary by day of the week. Public events – such as concerts – occur at specific times of day on known dates. A story from a news feed may include the time and date when it was published. Buses, trains and planes

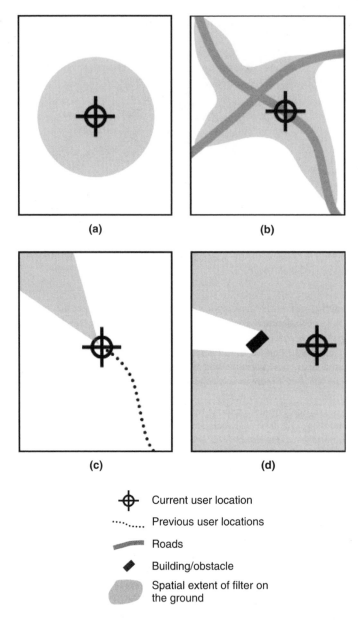

Current user location

Previous user locations

Roads

Building/obstacle

Spatial extent of filter on
the ground

Figure 6.8 Spatial filters for mobile search. Each filter can be thought of as defining a footprint on the ground representing the spatial extent of a query. Each filter is designed to satisfy information needs in a different scenario. (a) Nearby filter: Criteria: *Euclidean distance* Query scenario: *Which restaurants are found around here?* (b) Accessibility filter: Criteria: *Travel time* Query scenario: *Which restaurants can I reach within 30 minutes?* (c) Likely future location filter: Criteria: *Prediction of likely future location* Query scenario: *Which restaurants am I likely to pass in the next 30 minutes?* (d) Visibility filter: Criteria: *Places visible from current location* Query scenario: *Find reviews for the restaurants that I can see.* (Reproduced by permission of © David Mountain)

all operate according to predefined schedules. Equally, temporal cycles from the natural environment may be an important component of a mobile information need: plants may flower only at certain times of year; animal migration routes can take them to certain places at times which may be predicted in advance. In order to satisfy the immediate information needs of the mobile user, it may be necessary to take account of this temporal component to ensure that a mobile search service does not suggest a restaurant that is closed, a bus service that does not leave until tomorrow, or a good location to go whale watching, at the wrong time of year.

Mobile search systems that wish to ensure that the information they provide is timely must adopt some strategy for handling this temporal component. Simple strategies may make the assumption that newer information is more relevant, and such a strategy is employed effectively by news websites (and prior to this, newspapers), where the most visible content of the site consists of new information, and older materials may be removed or available only through archives. In a mobile context, users may wish to filter information to let them know about breaking news in their local area.

In other contexts however, the issue may be more problematic than simply representing the most up to date information. This requires that the temporal component associated with information is known, and may be represented in some way. For mobile navigation services based upon public transport, this may be represented by schedules which can be used to retrieve the most relevant options for a journey based upon when an individual intends to travel. Similarly mobile navigation services for private transport are increasingly taking account of congestion using real-time traffic information.

For other mobile information queries, it may be necessary to store the times and dates at which the information is relevant. Schedules may be effective in recording the times at which shops are open and services available. Some *events* may be considered relevant for a period of time running up to the time that they occur (for example, from tickets going on sale for a concert, until that concert occurs). Filtering information by time in this way can enable users of mobile search systems to retrieve more relevant information, for example, letting people know details of concerts, shows, plays or films that are available that evening, in the local area. An example of looking into time, location, and interest for events can be found in Bierig and Göker (2006).

6.3.2.3   Filtering information by user preferences

As well as filtering information to take account of the location of the user, or the time at which they made the request, the personal interests of individual users may be used to remove less relevant information and increase the precision of the retrieved set of results. These interests can be formalised as a *preference filter* and stored as part of a user's profile. Users are frequently asked to specify interests and preferences when they first use a system or service, and they can be refined, either manually or automatically, over time. The filter may be refined to take account of a user's behaviour, for example, learning the types of information that they are interested in, and using this to retrieve similar content; the recommender system StumbleUpon[5] is an example of such a filter. One approach to maintaining persistent filters is to allow users to rate the information that they access in terms of whether it was relevant to them. This allows a collaborative opinion to be formed on the overall quality of a particular website or source of information. It can also allow search systems to define the similarity in interests between users – potentially clustering them into groups – and use these associations to retrieve information for a user that was considered relevant by like-minded users. This promise of more relevant information gives incentive to users to provide feedback on the sites they visit.

6.3.3   Manual versus automatic filtering

When designing for mobile search, it is possible to manually pre-filter information to make it more relevant for mobile users in particular settings. We have already seen how this can be achieved when filtering information by time for online newspapers, by ensuring the most up-to-date information is

---

[5] http://www.stumbleupon.com/

available on most visible and accessible parts of a website. It is also possible to manually pre-filter content on embedded servers or mobile phones to ensure that information is relevant to the surrounding environment or context. Web designers can make *subjective* assessments using similar criteria to the filters described earlier. For example, a designer may choose to only include content that they consider to be relevant to the local area, or relevant to a particular theme.

A side benefit to manual pre-filtering is a reduction in the volume of information in the system. The key benefit, however, is that designers have more control of the user experience by adopting this manual strategy: news websites can order stories not just chronologically, but by how *important* they consider them to be for their readers – something that is difficult for a computer to gauge algorithmically. They can also consider the broader situation and context (see also Chapter 7), rather than just spatial location and time: they may consider that an embedded server's role is to provide information to travellers waiting at a bus stop, or visitors to a particular exhibit in an art gallery, context which may not be clear when considering spatial location alone. Hence pre-filtering can allow designers to create mobile search systems that are tailored for an intended purpose which serve a specific audience, but these search systems tend to be less flexible, and may not be able to satisfy more diverse queries.

The alternative to manually pre-filtering information is to automatically filter information at the time of query. Given the location and time associated with a user's query, and knowing the locations and times at which documents are considered to be relevant, the filters described in previous sections can sort information automatically. A temporal filter could remove old irrelevant information; a spatial filter can remove information considered not relevant, given the user's current spatial behaviour. Automatic filtering has the potential to make more diverse information sources relevant to the mobile user: it may be able to handle wider and less predictable information needs than a manual pre-filtering approach.

Filtering information offers an effective approach to ensuring that there is greater precision in the information retrieved, making it more relevant for mobile users who are less able to search manually through long lists of results to find the most relevant documents. The choice of whether to manually pre-filter information or to develop algorithms to filter at the time of query depends to a great extent upon the target audience and intended use of the mobile search system.

6.3.4   Using filters to push information

Filters can be applied to automatically *push* information considered relevant to users, without them having to construct a query directly. A mobile search system can keep track of a user's preferences, their location and the time, and use this to make a decision about whether to push information. Pushing information to users can be quite an invasive approach so such a system should be used very sparingly, and when information is pushed, it must be considered highly relevant by the recipients if they are to appreciate the intrusion. Many mobile operators are interested in push services to enable location-based advertising, however a service that indiscriminately tells users about special offers available for every shop they pass as they walk down a street is unlikely to be welcomed. Designers of mobile search must consider the sophisticated information needs of their users when constructing push services to avoid breeding resentment from their users. These kinds of user-centred challenges are discussed in further detail in the context of several experiments in Göker and Myrhaug (2008).

One technological approach to pushing information is to use intelligent personal agents to make a decision about what information to deliver, and when (Lech and Wienhofen 2005). These agents are developed to perform tasks on behalf of a user. Agents come in different forms. The presence of an agent may be obvious, it may be visualized as an animated character in two or three dimensions. They may speak natural language, they might be able to receive voice input, and some of them may look like a human face mimicking speech and emotions as it is having a dialogue with you. Alternatively, the agent may operate in the background, performing tasks that aim to improve your experience of a system, for example, adapting the user interface by concealing less frequently used functions. In both cases their role is to apply some 'intelligence' to improve the user experience.

Automatic push services can deploy agents with the aim of pushing only the most relevant information for mobile users, where the agents use space, time and interest filters as the basis for their reasoning. Changes in location, updated information resources or the passing of time may be the trigger for information to be pushed. For example, historians visiting London for the first time may wish to use a mobile visitor guide to let them know whenever they pass a site linked to a particular historical event, such as the Great Fire of London of 1666. Such a service could ensure that they do not pass by any significant location or monument without being made aware of its proximity to them. Employing a narrow spatial and interest filter in this way may reduce the information pushed to a desired level, and be seen as satisfying our persistent mobile information needs, rather than distracting us with unwanted information.

6.4   Case Studies

The following case studies describe two implemented mobile search systems, in two very different environments. They demonstrate how the filters described above can act as a framework in the design of mobile search systems, in both the built, and the natural environment.

6.4.1   Oslo airport – AmbieSense

One of seven demonstrations of mobile information systems that were developed as part of the EU funded AmbieSense EU-IST project[6] developed for use by air passengers at Oslo Airport, in which Oslo Airport itself was a partner. The system was evaluated within a larger iterative evaluation context (Göker and Myrhaug 2008).

6.4.1.1   User requirements

The needs of this specific user group are quite well defined, and assumptions can be made about the information required by the intended user group. The primary information need for passengers at air terminals is flight information. This includes information about the departure time of their flight and updates on any delays, information about the location and time of check-in, the departure gate number and whether the flight is yet boarding. An important secondary information need is the services on offer in the airport. Air passengers can experience a lot of waiting around and like to take advantage of places to eat, shop, and use facilities.

A distinct characteristic for air passengers is the pressure of time: passengers want to take advantage of the 'idle time' between checking in and departure as best as possible, but under no circumstances wish to miss their flight. A mobile search tool can allow passengers to keep up to date with their flight information, providing regular updates, but also inform them of things to do whilst they await their departure. This time pressure adds to the need for the device to return relevant information.

6.4.1.2   System functionality

The version of the system shown in Figure 6.9(a–b) was deployed on PDA devices that communicated with embedded servers via wireless connections (either Bluetooth or WLAN). In all, many wireless embedded servers were installed in the airport, designed to cover the full trajectory of the airport experience, from pre-check-in to departure.

The content was designed to fulfil the two distinct information needs of passengers at air terminals. Live flight information could be extracted from the Oslo airport database and information about the airport facilities (including shops, restaurants and bars) existed in digital form as it had been generated

---

[6] www.ambiesense.net and www.ambiesense.com

(a)   (b)

Figure 6.9 (a) The browser component for one of AmbieSense's mobile guides for Oslo airport (Reproduced by permission of © AmbieSense™ Ltd) (b) Relevant flight information retrieved from the OSL database and updated every minute (Reproduced by permission of © AmbieSense™ Ltd)

for the airport's website. This content was repurposed to be appropriate for a small screen device. An ontology (see Chapter 9 for further background) was created as the main organisational framework for filtering this information. The application homepage and the display of live flight information are shown in Figure 6.9(b).

The system applied a variety of filters to ensure that users received only the most relevant information, sacrificing recall for the sake of precision. Firstly information was filtered by user preferences: since a user's flight details were known, flight information was filtered to report just the one flight. A profile also recorded items of interest to a user (for example, preferences for food or shopping) and could be used to restrict the information retrieved by the system (see Figures 6.10 a–c). This information was filtered further by spatial proximity: the user accessed different repositories of information depending upon the embedded server they were connected to.

Intelligent agents were using filters to make decisions about if and when to push information to users. A change in the spatial location of a user (switching their connection from one wireless to another) acted as a trigger for agents to search information relevant given the user's declared preferences and the content. The results of this search were not returned in an intrusive manner, but just updated the list of recommended content as a user walked around the airport. Whilst the pushing of information about facilities was unobtrusive, information pushed concerning a change in flight information was more obvious. When a flight entered the 'boarding/gate closing' mode the agent pushed this information to the user, and they had to acknowledge this message before they could access any other information.

6.4.1.3   Evaluation

This version of the system, described above, was evaluated with 22 participants, as part of a series of large-scale user tests, who were all air passengers approached at Oslo airport. Göker and Myrhaug (2008) detail the iterative and incremental user-centred evaluation method and series of experiments on this indoor plus an outdoor scenario (see Chapter 2: User-centred Evaluation of Information Retrieval Systems for general background on user-centred evaluation). Testing with real air passengers at the airport ensured that the test was representative: field-testing of this kind is particularly essential for mobile applications, where lab-based studies may not be able to represent a user's environment and specific needs particularly effectively. The evaluation took the form of a demonstration by the team to show them how it worked, then let the participants alter their preferences and see different

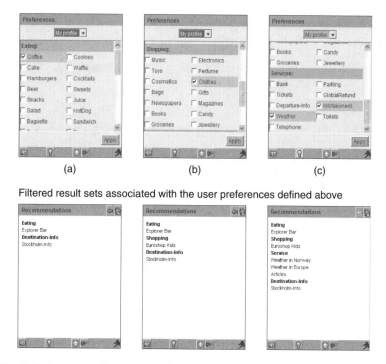

Figure 6.10   Defining user preferences in a profile (a) Eating preferences   (Reproduced by permission of © AmbieSense™ Ltd) (b) Shopping preferences   (Reproduced by permission of © AmbieSense™ Ltd) (c) Service preferences   (Reproduced by permission of © AmbieSense™ Ltd)

content popping up. The tests lasted between 20 and 45 minutes (walking around, and interviewing). Some users preferred to stay seated with their luggage while querying for airport information, whilst others walked round to see how content automatically changed depending on their spatial location and proximity to shops. After asking the user to use the system to conduct some tasks, feedback was generated in the form of an interview. The result led to the wider implementation of airport guides for mobile phones, since participants and respondents of a market survey wanted airport guides implemented for mobile phones instead of PDAs. The air passengers welcomed very much the automatic filtering of content while walking around in the airport.

6.4.2   The Swiss Alps – WebPark

As part of the EU funded WebPark project (Raper *et al.* 2007c), the Swiss National Park was chosen as a test-bed for a mobile search system, which offered a digital, location-aware alternative to paper guide books and maps for visitors to the park. The Swiss National Park Authority were themselves a project partner, and were actively involved in ascertaining user requirements, development, and the evaluation of the implemented system.

6.4.2.1   User requirements

Having an organisation representing end-users, that had practical experience in satisfying their infor-mation needs, acted as a constant reality check on development in the WebPark project, and was designed to ensure that the final system focused on and fulfilled the requirements of its users. The primary user group were visitors to the park, who would be using the tool in place of, or in addition

Figure 6.11 Interface for the Swiss National Park mobile search system. The images above show a homepage, customised search applications, an example of the mapping application (Reproduced by permission of © Camineo SAS. All rights reserved. Map data owned by © Rando Editions.)

to, traditional sources of information such as the park's information centre, information boards around the park, guidebooks, maps, and contact with the park rangers.

There were two approaches to eliciting user requirements: a paper questionnaire sent to a large number of potential users and a user shadowing exercise conducted with a handful of visitors. The questionnaire was sent to all subscribers of the National Park magazine and resulted in 1600 responses. Visitors were asked questions about how they acquired information relevant to the park (when in the field and outside the park), their level of experience in using Internet and mobile technology, whether they would be prepared to use such a system, and for what purposes. The results suggested that the type of information visitors considered important included information about maps for orientating oneself in the park and finding features of interest, information about flora and fauna and its distribution within the park, as well as other themed information (such as geology or history), and information about the length and difficulty of trails within the park. The user shadowing exercises encouraged visitors to raise every question that they had with a ranger, who accompanied them on their walk for that purpose and logged all questions raised. This exercise identified similar needs: most notably an interest in flora and fauna and its distribution relative to the user's location (Raper *et al.* 2007c; Mountain and MacFarlane 2007).

6.4.2.2 System functionality

Based on the user needs study it was decided that the content should include flora and fauna, recommended trails, the availability of facilities, and themed information based on the park's exhibitions (see Figure 6.11). The majority of this information could be adapted from the park's existing resources, disseminated via the web and CD-ROM. Functional requirements included a location-sensitive mobile search, and geography was identified as being the primary organisational framework for the system: the location or distribution of each item of information in the system should be known and can be used as a spatial filter when retrieving information. Two spatial filters were implemented and made available in the search tools: in addition to searching the 'whole park' (unfiltered information), users could search 'around me' (spatial proximity filter) or 'search ahead' (prediction filter). The system was deployed on a location-aware PDA, which stored information mostly at the portable tier (on the device itself), but could connect to remote servers via a wireless GPRS connection to access current information.

6.4.2.3 Evaluation

Three rounds of user evaluation were conducted during the lifetime of the project. The first two were conducted with 20–30 users and were intended to influence the development process, identifying the information and functionality that was well received and which required further work – as part of an iterative development. The final session was intended as a summative evaluation exercise, to gauge the success or otherwise of the final implemented system. By the time of this final evaluation exercise, the system was considered intuitive and robust enough to allow visitors to use it all day without supervision, after a short induction exercise, and provide feedback in the form of a questionnaire and interview at the end of the day. In this final round of testing, feedback was gathered from nearly 100 people, in areas including the general handling of the device, the ease of information provision and search tools, the quality of information content, the usefulness of specific applications, and the participant's willingness to pay for such a service. The system performed reasonably well in terms of ease of information provision, with 60% regarding this to be 'good' or 'very good' on the questionnaire. When broken down into the different types of search, filtered information was considered much more relevant than unfiltered information. Whilst only 54% of visitors considered unfiltered information to be either relevant or very relevant, roughly 90% considered information retrieved using one of the two implemented filters to be relevant. Of the two filters implemented, the 'around me' filter performed best, with two-thirds of participants regarding the information retrieved to be 'extremely relevant'. Using the 'search ahead' filter, roughly half of participants considered the information to be 'extremely relevant' (Mountain and MacFarlane 2007).

6.5   Summary

In this chapter we have focused on the factors that distinguish mobile search from desktop web search. Mobile search is a new and growing field, because of the increasing diversity and availability of mobile information environments. Mobile users now increasingly expect their individual information needs and interests to be satisfied at any time, from any place, and not just in front of a laptop or desktop computer.

We have investigated the distinctive nature of information for mobile search. One aspect of this concerns strategies for storing information at different tiers, and whether information is carried physically with us (portable tier), is made available on the basis of proximity in the physical world (embedded tier) or is made available over the Internet (remote tier). Another aspect of information concerns its ownership, and whether this is public, shared or personal to a particular individual.

Next we have considered approaches to providing mobile search that suit the distinctive characteristics of mobile users, and the need to filter results to help increase search effectiveness. Strategies for filtering information include filtering by spatial location, by time, and by personal preference. These filters can also act as triggers to push relevant information to mobile users. Finally we have seen two case studies, which describe the use of some of the concepts described in this chapter in two very different environments: the indoor built environment (Oslo Airport) and the natural environment (the Swiss Alps).

The issue of how best to meet the information needs of mobile users is likely to increase in importance for many years to come, and requires an interdisciplinary approach. As mobile devices become all-pervasive, the majority of queries may be submitted from mobile handsets.

For readers interested in research relating to mobile computing, and mobile search, we recommend the following journals:

- *Journal of Location Based Services*, ISSN: 1748-9733, Taylor & Francis.
- *Pervasive and Mobile Computing*, ISSN: 1574-1192, Elsevier.
- *International Journal of Mobile Information Systems*, ISSN: 1574-017X, IOS Press.
- *Wireless Communications and Mobile Computing*, ISSN: 1530-8669, John Wiley & Sons.

For those interested in further books on mobile media and design topics, we may recommend:

- *Multimedia Broadcasting and Multicasting in Mobile Networks*, by Iwacz, G., Jajszczyk, A., and Zajaczkowski, M., October 2008, John Wiley & Sons.
- *Mobile Interaction Design*, by Jones, M. and Marsden, G., December 2005, John Wiley & Sons.

- *Next Generation Wireless Applications: Creating Mobile Applications in a Web 2.0 and Mobile 2.0 World*, Golding, P., 2nd edition, John Wiley & Sons.
- *Mobile Advertising: Supercharge Your Brand in the Exploding Wireless Market*, Sharma, C. Herzog, J., and Melfi V., April 2008, John Wiley & Sons.

For those interested in more recent workshops on mobile information retrieval, have a look at:

- MobIR (2008) *Proceedings of the ACM- SIGIR 2008 Workshop on Mobile Information Retrieval*, 24 July 2008, Singapore.

Readers interested in geographic information systems and services, may consider:

- *Geographical Information Systems and Science*, by Longley P. A., Goodchild M. F., Maguire D. J. and Rhind D., 2005, W. John Wiley and Sons, Chichester.
- *Internet GIS: Distributed Geographic Information Services for the Internet and Wireless Networks*, Peng Z. R., and Tsou M. H., 2003, John Wiley and Sons, New York.

Finally, for current advances and observations in the mobile search area, with a strong user focus, readers may find the online knowledge resource below useful:

- Salz, P.A., Msearchgroove. *Blog*. www.msearchgroove.com [6 June 2009].

Acknowledgements. Hans Myrhaug and Ayşe Göker would like to acknowledge the valuable input of Stein Tomassen, Christopher Lech, Leendert W. M. Wienhofen, Marius Mikalsen, Tor Erlend Faegri, Borge Haugset, Jon Henrik Torshaug, Maj Irene Bye, Hannah Cumming, and Rob Engels in their contribution to the development of the specific version of the Oslo Airport system presented in this chapter. David Mountain would like to thank the WebPark team, the Department of Information Science at City University London, Placr Ltd, and Camineo SAS.

Exercises

6.1 Designing a mobile website with a search facility

This exercise has three parts (a), (b) and (c) that should be answered in this order.

Consider a leading website that you have used which offers a particular on-line service, for example this may be:

- A news site;
- A social networking site;
- An on-line auction site;
- Or any other on-line web service.

The site is designed for laptop use only and it has an on-line search facility. Imagine this business has just become one of your clients and they are requesting you to design their website for mobile phones, including the search. The planned mobile website is needed because it will be more convenient for the consumers to shop on their mobiles while commuting.

Assume the following design constraints for (a), (b) and (c):

- Mobile screens usually have a profile orientation with common measurements in the range of: 176 * 234, 240 * 320, or 480 * 640 pixels. Physical screen measurements can be as small as 34×45 mm or 45×60 mm. However, laptop screens have wide orientation with 1024 * 768 pixels or more, commonly with a 12–15 inch diagonal screen size.
- Vertical-scrolling-only is the chosen within-page-browsing mechanism for the mobile website.
- The website is currently accessible via Internet browsers, and your client would like as much of the existing content as possible to be reused for the mobile website.

6.1(a) Designing the mobile website

- Analyse the existing website in terms of website navigation, main display areas, and content (banners, product categories, and product descriptions).
- Establish a list of 10 navigational or graphical elements that are difficult to transfer to the mobile website without changes.
- Propose a new menu for navigation that is suitable for mobile phone use given the design constraints above. Describe what is similar and what is different with this new design.
- Analyse the main display areas of the web site and specify a sequence of web pages that together will be able to present the same information on a small screen. Explain which information will be similar, which information has been left out, and where in the new menu consumers can reach this content from.
- Identify whether the images and eventual tables are suitable for the mobile phones. Propose an optimal image size for the new mobile website. Propose an alternative way of presenting table content, i.e. without multiple columns.
- Propose a better way of presenting banners and product categories for mobile phones in terms of screen size and delivery time. Explain why you think the solution is more suitable for mobile phones.
- Redesign the product descriptions for mobile phone use.
- Summarise how the introduction of the mobile web site will affect the content creation and publishing process of the existing web site for laptop use.

6.1(b) The mobile search index and spatial filters

- What opportunities could this website exploit by implementing a service built upon local search?
- What spatial resolution of information is required for this service (national, regional, street level detail?)
- Which of the spatial filters listed in Section 6.3.2.1 are required or desirable for this local search service.

6.1(c) Ways to enable the mobile search

- Describe the search facility of the existing website in terms of layout, how many clicks it takes to find the search page, all the query parameters you think the search facility can take as input from the consumers. Decide which of them are more important for mobile phone use.
- Identify whether the search facility is using a traditional IR search index, or if it is based on an SQL database. In case there is an underlying SQL database, propose how you think the database structure may be organised in terms of entities and relations. In case it seems there is an underlying IR search index, try to identify which IR model may be in use, and which query operators the search facility is supporting.
- Design the new web page for mobile search, describe the similarities and differences between this new page and the page(s) of the existing search facility. Provide your design reasons for the similarities and the differences. What do you think would be the pros and cons of having a multiple-step versus single-step search page for the mobile search – in terms of navigation and visibility for the consumers?
- Design a personal settings page that enables the user to specify a minimum number of search queries/parameters over time. Explain why your design may improve the mobile search for the user over time, and how it is reached from the menu you proposed in (a).
- Summarise how the introduction of the mobile website with search page affects the indexing of the existing search facility for laptop use, and how the indexing method works when your client adds new, updates, or removes product descriptions.

References

Bierig, R. and Göker, A. (2006). Time, location and interest: an empirical and user-centred study. In *Proceedings of the 1st International Conference on Information Interaction in Context* (Copenhagen, Denmark, October 18–20, 2006). IIiX, vol. 176. ACM, New York, NY, 79–87. (DOI= http://doi.acm.org/10.1145/1164820.1164838)

Brimicombe, A. and Li, Y. (2006) Mobile Space-Time Envelopes for Location-Based Services. *Transactions in GIS*, **10**(1), 5–23.

Cairncross, F. (1999) The Death of Distance: How the Communication Revolution Will Change Our Lives. *American Quarterly*, **51**(1), 160.

dotMobi and AKQA Mobile (2008) Mobile Internet Usage and Attitudes Study. Available at http://mobithinking.com/best-practices/mobile-internet-usage-attitudes-study (accessed 6 June 2009).

Göker, A. and Myrhaug, H. (2008) Evaluation of a Mobile Information System in Context. *Information Processing and Management*, Special Issue on 'Evaluation in Interactive Information Retrieval', Elsevier, 2008, **44**(1), 39–65.

Göker, A., Watt, S., Myrhaug, H. I., Whitehead, N., Yakici, M., Bierig, R., Nuti, S. K. and Cumming, H. (2004a) An ambient, personalised, and context-sensitive information system for mobile users. In *Proceedings of the 2nd European Union Symposium on Ambient Intelligence* (Eindhoven, Netherlands, November 08–10, 2004). EUSAI '04, vol. 84. ACM, New York, NY, 19–24. DOI= http://doi.acm.org/10.1145/1031419.1031424

Göker, A., Myrhaug, H., Yakici, M. and Bierig, R. (2004b) A context-sensitive information system for mobile users. *27th Annual International ACM SIGIR Conference, Workshop on Information Retrieval in Context*, Sheffield, demos and demo paper, July 2004, 29–32.

Haller, R. (2004) Manager of the Swiss National Park. Personal communication.

Jiang, B. and Yao, X. (2006). Location-based services and GIS in perspective. *Computers, Environment and Urban Systems*, **30**(6), 712–725.

Krug, K., Mountain, D. M. and Phan, D. (2003) WebPark: LBS for mobile users in protected areas. *GeoInformatics*. March, 2003, 26–29.

Kwon, Y. J. and Kim, D. H. (2005) Mobile SeoulSearch: A Web-Based Mobile Regional Information Retrieval System Utilizing Location Information. *Lecture Notes in Computer Science*, 206–220.

Lane, G. (2003) Urban Tapestries: Wireless networking, public authoring and social knowledge. *Personal and Ubiquitous Computing*, **7**(3/4), 169–175.

Lech, T. C. and Wienhofen, L. W. (2005) AmbieAgents: a scalable infrastructure for mobile and context-aware information services. In *Proceedings of the 4th International Joint Conference on Autonomous Agents and Multiagent Systems* (The Netherlands, July 25–29, 2005). AAMAS '05. ACM, New York, NY, 625–631. DOI= http://doi.acm.org/10.1145/1082473.1082568

Mountain D. M. (2005) Exploring mobile trajectories: An Investigation of Individual Spatial Behaviour and Geographic Filters for Information Retrieval. PhD Thesis. City University, London. UK.

Mountain, D. M. and MacFarlane, A. (2007) Geographic information retrieval in a mobile environment: evaluating the needs of mobile individuals. *Journal of Information Science*, **33**(4), 515–530.

Myrhaug, H. and Göker, A. (2003) AmbieSense – interactive information channels in the surroundings of the mobile user, in *Proceedings of 10th International Conference on Human–Computer Interaction 2003*, 1158–1162.

Myrhaug, H., Fægri, T., Göker, A. and Lech, C. T. (2004) AmbieSense – a system and reference architecture for personalised and context-sensitive information services for mobile users. In *Proceedings of the 2nd International Symposium on Ambient Intelligence*, long paper, November 2004, Springer Verlag. www.eusai.net

Ostrem, J. (2003). *Palm OS User Interface Guidelines*. Palm OS User Interface Guidelines Document Number 3101-001-HW, February 24, 2003. Available at http://www.accessdevnet.com/docs/ui/UIGuide_Front.html (accessed 14 August 2008).

Portio Research (2008a) Slicing up the Mobile Services Revenue Pie. Market Report Summary at: [http://www.portioresearch.com/Slicing_Pie.html], Portio Research. 28 January 2008.

Portio Research (2008b) Mobile Data Services Markets 2008: A Complete region by region analysis of non-voice mobile services and revenues. Market Report Summary at: [http://www.portioresearch.com/MDSM08.html], Portio Research. 11 June 2008.

Portio Research (2008c) Worldwide Mobile Market Forecasts 2009-2013: Complete analysis of worldwide mobile technology, markets, and forecast subscriber growth. Market Report Summary at: [http://www.portioresearch.com/WWMMF09-13.html], Portio Research. 5 September 2008.

Purves, R. and Jones, C. (2006) Geographic Information Retrieval (GIR). *Computers Environment and Urban Systems*, **30**(4) 375–377.

Raper J. F. (2007a) Geographic Relevance. *Journal of Documentation*, **63**, 836–852

Raper, J. F., Gartner, G., Karimi, H. and Rizos, C. (2007b) A critical evaluation of location based services and their potential. *Journal Of Location Based Services*, **1**, 5–45.

Raper, J. F., Gartner, G., Karimi, H. and Rizos, C. (2007c) Applications of location based services: a selected review. *Journal of Location Based Services*, **1**, 89–111.

Reades, J., Calabrese, F., Sevtsuk, A. and Ratti, C. (2007) Cellular Census: Explorations in Urban Data Collection. *IEEE Pervasive Computing*, **6**, 30–38.

Rosenfeld, L. and Morville, P. (2006). *Information architecture for the World Wide Web: Designing Large-Scale Web Sites*. O'Reilly Media Inc, Sebastapol, CA.

Schwanen, T. and Kwan, M. P. (2008) The Internet, mobile phone and space-time constraints. *Geoforum*, **39**, 1362–1377.

Silva, M. J., Martins, B., Chaves, M., Afonso, A. P. and Cardoso, N. (2006) Adding geographic scopes to web resources. *Computers Environment and Urban Systems*, **30**(4), 378–399.

Stynes, K., Woolard, A. and Kahana, G. (2006) Market Research 2006: Industry Trends. Summary of Current Industry Trends related to Participate. London, BBC. Available from http://193.113.58.250/downloads/Participate_WP2.4_%20Market%20Research_Industry_Trends%202006.pdf (accessed 21 April 2009).

Sui, D. Z. (2008) The wikification of GIS and its consequences: Or Angelina Jolie's new tattoo and the future of GIS. *Computers Environment and Urban Systems*, **32**, 1–5.

Wikipedia (2009) Aldersgate, http://en.wikipedia.org/wiki/Aldersgate. (Last accessed 21 Jan 2009.)

7

Context and Information Retrieval

Ayşe Göker, Hans Myrhaug and Ralf Bierig

7.1 Introduction

Context information provides an important basis for identifying and understanding people's information needs. Utilising this should enable information retrieval systems better to meet these needs. 'Context' has also been known within the areas of 'user needs analysis' – in other words, what do users really want to know in terms of prior knowledge and what is really their information need.

Information is a crucial element of our daily lives. Today, the average person has access to more information than ever before. For example, the growth of the web is still measured on an exponential scale and other channels of information are also increasing. There were 172 million hosts on the web in August 2008 (Netcraft 2008), about 47 million more than in July 2007 and about 72 million more than at the end of 2006. This shows how important digital information has become and how critical it is to manage information effectively. We have a massive task of making more information accessible whilst ensuring it is still relevant and useful for our needs. There are many reasons why having contextually relevant information is important, including the timely delivery, and relevance of content. However, a major challenge is to enable this, even with the large amounts of content available.

We know that when using web search engines people on average type in about two words for their query[1]. On their own, this is quite minimal information to go by. Let us consider a recent popular search: the Olympics. It may be that you just wanted to find out the latest results and medal winners. It may be that you wanted to find out how the athlete representing your country performed. If you were in Beijing and doing the query in 2008, it may be that you wanted to know where the event was and how to get there. On the other hand, if you are in London there is much interest to find out about aspects of the planning for the games in 2012. If you also keyed in the name of a sport you are interested in, there could still be considerable variation in the breadth and depth of information that you would like – depending on whether you practice the sport or like only to watch it occasionally. Much depends on the context, the context of the query, and the user.

---

[1] Research on users' queries and weblogs indicate the average is usually just over two words (but less than three). Spink and Jansen (2004) describe and further point to a variety of analysis on this topic in their book.

Information Retrieval: Searching in the 21st Century   edited by A. Göker & J. Davies
© 2009 John Wiley & Sons, Ltd

The overall aim of this chapter can be viewed as twofold: firstly, to provide a background and understanding of context; secondly, to provide a framework for how context-aware[2] information system applications can be developed and evaluated.

Thus, in the first half of the chapter (i.e. Sections 7.2 and 7.3), we provide an overall introduction, and an understanding of the importance of context information in handling challenges of information search and retrieval. We describe relevant work on context in information retrieval (IR) and also from related fields, in order to help give a picture of the challenges in a wider sense.

Subsequently, in the second half, we discuss how context information can be captured and represented, along with examples and guidelines on how a context-aware information system or application can be developed. These are sections on context modelling and representation, and context and content. We also consider the related topics of personalisation and context, and mobility and context. We suggest considerations for how evaluation can be conducted in such context-aware information applications.

7.2 What is Context?

Numerous definitions of context exist. This is because of the multidisciplinary and rich nature of the topic itself. Later in this section, we provide some further background to these. However, in short, context can be defined as a description of the aspects of a situation (Myrhaug and Göker 2003; Myrhaug 2001). Figure 7.1 shows the relations between contexts, situations, and entities[3]. An entity or actor may be any person, system, or object involved in the situation.

The duration of context can range from being a very short moment to many years. The current context can depend on several criteria, e.g. the location, mental state, and so on. Capturing a person's context (or a user's context) can have a direct influence on the behaviour of a system. For example, a different choice of relevant information may be presented to different people.

Let us now consider a few definitions. Schilit *et al.* (1994) define context as 'where you are, who you are with, and what resources are nearby'. This might suggest that context is more focused on the user's surrounding as opposed to his/her inner states. Morse *et al.* (2000) describe context as 'implicit situational information'; Schmidt *et al.* (1999) had a more extended notion of context and defined it as 'interrelated conditions in which something occurs'. A more specific definition is provided by Dey *et al.* (2001, p. 5) in a special issue on context where they defined it as 'any information that

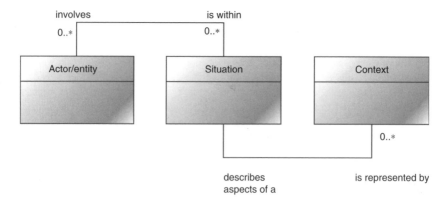

Figure 7.1 Actors, situations, and contexts (Reproduced by permission of © AmbieSense™ Ltd)

---

[2] Both 'context-aware' and 'context-sensitive' are phrases that are being used – though the first is more common. For consistency, we will use 'context-aware' in this chapter.
[3] The diagram is expressed in the standardised general-purpose modelling language called Unified Modelling Language (UML).

can be used to characterise the situation of an entity. An entity is a person, place or object that is considered relevant to the interaction between a user and an application, including the user and application themselves'. It becomes apparent from these more or less informal definitions that context generally 'suffers from the generality of the concept' (Schmidt *et al*. 1999, p. 3). Context as a term is used widely in everyday language[4] and also has a range of different meanings in information and computer science where it is also used to describe aspects of human–computer interaction and elements in natural language processing.

Context-aware information delivery methods use information about the user's current situation as a means to deliver relevant content to that situation. According to Salber *et al*. (1999), a system is context-aware if it uses context to provide relevant information and/or services to the user, where relevancy depends on the user's task[5] and other context information as well. For example, a mobile computer may exploit the user's task context better to adapt the information provided to his/her current situation. See Korkea-aho (2000) for a survey of the early context-aware applications.

Within IR, Ingwersen (1992, p. 195) early on said that a contextual IR theory might benefit the field because it could be a way of 'taking into account the uncertainty inherent in IR interaction'. He also refers to the pragmatic contexts, involving work tasks and (user) preferences (Ingwersen 1992, p. 196). We further will discuss the notion of context in IR shortly.

7.2.1 Whose context?

In general, one can speak of many different types of contexts – but why and for whom? The type of context depends on the entity or actor the context is intended for. In discussing context it is also useful to think of how context information may evolve over time (more on this later in the chapter). Below are some examples of context types according to the actor/entity that is involved:

- Contexts describing a user's situation should be named *user context*;
- Contexts describing a software's situation should be named *software context*;
- Contexts describing a document's situation should be named *document context*;
- Contexts describing a network's situation should be named *network context*.

User context will be discussed more in this chapter, but in the meantime let us consider, for example, a *document context*. The context of the document could mean information about the document such as: when it was created, who the author was, when it was accessed, in what ways or situations it was accessed (to the extent which this is possible to record), what media it is available in, where it is available, and other metadata about its content. In fact, much of this has already been investigated in IR without necessarily an explicit reference to the notion of context. However, if we are to develop context-aware information systems, it is useful to more clearly distinguish between aspects of a situation relating to a person with information need and the information we are trying to present.

7.3 Context in Information Retrieval

Within IR research, context information provides an important basis for identifying and understanding users' information needs. Cool and Spink (2002) in a special issue on context in information retrieval provide an overview of the different dimensions in which interest in context for information retrieval exists. They refer to an information environment level, information seeking level, information retrieval interaction level, and the query level. These categories are related and overlap. The first level is the information environment level. Here, Taylor's (1991) work on information need and information use environments is illustrative of the category of work. The second one is the information seeking level where, for example, Belkin's (1980) work on 'anomalous state of knowledge' reflects

---

[4] In many languages context means almost the same as a situation, but here we refine the concepts.
[5] Task should be understood here as goal-directed activity.

a problem-solving approach in information seeking. He states that users come to search tasks with varying degrees of knowledge about their subject when seeking information and that the anomalous state is a result of the gap between their information need and the internal resources to solve the problem or meet the need. The third level, information retrieval interaction, refers to user–system interactions, but from a cognitive perspective and is related to Ingwersen's (1992) cognitive communication model. The fourth of these dimensions is the query level which is exemplified in the myriad of work in TREC[6] (see also Chapter 2) on improving retrieval effectiveness (of a formal model, algorithm, or system) given a user query. Others have viewed context-aware retrieval more as a way of filtering results from normal retrieval techniques (Brown and Jones 2001).

Three theoretical models for *interaction* and *context* exist within IR. These are as follows:

- Cognitive model;
- Episode model;
- Stratified interaction model.

The *cognitive model* (Ingwersen 1992, 1996) views IR interaction as a set of cognitive representations and processes. Users interact according to their cognitive space that is defined along a set of different factors embedded in the user's personal context that closely resembles a user model. This cognitive space interacts with the social/organisational environment of that user that represents environment information. See Ingwersen and Jarvelin (2005, pp. 191–211) for more coverage on models for cognitive IR research.

The *episode model* (Belkin *et al*. 1995) views an 'information seeking episode [that] consists of a series of kinds of interactions (slices of time) structured according to some plan associated with the person's overall goals, problem, experience, . . .'. The model also includes aspects from an earlier model (Belkin *et al*. 1982) on users' anomalous state of knowledge (ASK). This episode model integrates the users' current contextual state as part of the information seeking process.

The *stratified interaction model* (Saracevic 1997) represents interactive information retrieval with a user and a system side that are connected through an interface. This model is influenced by human–computer interaction. Each side is divided into different levels or strata; the user side incorporates a strong contextual viewpoint that is divided into three different levels. The cognitive level deals with the ways users organise and structure information mentally, such as their state of knowledge and how they infer relevance from it. The affective level handles users' intentionality such as motivation, feelings and desires. The situational level represents the users' surrounding situation that triggers their information needs that are put forward to the retrieval system.

A survey of context work in IR can be found in Azzopardi (2002). However, more recently, a series of workshops on Context in Information Retrieval (e.g. Ingwersen *et al*. 2004; 2005) have provided a platform for the discussion of ideas and applications about context and information retrieval.

7.3.1   Context in the wider sense

The human mind is episodic in nature. We remember things, activities, and information easier if we can recall aspects of the situation where we experienced it. In fact, human discourse contains references to past, present, future, or imaginative situations. It helps us to remember, reason, and act (see Tulving 1972) for the distinction of episodic and semantic memory. Consider how we reflect upon photos that we take. We may recall them according to the picturesque sight, the friends and family we were with, an unusual event, and so on. Some of these 'triggers' can be difficult to capture and record. We present an example of how people choose to index their photos in a social arena shortly in this chapter.

Suchman's (1987) early work on 'situated actions' and the situatedness of most human behaviour is noteworthy. She highlighted the need to consider specific situations/settings of use and its importance

---

[6] For TREC (Text REtrieval Conference) run by NIST (National Institute of Standards and Technology) in the US see http://trec.nist.gov/

in moving beyond conventional system design. As Dourish (2004) also points out, her work is 'a common source for the idea that computer systems should respond to the settings within which they are used'. In the same paper, Dourish also gives a broad view of context and goes on to discuss the social and cultural aspects too.

Lieberman and Selker (2000, p. 617) make an interesting point within adaptive computer systems:

> ... a considerable portion of what we call intelligence in artificial intelligence or good design in human–computer interaction actually amounts to being sensitive to the context in which the artifacts are used. Doing the, 'right thing' entails that it be right given the user's current context.

In our view, this is an equally valid proposition for information retrieval systems.

A wide perspective of context has been discussed in some forums such as: within the fields of user modelling (Gross and Specht 2001a); information retrieval (Ingwersen *et al*. 2004); and in the European Information Society Technologies (EU-IST) Programme (cordis.europa.eu).

7.3.2   Perceptions of context in related fields

Several related fields also contain work on the notion of context. These include: user modelling; ubiquitous or ambient computing; human computer interaction; and artificial intelligence. We give a few perspectives from these here.

The importance of context, and its relevance in relation to user modelling, has also been identified for ubiquitous computing with a special issue by Jameson and Krueger (2005). Context within AI (artificial intelligence) usually has a more problem-solving focus. This has been addressed by a series of conferences on context (e.g. Bouquet *et al*. 1999; Akman *et al*. 2001), and onwards. A summary view can also be found in Brézillion and Turner (2002).

Work on context technologies can typically be found in the field of context-aware computing applications, see Dey *et al*. (2001) in a special issue on situated interaction and context-aware computing. In this field, a system is context-aware if it uses context to provide relevant information and/or services to the user where relevance depends on the user's task.[7] The definition has its roots in earlier work and definitions within the field (Schilit *et al*. 1994; Schilit 1995; Brown *et al*. 1997). Although these definitions typically involve the user and the environment, the field is technology-driven. The work in this field is typically about obtaining context and location information via sensors, and other digital equipment in the environment. There hasn't been a specific focus on the user's information needs.

Context-aware tourist applications have been developed (Abowd *et al*. 1997; Cheverst *et al*. 2000; Dunlop *et al*. 2003; Göker *et al*. 2004). See Chapter 6 for some related aspects. Considerable work in context-awareness has focused on location-based applications (Chen and Kotz 2000; Harter *et al*. 2002), devices, and networks (Ferscha and Mattern 2004). Also some work has been done on the interaction between the user and the device (CHI 2000). A broad perspective from this field is also given in Bradley (2005) on a user-centred design framework for context-aware computing.

7.3.3   Example: context and images

An example of where context information is provided for content is in Flickr[8], a public photo sharing website. The context information in this case is specified as user volunteered tags. This method of allowing users to tag their content with context information is currently a widely used method for

---

[7] Situation/context information has been used for several decades within artificial intelligence and machine learning applications. There are numerous intelligent systems using situation descriptions to describe the intermediate states when: solving complex problems, finding the best solution to a situation, planning, etc., with a current situation and the preferred goal situation to achieve, etc.

[8] See http://www.flickr.com

User tag	# Images *	Category
nikon	2864511	Camera
canon	2795794	
macro	1686691	
cameraphone	1230809	
film	1160898	
bw	1965754	Color
blue	1899819	
red	1785819	
green	1706871	
white	1557779	
wedding	7075247	Event
party	5565950	
travel	4991820	
vacation	3868691	
trip	3407057	
family	4474946	People
friends	3602993	
people	2336341	
me	2134245	
baby	1908037	

User tag	# Images *	Category
japan	4396587	Place
london	3776873	
california	3523133	
italy	3028034	
france	2886069	
summer	2515771	Time
night	2156441	
winter	1957041	
spring	1547115	
day	1319927	
beach	3714692	View
nature	3519273	
art	3079687	
music	2734043	
water	2527848	

*as per 5th May 2008
on www.flickr.com

Figure 7.2 Top 5 most popular tags according to the identified categories (Reproduced by permission of © AmbieSense™ Ltd)

enabling image and video search in many online services. Such user volunteered tags (keywords, categories, topics) that evolve over time is also referred to as a folksonomy where a group of people participate in evolving the needed vocabulary. On Flickr, if a user searches for an image using one or more words that in some way can be matched with the tags, a ranked list of relevant images will be returned. Once a tag term is established and linked to an image, the users can do a search. Each image can have several tag terms associated with it.

Flickr publishes the most popular[9] tags on their website. Figure 7.2 shows the list of these tags and the number of images they appear in. Additionally, we identified some common categories of tags. These include *place* (e.g. Japan, London), *people* (e.g. family, friends), *event* (e.g. wedding, party), *time* (e.g. summer, night), *view* (e.g. beach, nature), *color* (e.g. bw, blue), and *camera* (e.g. Nikon, Canon). The figure shows the top five tags in these categories.

Figure 7.3 shows the total number of images in these categories of 145 most popular tags. This figure illustrates that place, view/scenery, and event are highly used categories. These categories of tags can be said to represent some context information for the images uploaded on the site.

7.4 Context Modelling and Representation

A model can be described as a systematic description of an object or phenomenon that shares important characteristics with the object or phenomenon. It is, therefore, an explicit representation of important aspects of the object or phenomenon.

A context describes aspects of a situation seen from an actor/entity's point of view. From such a viewpoint, context modelling is the process of creating and capturing contexts as a means to help improve the effectiveness of an IR system. A systematic way of dealing with context information is important when developing an application. In this section, we cover background literature in the

---
[9] As per 5 May 2008

Millions of images in which
the 145 most popular
user volunteered tags
were in use.

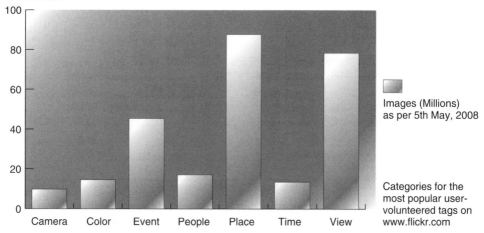

Figure 7.3   Most popular tags for context information in Flickr   (Reproduced by permission of ©
AmbieSense™ Ltd)

area of context models and present a model to help our understanding of context – especially when
designing systems.

7.4.1   Context modelling

7.4.1.1   Introduction

There has recently been increasing interest in context modelling (see Indulska and Roure 2004; Crestani
and Ruthven 2005; Ruthven *et al*. 2006; Borlund *et al*. 2008). Context modelling is motivated by a
general need for theory about context and its structures that will consequently help in building better
frameworks and more effective systems.

A number of surveys have explored existing context models such as Kaenampornpan and O'Neill
(2004), Strang and Linnhoff-Popien (2004), Baldauf *et al*. (2007). The latter two of these investigated
context models of various systems from a technical development viewpoint, i.e. architectures, data
formats, communication protocols and the use of standards. Additionally, work on various context
modelling toolkits exists (e.g. in Salber *et al*. 1999; Mikalsen and Pedersen 2005), as well as those
on context managing framework (e.g. Korpipaa *et al*. 2003).

The type of contextual information that a model contains (e.g. attributes and their relationships) is
perhaps more important than the technical structure of the model. The choice of the kind of attributes
and their values directly affect the performance of the system using it. The survey provided by
Kaenampornpan and O'Neill (2004) reviews context models based on the information they model and
relates them to each other.

7.4.1.2   Context models

Now, we will look more closely at some context models. The application they are designed to be
used for influences what is represented in the context model. In Schmidt *et al*. (1999) a context model
is defined by two categories: human factors (user information, social environment, task) and the
physical environment (location, available infrastructure, physical conditions). Lieberman and Selker

(2000) divide context into system, user and task. Whereas the system context is defined by the system implementation, the user context consists of the user state, history of past activities and preferences. The task context is defined by goals and actions. In Chalmers and Sloman (1999), a context model is defined along location, device characteristics, environment and the user activity. Gross and Specht (2001b) observed that most definitions of context had location, identity, time, and environment as dimensions. In Lucas (2001), context is categorised into physical context, device context and information context. For the purpose of map personalisation, Zipf (2002) identified relevant context attributes consisting of attributes about the user's physical condition, the weather, the user's task, user's cultural background and others. Reichenbacher (2007) defined context into the six categories of situation, user, user activity, physical environment, information and system.

7.4.1.3 Focusing on the user

The following *user context model* describes aspects of a situation seen from the user's point of view (Myrhaug and Göker 2003). This is proposed as a framework for exploiting user contexts within and across application domains. We believe this to be a comprehensive model that can be used with a wide range of applications involving user contexts. The structure is designed to enable effective matching and retrieval of contexts. The user context structure consists of five subcontexts:

- Environment context;
- Personal context;
- Task context;
- Social context;
- Spatio-temporal context.

Each subcontext should be considered as a container where an application developer can insert the important context attributes with values and types needed in the particular context-aware application, see Figure 7.4 below.

Environment context. This part of the user context captures the entities that surround the user. These entities may be (but are not limited to) things, services, temperature, light, humidity, noise and

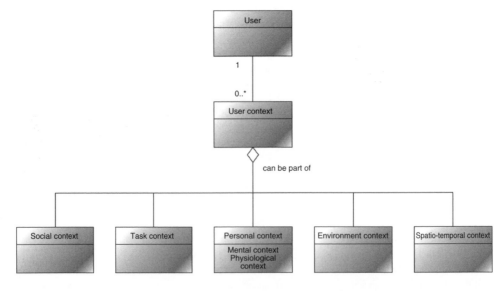

Figure 7.4 User context (Reproduced by permission of © AmbieSense™ Ltd)

persons. Information (e.g. text, images, movies, sounds) that is accessed by the user can be linked to the environment context. The various networks that are in the surrounding area can also be described in the user's environment context.

Personal context. This part of the user context consists of two sub contexts: the physiological context and the mental context. The physiological context contains information such as pulse, blood pressure, weight, glucose level, retinal pattern, and hair colour. The mental context contains information like mood, interests, expertise, anger, stress, etc.

Task context. This context describes what people are doing. The task context can be described with explicit goals, tasks, actions, activities, or events. Note that this can also include other persons' tasks that are within the situation. For example, considering a car with a driver and passengers, the situation can include the driver driving the car, passengers doing various activities such as reading, watching the car TV, and listening to music on the personal stereo. The task context of the driver and the passengers will be different.

Social context. This context describes the social aspects of the current user context. It can contain information about friends, neighbours, co-workers, relatives, and their presence. One important aspect in a social context is the role that the user plays in the context. A role can for instance be described with a name, the user's status in this role, and connections to the tasks in the task context. A role can, in addition, be played in a social arena. A social arena can have a name, such as 'at work'.

Spatio-temporal context. This context aspect describes aspects of the user context relating to the time and spatial location for the user context. It may contain attributes for time, location, direction, speed, shape (of objects/buildings/terrain), track, place, clothes of the user and so on (i.e. the physical extension of the environment and the things in it)[10].

Below, we relate others' work on context models within the model above.

- *Environment context* captures the entities around the user. It includes objects of the surrounding environment (e.g. buildings, outdoor facilities, infrastructure) and their state (i.e. temperature, light, humidity, noise). It also describes information – what Lucas (2001) calls information context – as part of the environment. Furthermore, devices that reside in the users' vicinity are also accounted for as part of the environment. Attributes of these devices define the extent with which personalisation can be performed (Chalmers and Sloman 1999). Examples include the processing power, screen size/resolution, colour support, sound capabilities and the kind and number of input devices. Similar to Göker and Myrhaug (2002), Schmidt *et al.* (1999) also categorise contextual information about device(s) as part of the physical environment.
- *Personal context* is equal to the user information that is stored in a typical user model (as also discussed later in following sections). It can be distinguished into user's physiological context (e.g. age or body weight) and user's mental context (e.g. interest). This is similarly described in Schmidt *et al.* (1999), which states that 'information on the user comprises for instance knowledge of habits, mental state or physiological characteristics'. Attributes that are represented by the mental context can generally be found in user models. Examples include user's identity, preferences, knowledge and skills as listed in Reichenbacher's context model (2007). User's interest is one of the most important attributes that is commonly modelled in most user models and has been widely applied in personalised information systems (Brusilovsky 1996).
- *Task context* contains information about what the user is doing or aiming for. It describes 'the functional relationship of the user with other people and objects' (Bradley and Dunlop 2004), including the benefits and constraints of this relationship. It can be modelled as explicit goals, actions and activities. User's activity is a common type of context in a number of context models such as the one described in Reichenbacher (2007) as well as Chalmers and Sloman (1999).

---

[10] Note that, in developing an application, we would need to decide whether an attribute such as location should be in the environment context or in the spatio-temporal context. This decision is largely influenced by what would be of more common sense given the application domain.

- *Social context* represents the social environment of a user. It may describe users' relationships to like-minded people that are connected to the user. Social filtering systems implement one special kind of social context modelling (Griffith and O'Riordan 2000). The recommendation output is solely based on a user being similar to other users who rated items previously. This social connection is then exploited to recommend more items. Social filtering systems have demonstrated good results for a range of relevant topics such as music (Shardanand and Maes 1995) and news (Resnick *et al.* 1994). Social context may model a user's list of friends or colleagues – perhaps explicitly expressed by that user or implicitly acquired through an email address book or buddy list on a website.
- *Spatio-temporal* context represents physical space and time. The *spatial* aspect represents physical space. Location is the most common aspect of a spatial context. Mobile guides such as Cyberguide (Abowd *et al.* 1997), GUIDE (Cheverst *et al.* 2000) and the CRUMPET system (Zipf 2002) belong to a special class of applications that are commonly referred to as location-based services. Based on its relevance, location modelling has emerged as one research branch in location-based services; a comprehensive overview is provided in Jiang and Yao (2006). Many location-based services use location information for the personalisation of geographic maps. Location can be represented either geographically or semantically as described in Beigl (2002). The geographic representation exhibits locations by its position, i.e. coordinates provided from the Global Positioning System (GPS). On the other hand, the semantic representation describes locations in a more descriptive and humanly understandable way, yet still able to be processed by a computer. The comMotion system described in Marmasse and Schmandt (2000), for example, learned meaningful locations semantically by analysing users' GPS logs over time. Besides location, spatial context also models attributes such as the direction of movement, the viewing direction and the speed of movement.
- The *temporal* aspect, of the spatio-temporal subcontext, refers to time and can be represented as an absolute measurement or in a more relative manner (e.g. 'in the evening' or 'before a meeting'). In Hull *et al.* (1997), time is part of the users' environment. Similarly, Reichenbacher (2007) views it together with location as part of a situation. In Schmidt *et al.* (1999), temporal context is related to all other context attributes representing the contextual change of those attributes over time. In Bradley and Dunlop (2004), temporal context is described as being 'embedded within everything, and is what gives a current situation meaning … '.

Overall, although there is some empirical work (Bierig and Göker 2006) that investigates more closely a context model and its attributes, there is generally a lack of empirical work for investigating the relation between various context attributes.

7.4.2   User models and their relationship to context

In scanning related work, we often come across the concept of a *user model*. Kobsa (1995) defines user models as 'collections of information and assumptions about individual users (as well as user groups), which are needed in the *adaptation* process'. User modelling can be used to understand humans (both as individuals and in groups). Every human is able to create implicit forms of such (user) models that are essential for daily tasks especially those involving communication (Rich 1979). A user model is a prerequisite for adaptive information systems in general and personalised information systems in particular. Rich (1979, p. 331) also identified the need for user models as a means for personalisation.

Another definition of what constitutes a user model is provided at a user modelling conference:

A user model is an explicit representation of properties of individual users or user classes. It allows the system to adapt its performance to user needs and preferences (Brusilovsky *et al.* 2003)

Work on user modelling started around the beginning of the 1980s with the work of Allen, Cohen and Perrault, as described in Kobsa (2001). Whereas early systems had no clear distinction between an application and user model, systems developed after the mid-1980s started to separate the two.

The focus was set to abstract the concept of user modelling and to introduce reusability for future applications and projects. User modelling mostly applied techniques for the automatic detection of users' prepositional characteristics expressed as *user properties* (such as interests or pre-existing knowledge) based on users' past interactions with the system. According to Kobsa (2001), the types of user properties and the way they are structured in a user model are largely based on intuition and experience. As part of a larger survey paper Kobsa *et al*. (2001) provide a list of user properties as a potential guideline for the creation of a user model. These include: demographic data, information about level of domain knowledge or experience, goals and plans with short/long-term intentions, and interests/preferences. Interest is a frequent user attribute (Brusilovsky 2001) and is a key property for many recommender systems.

From an IR perspective, both user modelling and context modelling are relevant because they can be concerned with the creation and capturing of information relevant to formulate a query, retrieve, and filter information. They have their focus on different subcontexts in the user context model described above (see Figure 7.4). The field of user modelling has mainly been concerned with the part we have named personal context, and task context. Whereas, the work in context-aware ubiquitous computing has mainly been concerned with the spatio-temporal context and the environment context. It is worth noting that there is a trend to integrate elements from each type of model. On one hand, many user modelling systems have started to model aspects from the surroundings of the user and therefore widened their initial approach. On the other hand, context models have expanded from their technical, measurement and sensor-based viewpoint (i.e. device and positioning information) to a viewpoint that also includes the user as part of the surroundings – particularly in mobile and ubiquitous information access.

7.4.3   Past, present and future contexts

In this section we present one way of thinking of context information that evolves over time. The *context space* is a viewpoint into a user's situations, where the information is linked to *past, present* and *future* contexts. There are past contexts (i.e. history), the current context, and possible future contexts (Myrhaug 2001). In other words, we can reuse past contexts if they are relevant to current user queries or to help predict future ones. The UML model of the context space is depicted in Figure 7.5.

The *context history* contains all past contexts. For example, a person takes a picture X in context A, and this association is soon after stored in the context history. Two days later when the person is in a context B, quite similar to A, the system could present picture X to the user. The *current context* describes aspects of the current situation, i.e. now. Various context sources can feed information into the current context (e.g. people, sensors, software, etc.).

Future contexts are contexts that systems or users predict. As with past and present contexts, these also can be created and explicitly stored. For example, planned activities in a personal calendar can all be thought of as examples of future contexts. A system might also predict some future contexts when a user subscribes to a service. An example is when a user is travelling from one city to another. He/she encounters various contexts such as 'in the taxi', 'in the airport', 'at the gate', 'on board' 'at arrivals', 'baggage claim' and so on. Different content can be associated with future contexts, such as an e-ticket, a city map, shopping content for the airport and so on.

Context can change rapidly. When users travel, for instance, the change of context is likely to occur more frequently than if the user stays at the same place. However, one can easily argue that the situation changes, even if a person remains in the same room, due to activities nearby – and most importantly as time passes. For example, the situations can be 'in the morning', 'at breakfast', 'at noon', 'in the afternoon', 'dinner', 'supper', and 'night'. At breakfast, the content associated with the context might be the wake-up melody, at breakfast the local newspaper, at noon the CNN or BBC news, in the afternoon the latest stock prices, at dinner time shopping ads, and so on. In other words, contexts can be planned to occur at recurring times during the day, week, month, and year. When the current context differs sufficiently from the previously stored context, it may automatically be stored as a unique context.

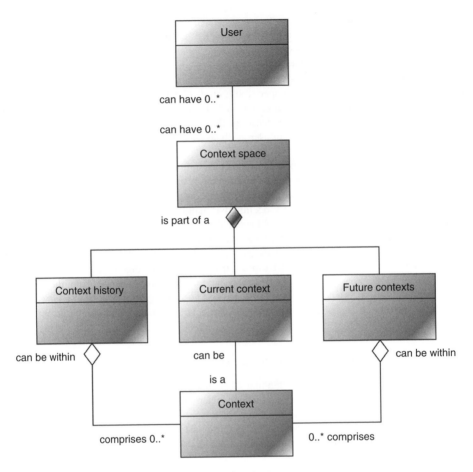

Figure 7.5 Context space with past, present, and future contexts (Reproduced by permission of ©
AmbieSense™ Ltd)

7.4.3.1 Context space

We can think of context space as a persistent storage area for the contexts encountered over time.
There can be only one current context for an actor/entity, but there can be zero, one or many contexts
as part of the context history or the future. It can be useful to be able to query for contexts that
contain certain attributes and values. It can also be beneficial to avoid matching a context against all
other contexts, because contexts can be of different types. We advise that different types of contexts
should not, in practice, normally be matched with each other. They are normally made for different
application domains, even if the names of two attributes in two different types are the same. For
instance, an attribute called 'location' can refer to a geo-code in a city guide application, but also to
a place on the human body in a healthcare application.

In discussing context space, both context management and context trace are important. *Context
management* is about the creation of new contexts, determining the current context; the storing,
searching, matching, retrieval of, and sharing of contexts with other people. Content may be linked
to contexts either explicitly or with the use of retrieval algorithms. In fact, context technology is
only useful for users once contexts are linked either directly to content, or indirectly via information
retrieval and extraction mechanisms (i.e. search functions). A context without relevant content is
useless, whilst relevant content without context would still be useful.

A *context trace* is a subset of user contexts in the context history. They emerge as associations between contexts and content. For example, a context trace regarding your visits to the Olympic stadium brings together a small photo album from the stadium, and a context trace regarding what news you have read on the train may bring together a trail of related news items. More on the origin of context traces can be found in Rahlff *et al*. (2001).

7.5   Context and Content

In this section, we provide a framework for thinking about context and content when developing context-aware information systems. If we are to develop context-aware systems, it is useful to be able to deal with the concepts of context and content separately. We discuss representations of context, how it can be captured, the context and content matching process, and the role of context templates. The focus is on aspects that are independent of any specific algorithm or method, and therefore is aimed at a more generic level.

7.5.1   *Representation of context*

Broadly speaking, we can have two main approaches to representing context information: structured and unstructured. We can think of unstructured context information as a set of varied context attributes/features that have not been categorised. For example, 'in the car, yesterday, music, track 14'. The structured approach to context information reflects a view that some consistency or stability in representation, categorisation, and expression of relationships between entities is desirable. Various forms of structured representations are tree structures, semantic networks, graphs and so on. Structures can be flexible or firm.

Here, we consider context as a hierarchical tree-based representation of aspects of real-world situations. Figure 7.6 shows context, and it has attributes and its relationship with content. The diagram focuses more on context than content. A *context* has:

- Name;
- Type;
- Attributes.

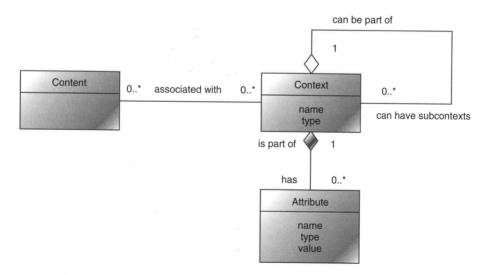

Figure 7.6   Context and content   (Reproduced by permission of © AmbieSense™ Ltd)

A context can have subcontexts (environment, task, personal, etc.). We can read the diagram as follows: a context can be part of another context, and a context can consist of several subcontexts. In general, a context should always contain one or more attributes. Each attribute of a context has a name, type, and value. The type of an attribute is a data primitive (e.g. Boolean, integer, float, double, string ...). However, when a context consists of several subcontexts, it is also possible to make meaningful contexts without including any attributes at the top-level context.

The value of an attribute can be constrained in some applications. This can be achieved by using finite value sets or specific value ranges. For example, an attribute with name = 'Country', type = string, can be constrained to the values = {'Australia', 'Brazil', 'China', 'Germany', 'Italy', 'Spain', 'Turkey', 'United Kingdom', 'USA'}.

Some examples of value ranges are:

Value range A = [10, 1000], type = integer
Value range B = [0.1, 0.12 >, type = real
Value range C =< −1.0, 1.0 >, type = complex
Value range E = [false, true], type = Boolean
Value range F = [June 1 2010, Aug 31 2010], type = date

Some examples of value sets are:

Value set 1 = {'Red', 'Blue', 'Green'}, type = string
Value set 2 = {'Little', 'Average', 'Much'}, type = string
Value set 3 = {'12', '14', 'N-12', 'N-14'}, type = string
Value set 4 = {1, 10, 100, 1000}, type = integer
Value set 5 = {false, true}, type = Boolean
Value set 6 = {May 1 2011, Jan 1 2011}, type = date

7.5.2   Capturing context

Context information can be captured either automatically by the system or manually by user input. Some context information is quite static while others are more dynamic. This means that some attributes may never change; others change once a year, or even every millisecond. Since an actor/entity encounters new situations, the context can be updated, or replaced by a completely new context. This section is about capturing context and sources of context information.

7.5.2.1   Capturing context by system and user input

Context can be captured automatically by systems or through user input. People generally prefer to have automated methods to fill in as much of the context information as possible. For example we can define a spatio-temporal context, made to enrich or fill in the user context as automatically as possible when the user encounters a new situation in the vicinity of a place.

Imagine a professional photographer taking pictures where the camera automatically notes the date, time, the user-id of the photographer, the geo-position when the picture was taken and even the direction of the lens.

On the other hand, the manual approach of inputting context information is also sometimes helpful in providing clues for future retrieval. For instance, the professional photographer may label digital pictures taken at sports events by specifying context information that would be useful in retrieving individual pictures for his needs. These can be information such as the name of the sport, the match, any highlight, who was in the picture, something about the originality of the picture, and perhaps an indicative price. Hence the individual information need can be a strong motivator for users to manually input context information.

7.5.2.2   Sources of context information

Context information can be found in a variety of sources. These can be as follows:

- Features in content items or the items themselves, these can include metadata about the content item;
- Index terms;
- Structure of a subject area as represented by a classification scheme, thesaurus or ontology;
- User searches, user logs (containing queries, click behaviour, timings and so on);
- Comments on content items (from individual users, groups, experts, etc.);
- Sensor information about the environment (location, temperature, time, etc.);
- Actuator information such as a remote control mechanism or a switch that records its state when someone has pressed it.

The above is a list of some of the possibilities at present. It is conceivable for other sources of context information to be available in the future.

7.5.3   Searching with context information

In discussing searching with context information, this section looks at how context information can affect queries, some aspects of associating context with content, and also the potential use of associating contexts with each other in order to retrieve useful content.

7.5.3.1   Context and queries

Context information can affect a query in the following ways:

- Context as the query;
- Query extraction from context;
- Query amplification with context;
- Query replacement due to context.

The first approach uses the context as the query itself. The second one involves extracting a query from the context information available. The third approach adds context information to enrich a query. The fourth is when the occurrence of a particular context results in a query being completely replaced with another. For example, if the query was 'sandwich' and the context 'lunch', a replacement could be 'lunch pack deals'. The third and fourth items above relate to query expansion – often referred to as a reformulation of a query in IR.

7.5.3.2   Associating context with content

Associating *context* with *content* can be done by humans or automatically through computerised systems. Broadly speaking, context and content can be associated with each other in three ways:

- *Content can contain context information*, such as with metadata associated with content items;
- *Context information can contain explicit links or pointers to content items*, such as deliberately linking specific photographs to a travel situation;
- *Context information is separate from the content items*, i.e. a less embedded approach.

In associating a context with content items, at some point we need to go through a matching process. This involves identifying the appropriate context, the collection of content items, and having a method to match between these. There is plenty of scope for variation in the matching of content to context methods just as there is scope for variation in assigning importance/weights to context attributes.

When associating content with context, we can identify possibilities at different levels of granularity. A specific value of a context attribute can be used as the starting point for the association process, alternatively the name of an attribute can be used, or thirdly a higher level such as the name of a context can be used.

7.5.3.3 Associating contexts with each other

Associating contexts with each other is an indirect means to associating context with content. This can involve looking at the closeness of contexts from different users, such as in a recommender system, or looking at those of the same user for his/her past or predicted future contexts. This can, for example, be useful in long-term interests or shared interests.

With this approach, content can be found by first looking for similar contexts. For example, a user specifies a context or a query like 'find content of similar situations like this'. This is matched with other contexts for the system to retrieve a set of similar contexts, and the results are then presented. Similarity thresholds can be used in helping decide on the set of similar contexts.

However, not every context should be compared with literally any context, because there would be a risk of: (1) slowing down the context retrieval method, or (2) comparing contexts that are inappropriate to compare with each other. The result may be a slow response or a search that returns irrelevant content.

7.5.4 Context templates

We believe the use of a ready-made mould or template for capturing context information can be useful for acquiring contexts, and also for the context management process. Such a *context template* helps standardise the context structure and makes it a firm pattern that can enable efficient matching of similarities and differences between contexts. A context template can be used for describing typical situations or covering more general aspects of a situation that an actor/entity can encounter.

The sources of information for creating context templates are the same as those for contexts, as covered earlier. However, it can be important to carefully analyse and understand the various types of situations in which specific context attributes reoccur, can be captured, and when content would be useful to appear.

Contexts can be created or instantiated from context templates. Templates can be used in order to standardise the creation and exchange of contexts both within and between applications. By developing a context template for an application we can achieve the following:

- A shared understanding and vocabulary of the context as an entity;
- Ensure that contexts are comparable information structures.

In order to illustrate the use of attributes, value sets, and value ranges for context templates, an example is provided below.

Example application. A photo album application helps the user to manage their personal digital image collection. The user takes the pictures with a digital camera and annotates a context for each of the pictures in order to make the photo album searchable. A single 'user context template' was made, to which each 'user context' created by the user must adhere.

The attributes of the 'user context template' are:

Attribute 1: name = 'Title', type = string
Attribute 2: name = 'Persons', type = string
Attribute 3: name = 'Location', type = string
Attribute 4: name = 'When', type = date, value range = [1 Jan 1900, 31 Dec 2099]
Attribute 5: name = 'Category', type = string, value set = {'portrait', 'landscape', 'people'}
Attribute 6: name = 'My rating' , type = string, value set = {'favourite shot', 'normal shot'}

For simplicity of the example, we have not introduced subcontexts here.
An instantiated photo context from the template above can be:

Attribute 1: name = 'Title', type = string, value = 'Grandma 2008'
Attribute 2: name = 'Persons', type = string, value = 'Grandma and me'

Attribute 3: name = 'Location', type = string, 'In the garden at home'
Attribute 4: name = 'When', type = date, value = '21 Aug 2008'
Attribute 5: name = 'Category', type = string, value = 'people'
Attribute 6: name = 'My rating', type = string, value = 'favourite shot'

The user can fill in the values, given the template, for the six fields for each picture. The user can search and sort the photos according to attributes and values of created contexts.

In short, context templates shape contexts with regard to both choice of subcontexts, attributes, and attribute values.

7.6   Related Topics

In discussing context in information retrieval, the topics of personalisation and mobility often emerge. In this section, we briefly cover how they are related to context.

7.6.1   *Personalisation and context*

Personalisation is about tailoring products and services to better fit the user. There are several ways of achieving this. We advocate user context as a means of capturing all these. Personalisation can be achieved by tailoring products and services either to large user groups, smaller interest groups, or the individual user. An example of personalisation can be in car purchasing: once you have selected the car model, you can tailor it with extra equipment, colour, dashboard interior, seat textile, the engine, and special wheels. Despite the popularity of search engines, a survey of 60 search engines (Khopkar *et al.* 2003) found that few of these provided personalisation features and that generally there was a minimalist approach. It would appear that the personalisation features are difficult to access and use.

A number of systems have been developed that provide personalised search services in desktop/laptop environments. Pretschner and Gauch's (1999) OBIWAN system is an interface for personalised web search that uses ontology-based user profiles as concept hierarchies and re-ranks and filters results from common Internet search engines. PowerScout (Lieberman *et al.* 2001) more proactively offers its users website recommendations based on users' past browsing history from submission of queries to public search engines. The Outride system (Pitkow *et al.* 2002) operated as a browser sidebar, and delivered personalised search output based on query modification and the filtering and re-ranking of search results. They showed that personalised information retrieval can help users to reduce their search effort on the web by more than 50%. Billsus and Pazzani's (2000) PDA edition of the Daily Learner allowed users to access news from nine different categories based on recommendation. Re-ranked results are provided based on users' past interaction with the system.

At query time, some personalised information retrieval systems use query modification. For example, the WebMate browsing and searching agent (Chen and Sycara 1998) assisted users' search by expanding search queries with learned keywords based on a model of correlated word pairs. The Outride system, mentioned earlier, did that in a similar fashion but additionally used implicit feedback; as users also browsed for content using an ontology, Outride used its category information to augment subsequent queries. Göker and He (2003) identified topical session boundaries using temporal and query term information from online queries to help personalise a ranking order.

Re-ranking is the reordering of an existing list of search results depending on the information provided in a user and/or context model. The OBIWAN system, mentioned earlier, used a publicly available search engine and modified its generic search result rank into a personalised search rank based on the user's interest in a number of topical categories. The weights of these categories were constantly adjusted along page content and the user's viewing time. Re-ranking in the Outride system, on the other hand, was based on the correlation between the content (titles and metadata) and the user profiles using the vector space model. In PResTo! (Keenoy and Levene 2005), re-ranking was also performed along a vector space model taking into account the URL structure as well as temporal information. Temporal information included information about the hit frequency (number of times this URL has been found before), lookup frequency (number of times this URL has been accessed) and the age of the URL in the user model.

Filtering, in the sense of removing non-relevant content from a result list, has also been used. Pitkow *et al.* (2002) suggested removing search results that the user has seen before, to support focusing on new results. This approach was challenged in Bradley and Smyth (2003) in support for re-ranking as being generally advantageous over filtering in situations that require a system to deliver a good coverage of available information. Filtering actively suppresses content that might have been useful for the user – though this may be justified in situations where there is a constant information stream or push that does not allow users to fully examine their content.

7.6.2   Mobility and context

The increase in mobile computing is perhaps one of the most significant and rapid changes in recent years regarding how people use information systems. Users are mobile and want to use hand-held devices for daily tasks while being away from their normal office and home environments or on the move between these environments (see also Chapter 6). This new way of using information systems imposes new challenges on existing information retrieval systems that so far have predominantly been introduced for desktop/laptop environments. The mobile information environment presents situations where the context is a strong trigger for information needs and where the context shift is more likely and rapid.

In the past decade, context-aware systems were often developed in the form of mobile guides. Following this, several frameworks for context management were established that supported the development of systems and the modelling of context from a certain technical perspective. Despite the technical development focus, there are few studies that investigate these context models and their attributes in more detail. Bierig and Göker (2006) investigated the effects of three context attributes – time, location, and interest – in a user study showing profound contextual effects on users' perception of usefulness both for individual attributes as well as in interaction[11]. There is a need for further work that critically evaluates the impact of context. Individual context attributes need to be understood in much greater detail relative to how they affect people's search behaviour.

7.7   Evaluating Context-aware IR Systems

Evaluating any interactive information retrieval system is a challenge. Evaluating a context-aware information system is equally so, if not more. What works under laboratory test conditions may not work in a real interactive setting. The origins and challenges of evaluation are discussed in Chapter 2. Here, we outline a framework for the evaluation of context-aware IR systems.

The overall purpose and intent in evaluation is key. Saracevic (1995) stated that many approaches can be used in evaluation – but that the choice depends on the intent. He states that both system and user-centred approaches are needed and should 'work together and feed on each other'. To evaluate these systems in their natural setting is equally important as is also stated by Borlund and Ingwersen (1997) in emphasising the need for realism in experiments. Examples of evaluation of context-aware IR systems in their natural settings can be found in Göker (1994; 1997), Göker and Myrhaug (2008) and also in the system described in Raper *et al.* (2007). The first two of these describe an example for a bibliographic IR system (Okapi), and the last two were for mobile information systems.

When evaluating context-aware information systems it is useful to follow two themes of discussion. One on *formative* vs *summative* evaluation and the other on *quantitative* vs *qualitative* methods. Overall, formative studies are intended to inform us on how to improve the system, and summative studies (typically quantitative) are intended to provide information on whether the 'product' is doing

---

[11] The work focused on context-awareness for event information.

what it is meant to do. Formative evaluation is often conducted more than once (Scriven 1991; for earlier work see Scriven 1967). Summative evaluation is a process of identifying larger patterns and trends in performance and judging these summary statements against criteria to obtain performance ratings, for example. When conducting studies in situ, it can be useful to consider the possibility of these distinctions and where the transition from one to the other may occur. See also Chapter 2 for some relevant discussion relating to the above and systems/user evaluation approaches.

The second theme is about quantitative and qualitative methods of evaluation. This encompasses discussions on reliability, validity, rigour and quality in qualitative research and how this can be achieved for trustworthiness (see, for example, Lincoln and Guba 1985; Golafshani 2003).

We believe an iterative approach to developing both the system and the information to improve the quality of the system, the information, and the interaction is important. It is also important to conduct iterative evaluations of these. We outline a method detailed in Göker and Myrhaug (2008) in a Special Issue on Evaluation of Interactive Information Retrieval Systems (Borlund and Ruthven 2008).

7.7.1   Principles of methodology

There are two main principles in the methodology described in detail in Göker and Myrhaug (2008). The principles are:

- Conduct user studies in context;
- Have iterative development of the system, the information, and the evaluation.

The first of these, *user studies in context*, means to ensure there is a contextual match between the study and the application of the system in the situation or intended use. The guidelines for this principle are as follows:

- Involve the right participants;
- Choose the right situations;
- Set relevant user tasks;
- Document results whilst in the situation;
- Document the context;
- Use relevant evaluation approach and measures.

For further clarity of the latter items above, *documenting the context* is about documenting/recording aspects of a situation during user studies so that comparisons between findings are made easier and are also more illuminative. Using a relevant evaluation approach and measure is to recognise that at each stage of an iterative development and evaluation there may be slightly different subgoals as part of an overall objective. For example, a three-way judgement on *relevance* (relevant, partially relevant, non-relevant) can be more cost-effective in earlier studies. A more granular graded scale[12] can be used for a more close analysis of relevant items in comparing different retrieval algorithms in later studies. Like Saracevic (1995) and Scholtz (2006), we see different evaluation approaches and measures being appropriate for different stages of an iterative process.

The second principle is *iterative development* of the system, the information, and the evaluation. The guidelines for this principle are as follows:

- Iterative development;
- Iterative studies;
- Scaling up the number of participants;
- Shortening the time spent per participant;
- Specifying a user test plan.

---

[12] A graded scale for relevance judgments was introduced by Kekalainen and Jarvelin (2002).

To expand on these briefly, *iterative development* refers to the improvement of both the information and the system progressively. It includes considering information/content access and navigation issues in addition to considering effectiveness of information retrieval algorithms. *Iterative studies* (or experiments) refer to improving the studies iteratively towards an overall goal. This helps manage the complexities of evaluating context-aware systems in their natural setting and is also useful in risk management. *Scaling up the number of participants* between the experiments to help progress from formative to summative evaluation, and gather more evidence for conclusions. At the start even a small number of participants in the first study can be quite useful in at least giving indications as to what is not working well. It is important to be able to *shorten the time spent per participant*, through practice and experience, for scaling up. As we scale up, since each experiment reveals new problems and determines improvements on old ones, we need to specify what to test for and how to test it, i.e. to *specify user test/study plans* – where the test plan is the driving document for conducting studies.

7.8 Summary

In this chapter, we have described the background to context and ways in which we may represent and model it in developing context-aware IR systems. The background work covered literature on context in IR, but also considered the wider literature from related fields. We then presented various context models. We took one widely applicable framework and showed how context information can be represented and used. We also briefly discussed two related topics of personalisation and mobility. Lastly, we presented evaluation approaches for context-aware IR systems.

For further reading on context we recommend following the references in the early section on Context in IR. The names of the subsections indicate wider coverage areas.

Additionally, to follow recent trends and work in context, the following conferences and forums include a variety of aspects of context:

- IIiX (International Symposium on Information Interaction in Context);
- CONTEXT (International and Interdisciplinary Conference on Modeling and Using Context);
- UBICOMP (International Conference on Ubiquitous Computing);
- UM (International Conference on User Modeling).

The more general IR conferences such as the ACM SIGIR (Special Interest Group on Information Retrieval Conference) and the BCS-IRSG ECIR (European Conference on Information Retrieval) sometimes have papers on context in IR. The journals listed below either generally contain papers around the theme of context or from time to time have special issues addressing the topic:

- *Journal of Information Processing and Management* (Elsevier)
- *Journal of the American Society of Information Science and Technology* (Wiley)
- *Journal of Location Based Services* (Taylor & Francis)
- *Journal of Personal and Ubiquitous Computing* (Springer)
- *Journal of User Modeling and User-Adapted Interaction* (Kluwer)

Acknowledgements. We would like to acknowledge the wider efforts of several people who worked on context with us in related projects. These include: Marius Mikalsen, Stein L. Tomassen, Odd-Wiking Rahlff, Murat Yakici, Stuart Watt, Hannah Cumming, Tor Erlend Faegri, Erik G Nilsson, and Juergen Pollich.

Exercises

7.1 *An investigation into context dimensions*. Choose a popular web search engine that also displays information about user query statistics. Consider the top 5 most frequently used queries by people. Pick one and try to imagine different scenarios for its interpretation. How many such different situations can you think of and why? Try to see if these can map on to the user context, described diagrammatically, in the chapter.

7.2 *Context information for an image-sharing website*. Choose an existing picture sharing website. Imagine the owner of this website consulted you to improve the search service of the site. The request is based on repeated negative user feedback about the site's search facility – describing it as random and/or poor. The business owner fears that this might put off users and make them migrate to competitors. For that reason you have decided to enhance the search facility with context information.

(a) *Existing context information for publishing and search*
Analyse the page where you can upload/publish images to see what context information users can specify when publishing their images.

    i. Try to identify, on this page/site, what kind of contextual features are possible to attach to a picture when it is being uploaded and published. How can you include these features e.g. menus, text input, etc.?

    ii. Publish five images of your choice in the web service, but remember to save/upload all of the context information you specify for each image.

    iii. Search for the five images by formulating queries from the context information you saved. Describe your observation for each query.

    iv. Describe if and how the existing context information is being used in the search index, and conclude which of the context information attributes are in use, not in use, and if an image provider can contribute to the term vocabulary with new categories (tags) or not.

    v. Summarise if it is possible to provide free text only, predefined categories (tags) only, or a combination of both free text and predefined categories (tags) as context information when publishing images. Consider also if features such as colour, texture, shape are automatically included by the system. Conclude which context information is certainly in use, and how you think this context information is included in the search index.

(b) *Improved context information for publishing and search*
You need to create a context information model suitable for both image publishing and search, and describe it according to Figure 7.4 (user context) and Figure 7.6 (context and content) earlier in the chapter:

    i. Make a table called Current Context Attributes with the relevant names, value, and type. Consider these also for the five subcontexts described in User Context section.

    ii. Were any of the attributes difficult to classify? Ask yourself if the attribute is information about the picture (context), or about something inside the picture (content).

Whether the information is more context or more content, try anyway to propose a new context attribute for the information. Revise if any of the identified attributes capture the same context information, and merge any similar attributes into one attribute.

iii. Make a similar Competitor's Context Attributes table for a competitor's web site on the web.

iv. Propose an Improved Context Attributes table.

v. Draw a UML class diagram in (Unified Modelling Language) that represents the User Context with attributes as part of the different subcontexts. Look at the Improved Context Attributes table for this. Create an XML-schema (eXtensible Mark-up Language) representing the UML class diagram.

vi. Draw or sketch a new web page for publishing images, discuss how context information represented by your context XML-schema, can be used to update and maintain the search index.

vii. Discuss the approaches: (a) Context information containing only one attribute with any text input, versus (b) Context information with 10 attributes with predefined values. Consider, in your discussion, the original feedback given by users that prompted the context-aware IR approach.

References

Abowd, G. D., Atkeson C. G., Hong, G., Long S., Cooper R. and Pinkerton M. (1997). Cyberguide: a Mobile Context-Aware Tour Guide. *Wireless Networks*. 3 : 5. Special issue: Mobile computing and networking: selected papers from MobiCom '96., 421–433, 1997.

Akman V., Bouquet P., Thomason R. H. and Young R. A. (Eds.) (2001): Modeling and Using Context, 3rd International and Interdisciplinary Conference, CONTEXT, 2001, Dundee, UK, July 27–30, 2001, Proceedings. *Lecture Notes in Computer Science*, **2116** Springer 2001, ISBN 3-540-42379-6.

Azzopardi, L. (2002). *Finding Order in the Chaos of Context: Towards a Principled approach to Contextual Information Retrieval, Six Month Report*, University of Paisley, 2002. http://cis.paisley.ac.uk/azzo-ci0/download/azzopardi2002report.pdf.

Baldauf, M., Dustdar, S. and Rosenberg, F. (2007). A Survey on Context-Aware Systems. *International Journal of Ad Hoc and Ubiquitous Computing* **2**(4): 263–277.

Beigl, M. (2002). Special Issue on Location Modeling in Ubiquitous Computing. *Personal and Ubiquitous Computing* **6**(5–6): 311–357.

Belkin, N. J. (1980). Anomalous States of Knowledge as a Basis for Information Retrieval. *Canadian Journal of Information Science* 5: 133–143, 1980.

Belkin, N., Cool, C., Stein, A. and Thiel, U. (1995). Cases, Scripts, and Information-Seeking Strategies: On the Design of Interactive Information Retrieval Systems. *Expert Systems with Applications* **9**(3): 379–395.

Belkin, N., Oddy, R. N. and Brooks, H. M. (1982). ASK for Information Retrieval: Part 1. Background and Theory. *Journal of Documentation* **38**(2): 61–71.

Bierig, R. and Göker, A. 2006. Time, Location and Interest: An Empirical and User-Centred Study. In *Proceedings of the 1st International Conference on Information Interaction in Context*, Copenhagen, Denmark, October 18–20. IIiX, vol. 176. ACM, New York, NY, 79–87.

Billsus, D. and Pazzani, M. (2000). User Modeling for Adapative News Access. *User Modeling and User-Adapted Interaction* **10**(2–3): 147–180.

Borlund, P. and Ingwersen, P. (1997). The Development of a Method for the Evaluation of Interactive Information Retrieval Systems. *Journal of Documentation* **53**(3): 225–250.

Borlund, P. and Ruthven, I. (Eds). (2008). Special Topic Issue on: Evaluation of Interactive Information Retrieval Systems. In: *Information Processing and Management* **44**(1).

Borlund, P., Schneider, J. W., Lalmas, M., Tombros, A., Feather, J., Kelly, D., de Vries, A. and Azzopardi, L. (2008). Information Interaction in Context. *Proceedings of the 1st IIiX Symposium on Information Interaction in Context*, 14–17 October 2008, London, UK ACM, New York, NY, USA.

Bouquet, P., Serafini L., Brézillon P., Benerecetti M. and Castellani F. (Eds.) (1999) *Modeling and Using Context, 2nd International and Interdisciplinary Conference, CONTEXT'99*, Trento, Italy, September 1999, Proceedings. *Lecture Notes in Computer Science*, Springer 1999, ISBN 3-540-66432-7.

Bradley, N. A. (2005). *A User-centred Design Framework for Context-Aware Computing*. PhD Thesis. Department of Computer and Information Sciences. Strathclyde University.

Bradley, N. A. and Dunlop, M. D. (2004). Towards a User-Centric and Multidisciplinary Framework for Designing Context-Aware Applications. *6th International Conference on Ubiquitous Computing (UbiComp), 1st International Workshop on Advanced Context Modelling, Reasoning And Management*, Nottingham, UK, 2004.

Bradley, K. and Smyth, B. (2003) Personalized Information Ordering: A Case Study in Online Recruitment. *Knowledge-Based Systems*, **16**(5–6) 269–275.

Brézillon, P. and Turner R. (2002). Modeling and Using Context in AI. *ECAI'01 Tutorial presented at the European Conference on Artificial Intelligence*, Lyon, France, July 22, 2002.

Brown, P., Bovey, J. and Chen, X. (1997). Context-Aware Applications: From the Laboratory to the Marketplace. *IEEE Personal Communications* **4**(5): 58–64.

Brown, P. J. and Jones, G. J. F. (2001). Context-Aware Retrieval: Exploring a New Environment for Information Retrieval and Information Filtering. *Personal and Ubiquitous Computing* **5**(4): 253–263, 2001.

Brusilovsky, P. (1996). Methods and Techniques of Adaptive Hypermedia. *User Modeling and User-Adapted Interaction*, **6**(2–3) 87–129.

Brusilovsky, P. (2001). Adaptive Hypermedia. *User Modeling and User-Adapted Interaction (Ten Year Anniversary Issue)*, **11**(1–2) 87–110.

Brusilovsky, P., Corbett, A. and Ross, F. D. (2003). *Preface of Proceedings. 9th International Conference on User Modeling (UM)*, Johnstown, PA, USA, Springer Verlag, 2003.

Byun, H. E. and Cheverst, K. (2001). Exploiting User Models and Context-Awareness to Support Personal Daily Activities. *8th International Conference on User Modeling (UM), Workshop on User Modeling for Context-Aware Applications*, Sonthofen, Germany, 2001. Citation

Chalmers, D. and Sloman, M. (1999). QoS and Context Awareness for Mobile Computing. *1st International Symposium on Handheld and Ubiquitous Computing*, Karlsruhe, Germany, Springer Verlag, 380–382, 1999.

Chen, L. and Sycara, K. (1998). WebMate: A Personal Agent for Browsing and Searching. In *Proceedings of the 2nd International Conference on Autonomous Agents*, Minneapolis, Minnesota, United States, May 10–13, 1998. K. P. Sycara and M. Wooldridge (eds). AGENTS '98. ACM, New York, NY, 132–139.

Chen, G. and Kotz, D. (2000). *A Survey of Context-Aware Mobile Computing Research*. Hanover, NH, USA, Dartmouth College, 2000.

Cheverst, K., Davies, N., Mitchell, K., Friday, A., and Efstratiou, C. (2000). Developing a Context-Aware Electronic Tour Guide: Some Issues and Experiences. *Conference on Human Factors in Computing Systems (CHI)*, The Hague, The Netherlands, ACM Press, 17–24, 2000.

CHI (2000). Workshop on Situated Interaction in Ubiquitous Computing. *Conference on Human Factors in Computing Systems (CHI)*, The Hague, The Netherlands, ACM Press.

Cool, C. and Spink, A. (2002). Special Issue on Context in Information Retrieval. *Information Processing and Management* **38**(5) 605–611.

Crestani, F. and Ruthven, I. (Eds.) (2005). Context: Nature, Impact, and Role. *5th International Conference on Conceptions of Library and Information Sciences (CoLIS 2005)*, Glasgow, UK, Springer Verlag, Proceedings. Lecture Notes in Computer Science Series, vol. 3507, June 2005.

Dey, A. K. (2001). Understanding and Using Context. *Personal and Ubiquitous Computing* **5**(1) 4–7.

Dey, A., Kortuem, G., Morse, D. and Schmidt, A. (2001). Special Issue on Situated Interaction and Context-Aware Computing, *Journal of Personal and Ubiquitous Computing*, **5**, 1.

Dourish, P. (2004). What We Talk About When We Talk About Context. *Personal and Ubiquitous Computing* **8**(1) 19–30.

Dunlop, M., Morrison, A., McCallum S., Ptaskinski P., Risbey, C. and Stewart F. (2003). Focused Palmtop Information Access Combining Starfield Displays with Profile-Based Recommendations. Mobile and Ubiquitous Information Access, *Lecture Notes in Computer Science* **2954**, 79–89, Springer-Verlag.

Ferscha, A. and Mattern, F. (Eds.) (2004). *2nd International Conference on Pervasive Computing*, Linz-Vienna, Austria, Springer Verlag. Proceedings, Lecture Notes in Computer Science Series, 2004.

Göker A. (1994). *An Investigation into the Application of Machine Learning in Information Retrieval*. PhD Thesis. Department of Information Science. City University, London, UK.

Göker, A. (1997). Context Learning in Okapi. *Journal of Documentation* **53**, 80–83.

Göker, A. and He, D. (2003). Personalization via Collaboration in Web Retrieval Systems: A Context Based Approach. *Proceedings of the American Society for Information Science and Technology*, **40**(1), 357–365.

Göker, A. and Myrhaug, H. (2008) Evaluation of a Mobile Information System in Context. *Information Processing and Management*, Special Issue on 'Evaluation in Interactive Information Retrieval', Elsevier, **44**(1), 39–65.

Göker A., Watt. S., Myrhaug H. I., Whitehead N., Yakici M., Bierig R., Nuti S. K. and Cumming H. (2004). An Ambient, Personalised, and Context-Sensitive Information System for Mobile Users. *ACM International Conference Proceeding Series; Proceedings of the 2nd European Union symposium on Ambient intelligence* Eindhoven, **84**, 19–24. 2004. ISBN:1-58113-992-6.

Golafshani, N. (2003). Understanding Reliability and Validity in Qualitative Research. *The Qualitative Report* **8**(4) 597–607.

Griffith, J. and O'riordan, C. (2000). *Collaborative Filtering*. Galway, Ireland, Department of Information Technology, National University of Ireland.

Gross, T. and Specht, M. (2001a). Workshop on User Modelling for Context-aware Applications. *10th International Conference on User Modelling (UM 2001)*, Sonthofen, Germany. http://mc.informatik.uni-hamburg.de/konferenzbaende/mc2001/V33.pdf.

Gross, T. and Specht, M. (2001b). Awareness in Context-Aware Information Systems. *Mensch & Computer* 2001. H. Oberquelle, R. Oppermann, J. Krause (eds) Fachübergreifende Konferenz. Stuttgart, B.G. Teubner, 2001, 173–181.

Harter, A., Hopper, A., Steggles P., Ward A. and Webster P. *et al*. The Anatomy of a Context-Aware Application. *Wireless Networks* **8** 187–197.

Hull, R., Neaves, P. and Bedford-Roberts, J. (1997). Towards Situated Computing. *1st International Symposium on Wearable Computers*, Cambridge, MA, USA, IEEE Computer Society Press, 1997, p. 146.

Indulska, J. and Roure, D. D. (2004). Workshop on Advanced Context Modelling, Reasoning And Management. *6th International Conference on Ubiquitous Computing (UbiComp)*, Nottingham, UK. 2004.

Ingwersen, P. (1992). *Information Retrieval Interaction*. London, Taylor Graham, 1992.

Ingwersen, P. (1996). Cognitive Perspectives of Information Retrieval Interaction: Elements of a Cognitive IR Theory. *Journal of Documentation* **52**(1) 3–50.

Ingwersen, P. and Jarvelin, K. (2005). *The Turn: Integration of Information Seeking and Retrieval in Context*. The Netherlands, Springer. ISBN:140203850X, 2005.

Ingwersen, P., Jarvelin, K. and Belkin, N. (2005). Workshop on Information Retrieval in Context. *28th Annual International ACM SIGIR Conference on Research and Development in Information Retrieval*, Salvador, Brazil, Royal School of Library and Information Science, Copenhagen, Denmark, 2005.

Ingwersen, P., van Rijsbergen, C. J. and Belkin, N. (2004). Workshop on Information Retrieval in Context (IRiX). *27th Annual International ACM SIGIR Conference on Research and Development in Information Retrieval*, Sheffield, UK, 2004, http://ir.dcs.gla.ac.uk/context/.

Jameson, A. and Krueger, A. (2005). Special Issue on User Modeling in Ubiquitous Computing. *User Modeling and User-Adapted Interaction*: 15(3–4): 193–338.

Jiang, B. and Yao, X. (2006). Location-Based Services and GIS in Perspective. *Computers, Environment and Urban Systems*. **30**(6) 712–725.

Kaenampornpan, M. and O'Neill, E. (2004). An Integrated Context Model: Bringing Activity to Context. *6th International Conference on Ubiquitous Computing (UbiComp), 1st International Workshop on Advanced Context Modelling, Reasoning And Management*, Nottingham, UK, 2004.

Keenoy, K. and Levene, M. (2005). Personalisation of Web Search. *Intelligent Techniques for Web Personalisation*, Springer Verlag, 201–228.

Kekalainen, J. and Jarvelin, K. (2002). Using Graded Relevance Assessments in IR. *Journal of the ASIST* **53**(13) 1120–1129.

Khopkar, Y., Spink, A., Giles, C. L., Shah, P. and Debnath, S. (2003). Search Engine Personalisation: An Exploratory Study. First Monday, 8(7), July 2003, http://firstmonday.org/htbin/cgiwrap/bin/ojs/index.php/fm/article/view/1063/983, accessed 21 June 2009.

Kobsa, A. (1995). Editorial. *User Modeling and User-Adapted Interaction*, **4**(2) iii–v.

Kobsa, A. (2001). Generic User Modeling Systems. *User Modeling and User-Adapted Interaction* **11**(1–2) 49–63.

Kobsa, A., Koenemann, J. and Pohl, W. (2001). Personalized Hypermedia Presentation Techniques for Improving Online Customer Relationships. *The Knowledge Engineering Review* **16**(2) 111–155.

Korkea-aho, M. (2000). Context-Aware Applications Survey. Technical report, Internet Working Seminar (Tik-110.551), Helsinki University of Technology, 2000.

Korpipaa, P., Mantyjarvi, J., Kela, J., Keranen, H. and Malm, E. J. (2003). Managing Context Information in Mobile Devices. *Pervasive Computing*, **2**(3) 42–51.

Lieberman, H., Fry, C. and Weitzman, L. (2001). Exploring the Web with Reconnaissance Agents. *Communications of the ACM* **44**(8) 69–75.

Lieberman, H. and Selker, T. (2000). Out of Context: Computer Systems That Adapt To, and Learn From, Context. *IBM Systems Journal* **39**(3/4) 617–632.

Lincoln, Y. S. and Guba, E. G. (1985). *Naturalistic Inquiry*. Newbury Park, California, US, Sage Publications, 1985.

Lucas, P. (2001). Mobile Devices and Mobile Data – Issues of Identity and Reference. *Human–Computer Interaction* **16**(2, 3\4): 323–336.

Marmasse, N. and Schmandt, C. (2000). Location-aware information delivery with comMotion. *2nd International Symposium on Handheld and Ubiquitous Computing (HUC)*, Bristol, UK, Springer Verlag, 157–171, 2000.

Mikalsen, M. and Pedersen, A. (2005). Representing and Reasoning about Context in a Mobile Environment, *Revue d'Intelligence Artificielle*, **19**(3) 479–498, 2005, Lavoisier, France.

Morse, D. R., Armstrong, S. and Dey, A. K. (2000). The What, Who, Where, When, Why and How of Context-Awareness. In *CHI '00 Extended Abstracts on Human Factors in Computing Systems*, The Hague, The Netherlands, April 1–6, 2000). CHI '00. ACM, New York, NY, p. 371.

Myrhaug, H. I. (2001). Towards Life-Long and Personal Context Spaces. *Workshop on User Modelling for Context-Aware Applications*, Sonthofen, Germany, 2001.

Myrhaug, H. I. and Göker, A. (2003). AmbieSense – Interactive Information Channels in the Surroundings of the Mobile User. *2nd International Conference on Universal Access in Human – Computer Interaction*, Crete, Greece, Lawrence Erlbaum Associates, 1158–1162, July 2003.

Netcraft (2008). Netcraft report. http://www.netcraft.com, accessed 15 September 2008.

Pitkow, J., Schuetze, H., Cass, T., Cooley, R., Turnbull, D., Edmonds, A., Adar, E. and Breuel, T. (2002). Personalized Search. *Communications of the ACM*, **45**(9) 50–55.

Pretschner, A. and Gauch, S. (1999). Ontology Based Personalized Search. In *International Conference on Tools with Artificial Intelligence (ICTAI)*, Chicago, USA, 391–398,.

Rahlff, O., Kenneth Rolfsen, R. and Herstad, J. (2001). Using Personal Traces in Context Space: Towards Context Trace Technology. *Personal Ubiquitous Computing* **5**(1) 50–53.

Raper, J., Gartner, G., Karimi, H. and Rizos, C. (2007) Applications of Location Based Services: A Selected Review. *Journal of Location Based Services*, **1**, 89–111.

Reichenbacher, T. (2007). Adapation in *Mobile and Ubiquitous Cartography. Multimedia Cartography*. W. Cartwright, M. P. Peterson and G. Gartner (eds). Berlin, Springer Verlag: 383–397, 2007

Resnick, P., Iacovou, N., Suchak, M., Bergstrom, P. and Riedl, J. (1994). GroupLens: An Open Architecture for Collaborative Filtering of Netnews. *5th Conference on Computer Supported Cooperative Work*, Chapel Hill, NC, USA, ACM Press. 175–186, 1994.

Rich, E. (1979). User Modeling via Stereotypes. *Cognitive Science* **3**(4) 329–354.

Ruthven, I., Borlund, P., Ingwersen, P., Belkin N. J., Tombros A. and Vakkari P. (Eds.) (2006). Information Interaction in Context. *Proceedings of the 1st IIiX Symposium on Information Interaction in Context*, 18–20 October 2006, Copenhagen, Denmark. vol. 176. ACM, New York, NY, USA.

Salber, D., Dey, A. K. and Abowd, G. D. (1999). The Context Toolkit: Aiding the Development of Context-Enabled Applications. *Conference on Human Factors in Computing Systems (CHI)*, Pittsburgh, PA, USA, ACM Press, 434–441, 1999.

Saracevic, T. (1995). Evaluation of Evaluation in Information Retrieval. *18th Annual International ACM SIGIR Conference on Research and Development in Information Retrieval*, Seattle, WA, USA, ACM Press, 138–146, 1995.

Saracevic, T. (1997). The Stratified Model of Information Retrieval Interaction: Extensions and Applications. *Proceedings of the American Society for Information Science (ASIS) Annual Meeting*, 313–327, 1997.

Schilit, B. (1995). *System Architecture for Context-Aware Mobile Computing*. New York, Columbia University, 1995.

Schilit, B., Adams, N. and Want, R. (1994). Context-Aware Computing Applications. In *Proceedings of the Workshop on Mobile Computing Systems and Applications*, 85–90. IEEE Computer Society, Santa Cruz, CA, December 1994.

Schmidt, A., Beigl, M. and Gellersen, H. W. (1999). There is more to context than location. *Computers and Graphics Journal* **23**(6) 893–902.

Scholtz, J. (2006). Metrics for Evaluating Human Information Interaction Systems. *Interacting with Computers* **18**(4) 507–527.

Scriven, M. (1967). The Methodology of Evaluation. *Perspectives of Curriculum Evaluation*. R. W. Tyler, R. M. Gagne and M. Scriven (Eds). Chicago, IL, US, Rand McNally. 39–83.

Scriven, M. (1991). *Evaluation Thesaurus*. Newbury Park, CA, US, Sage Publications.

Shardanand, U. and Maes, P. (1995). Social Information Filtering: Algorithms for Automating 'Word of Mouth'. *Conference on Human Factors in Computing Systems (CHI)*, Denver, CO, USA, 1995, 210–217.

Spink, A., and Jansen, B. J. (2004). Web Search: Public Searching of the Web, Series: Information Science and Knowledge Management, Vol. 6, 199 p., 2004. Hardcover ISBN: 978-1-4020-2268-5.

Strang, T. and Linnhoff-Popien, C. (2004). A Context Modeling Survey. *6th International Conference on Ubiquitous Computing (UbiComp), 1st International Workshop on Advanced Context Modelling, Reasoning and Management*, Nottingham, UK, 2004.

Suchman, L. (1987). *Plans and Situated Actions: The Problem of Human – Machine Communication*. Cambridge, Cambridge University Press.

Taylor, R. S. (1991). Information Use Environments. *Progress in Information Sciences*. B. Dervin and M. Voight (eds). Norwood, NJ, Ablex, Vol. 10, 217–255.

Tulving, E. (1972). Episodic and Semantic Memory. In E. Tulving and W. Donaldson (Eds.), *Organization of Memory*, pp. 381–403. New York, Academic Press.

Zipf, A. (2002). User-Adaptive Maps for Location-Based Services (LBS) for Tourism. In *9th International Conference for Information and Communication Technologies in Tourism (ENTER 2002)*, 328–338, Innsbruck, Austria. Springer Verlag.

8

Text Categorisation and Genre in Information Retrieval

Stuart Watt

8.1 Introduction: What is Text Categorisation?

Searching is perhaps the most well known application of information retrieval, it is far from being the only one. This chapter will consider a different application: text categorisation. Text categorisation, put simply, is the process of classifying or labelling documents in some way, to make the documents easier to manage. At first glance, this may seem one of the easier parts of information retrieval. After all, there is often no need to worry about gigabytes of text, or to find and return the very best document within those gigabytes of text. Text categorisation is often used with smaller corpuses, such as an individual person's email, or a library catalogue. The complexity lies in the categories and labels used: these are often semantic, corresponding to hidden aspects of the documents, and to social conventions shared among users of the documents.

As with other methods of information retrieval, there are many different techniques for text categorisation, for example, looking for clusters of similar texts, or using statistical estimates of the probability of category membership, given the evidence presented by features. Text categorisation techniques almost exclusively derive directly from traditional classification techniques, using information retrieval document measures. Although these approaches can generate classifications effectively, they may fail to capture some of the wider context of text categorisation. This chapter will look at text categorisation in a wider context – helping people to complete tasks effectively – and use the psychology of categorisation, which explores different kinds of category and feature representations, and which integrates perception and decision-making into the categorisation process. The chapter uses this to map out approaches to text categorisation in information retrieval that draw together the work from information retrieval and cognitive psychology, and provides a framework of possible approaches to further work on text categorisation in information retrieval.

Why is text categorisation important? The main difference between text categorisation and other kinds of information retrieval is that text categorisation involves a qualitative decision. Documents are not just ranked: some are labelled and some are not. This is done for a reason, typically to filter or route documents according to when and how they are likely to be needed. This argument places text categorisation within a wider problem-solving framework, just as Belkin's (1980) ASK model

Information Retrieval: Searching in the 21st Century   edited by A. Göker & J. Davies
© 2009 John Wiley & Sons, Ltd

does. Text categorisation is not – or should not be – treated as a mathematical problem, but as a tool that assists in human problem-solving. Characteristics of the problem domain therefore begin to play a more important role in the categorisation process.

The aim of this chapter is to provide background on the purpose of categorisation, how to build a categorisation system, and discuss the evaluation of text categorisation to define the relationship between categorisation and genre.

8.1.1   Purpose of categorisation

Before pushing the argument mentioned above about treating categorisation as a mathematical problem, it may be helpful to provide a backdrop, by looking at the purposes of categorisation for people. People are natural categorisers – they can tell at a glance whether or not an object looks like a chair or a table, a dog or a cat. The features that distinguish these objects can be remarkably subtle, and the actual categories themselves are not crisply or clearly defined. If you happen to be out walking in the hills, a large rock may become either a chair or a table. This is something we learn very early in childhood – although we do not learn it without making some mistakes along the way. Young children may over-generalise the 'cat' category, and see dogs as 'cats', until they have experienced enough cats and dogs to be able to tell the difference.

There is one very important difference between natural categories, such as chairs and cats, and information categories as represented in documents. This difference is that information categories are far more malleable, and can be formed in very short periods. A good example, and one that is the laboratory rat of the text categorisation world, is email spam filtering. Commercial spam email first started in 1994, and has reached large-scale proportions only in the last few years; today more than three-quarters of all email messages are typically spam of one kind or another. Technology has been undergoing development to automatically filter out the junk email, and although it is irritating, people generally have little difficulty in recognising and ignoring spam email messages.

The typical view of text categorisation in information retrieval is that it involves labelling documents according to a set of categories (Sebastiani 2002). For example, a classification might tag news articles according to whether they are politics, sport, science, and so on. Classification involves deciding, for any document, which categories it belongs to. The problem is not just one of texts; images and videos, for example, all involve similar classification issues. More formal descriptions of categorisation in information retrieval (e.g., from Sebastiani 2002) often make use of a number of further assumptions, such as:

- Assumption A1 – The categories are normally assumed to be independent.[1]
- Assumption A2 – The meaning of the categories is not defined.
- Assumption A3 – No additional knowledge, apart from that attached to the documents themselves, is available.

While these assumptions may make the implementation of a classifier system easier, clearly they are all false for human categorisers. Although additional semantic knowledge can make it harder to move between corpuses (as the semantic knowledge would need to change to reflect this move), it can still be very useful.

So far, we have been fairly agnostic about the actual classifications used. Most commonly, categories are considered independent and 'flat', that is, each category label is assigned without assuming any relationships to any other category label. This is unlikely to be the case in practice: Figure 8.1 depicts a

---

[1] This is a handy convenience – assuming that the categories are independent allows Bayesian statistical techniques to be used in devising classification techniques. In practice, most machine learning approaches to text categorisation assume: (a) that classification is a binary labelling, a document is either in a given category, or it is not; and (b) that categories are independent of one another. As we shall see later, there is good evidence to show that assumptions (a) and (b) are both false.

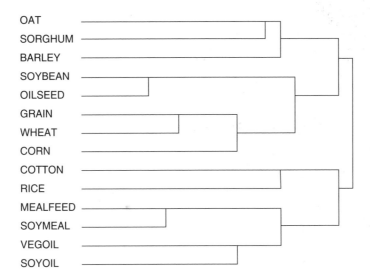

Figure 8.1 Part of a dendrogram by category association within Reuters-21578

section of the categories in the Reuters-21578[2] collection as a dendrogram, showing which categories are most strongly associated by co-occurrence in texts. Some categories are clearly closely associated (e.g., wheat and grain, soy oil and vegetable oil). However, these associations are not consistent; they reflect geographical as well as functional associations. For example, zinc and lead co-occur (they are associated minerals) as do copra and plywood (they frequently come from the same countries, despite the different industries). Categories are not independent: they reflect (often hidden) knowledge within the collection and the task, and often form fairly clear hierarchies.

The hierarchical model also has its problems. For example, faceted classification is important in library and information science. Faceted classification systems use several labels for each document – with facets playing structured roles. So, while a restaurant review might topically be hierarchically classified either under 'restaurants' or 'reviews' (but not both), a faceted classification system could classify it under both, as well as associating it with location, price, style of cuisine, and so on. The structure provided by a faceted classification system can capture all the different aspects of why people might want to use the document, allowing a much richer search, in the manner of a knowledge-based ontology (Domingue and Motta 2000).

This label-oriented view of text categorisation is common. For example, Sebastiani (2005) regards 'topic spotting' as identical in meaning to text categorisation. Although topic spotting is one type of text categorisation, many categories (such as junk email) are not really topics, but 'genres' (Bakhtin 1986) – socially-formed common forms of communication, irrespective of the actual topic. Genre, as an important area of new research in text categorisation, (Dewdney *et al.* 2001; Freund *et al.* 2005; Finn and Kushmerick 2006) will be discussed in a later section.

Text categorisation, therefore, is part of a wider context of use that differs from some other areas of information retrieval. The main points can be summarised as follows:

1. Text categorisation is intended to make a set of documents easier to manage in some way.
2. Effective management of documents depends on its intended purpose, for both sender and recipient.
3. The intended purpose of a document is reflected in its structure and layout, as well as in its use of language.

---

[2] The Reuters-21578 collection, strictly 'Reuters-21578, Distribution 1.0' is widely used to evaluate text classification systems and techniques. It is freely available for research use at: http://kdd.ics.uci.edu/databases/reuters21578/reuters21578.html.

Another closely related topic is information filtering, which is like classification, except that instead of labelling the documents, there is an autonomous decision about who they should be shown to. A typical information filtering system will have a representation of a user's interests, and may filter incoming news (for example) to reflect their interests, hiding everything else. A good example was InfoFinder (Krulwich and Burkey 1997) – this used a decision tree to learn people's interests, automatically sending them information which might be interesting. In practice, InfoFinder used a kind of relevance feedback system to train the decision tree, which meant that it could become committed to an invalid model.[3]

Although, at first glance, the interaction in information filtering seems very different, in practice they are not so very different. In both cases, categorical decisions are being made, and often (not always) categorisation is being used to implement the decision-making process needed for information filtering. Also, the approaches to evaluation are similar, both in the forms used, and in the need for different weightings of precision and recall, depending on the context of use.

8.2 How to Build a Text Categorisation System

There are several parts to a typical text categorisation system, as shown in Figure 8.2. First of all, there is a component that can take a training set of document texts and can transform them into some category representation. An important part of this is feature selection, which focuses the efforts of the classification system by selecting which features are likely to be useful in distinguishing between categories. The same feature set and category representation are used in the classifier itself. The output from the feature selection is a document representation that is used typically for two different purposes.

First, a large batch of preclassified documents is used to train a machine learning algorithm, which generates a set of category representations. The actual nature of these category representations varies widely, depending on the technique. For some algorithms (e.g., Rocchio) it may be as simple as one 'typical' document; for others (e.g., neural networks, latent semantic indexing), it is a weighted set of associations between features and categories. In practice, the actual nature of these representations does not matter, so long as it can be used by a classifier component to decide how to categorise new, unseen, document representations.

As the diagram in Figure 8.2 shows, the classification system and the feature selection technique are well bounded, and while both of these contribute to the effectiveness of the overall process, they can be explored independently. The training side, of course, can also be enhanced manually – additional heuristics for feature selection can be built in, or even the entire classification system built from scratch by hand.

It might be helpful to compare this with Figure 8.3, which shows a similar diagram for a basic text retrieval algorithm. There are a few important differences: four in particular:

- First, text categorisation is often done without a user being immediately present; it is often used to filter, index, or organise information in some way for the benefit of the user, frequently in a setting where the user is not using the system interactively. In other words, the role of the system can be quite varied (Masterton and Watt 2000).
- Second, text categorisation systems do not use queries; instead, documents are used directly, and corresponding labels are produced for use.
- Third, there is no direct equivalent to a category representation in a text retrieval system; the retrieval systems generally use a representation which preserves, at least in part, the original document structure.
- Finally, the feedback systems in text retrieval are often more complex, or at least more dynamic. A user will almost certainly revise a query and re-use it to refine the results. However, the category representations may be refined, as in the case of email filtering.

---

[3] In one trial, this resulted in InfoFinder deciding that we were interested in Olympic skiing events, somehow, which happened not to be the case, and it seemed to be impossible to further refine its model to overcome this commitment.

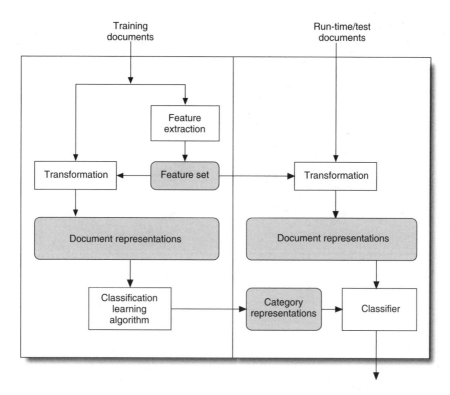

Figure 8.2   Diagrammatic overview of a typical text categorisation system

Given the outline structure in Figure 8.2, it is worth looking at each of the main parts of a text categorisation system in more detail. Rather than starting with the input documents, it is best to start with the classifier itself, as this influences the kinds of feature extraction that might be needed. In the rest of this section, then, we will focus on the classifier component, the machine learning component, and the feature extraction component in turn.

8.2.1   The classifier component

The job of the classifier is a deceptively simple one. Given a set of categories and a document, it needs to decide which categories the document belongs to. There are many different ways of doing this, and they have different advantages and disadvantages.

The diagrams in Figure 8.4 show a very simplified version of some of the alternatives. The diagrams show two different documents, D1 and D2, in an imaginary space of possible documents, and how different classifiers might make a decision about whether these belong to two categories, A and B. Of course, reality is a lot more complicated than this, as these diagrams are reducing rich document representations down to two dimensions.

(a) *Defining features.*   Known in philosophy as the 'classical view' of categories, these are the oldest categorisation schemes, dating back to Plato. They assume there is a (finite) set of characteristics, which something *must* have to belong to the category. Things that are missing even one characteristic fall outside. Rule-based classifiers use this approach; visually, the defining features represent areas of the space of documents with a well-defined boundary. Documents inside the boundary (e.g., D1) are in a category; those outside (e.g., D2) are unclassified.

(b) *Spatial boundaries.*   A second way of using a rule-like approach is to cut the space of possible objects into sections, with objects on one side of the boundary falling into the category, and those

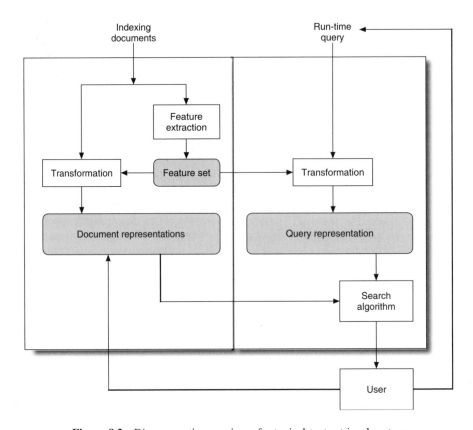

Figure 8.3   Diagrammatic overview of a typical text retrieval system

on the other side falling outside. This has some of the same shortcomings as the classical view, but they are easy to calculate, and can be effective at teasing out approximate structures when the number of known examples is relatively small.

(c) *Prototypes.*   In both philosophy (e.g., Wittgenstein 1953) and psychology (e.g., Rosch 1978; Rosch and Mervis 1975) the classical view has failed to deliver; for example, there is no single set of features which defines a game (Wittgenstein 1953). One alternative is to use prototypes – some representation of a 'typical' member of the category. If an object is sufficiently close, it falls in the category; otherwise it is unclassified. This has some advantages: the tolerance can be restricted or relaxed according to context. However, for some categories, such as news stories, although 'typical' news stories are easy to recognise, there can be enormous variation that is hard to capture in a coherent prototype.

(d) *Probabilistic models.*   At first glance, probabilistic models look like a fuzzy version of the classical view. This is deceptive. As with information retrieval (e.g., Sparck Jones *et al.* 2000), they represent a different way of thinking about the whole problem. The classifier now combines evidence from different sources to establish a document's category. When used under Bayes' rule, assuming independence between the sources of evidence, it does come closer to the classical view; however, they also have some of the flexibility of prototypes, as their threshold of confidence can be adjusted according to need. As the diagram shows, this does not depend on similarity within a space of possible documents.

(e) *Exemplars.*   One solution to the problem of prototypes is simply to remember absolutely every-thing! To classify an object, simply find the most similar, and use its category. Alternatively, look for the nearest three or so, and let them vote on it. As the diagram shows, this can produce different

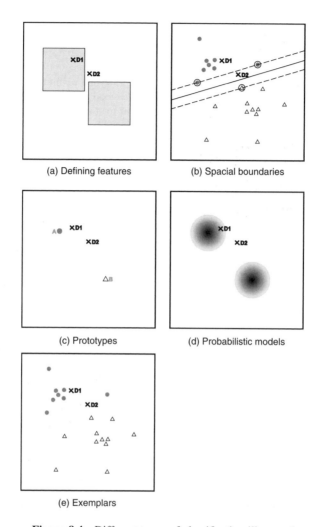

Figure 8.4 Different types of classification illustrated

results when categories are dispersed. When many known examples are available, exemplars can be very effective; they become less useful when there are gaps in the space of possible documents.

(f) *Hidden dependency models.* Some classifiers are hard to show visually, because there is a more complex tangling of features in the classification process. One common approach is to use language modelling, to build a statistical model of the relationships within the language, and using these to predict whether a new text plausibly fits with the category. How this is done may vary widely, from language models (e.g., Teahan and Harper 2001) through to knowledge-based extensions to similarity, such as case-based reasoning (e.g., Hahn and Chater 1997).

The differences between these interpretations of categories, as shown in the examples in Figure 8.4, bring out significant differences in the classification of boundary cases. It really is not clear how D2 should be classified – it is not especially close to any prior examples or a prototype. Under circumstances like these, the best we can do is to use techniques that try to establish a 'deeper' model of the nature of a category, looking at the relationships between features and weighting them appropriately, and using this additional information in the classification systems. Besides, a text categorisation system needs to be able to build these category representations effectively, and this is a very different challenge. This is where machine learning comes in.

8.2.2   *The machine learning component*

The task of the machine learning component is to put together an appropriate set of category representations for the classifier. Obviously, the task depends on the category representation that seems best for the task in hand. This may even depend on factors beyond performance, such as the extent to which the category representations can be read and understood by users (Pazzani 2000).

Of course, in principle there is no absolute requirement to use machine learning in a text categorisarion system. The category representations could be hand-built – especially if they are easy to understand, as is typically the case for defining features, rules, or even prototypes. However, hand-built classifiers are difficult (and expensive!) to produce, hard to maintain, and awkward to extend to new topics or problems. Machine learning – 'training' a system to build the category representations from a set of known examples – makes it all quicker, cheaper, and more adaptable.

A wide range of machine learning algorithms have been used to build text categorisation systems, although the most common are Bayesian probabilistic learning algorithms, support vector machines, nearest-neighbour exemplar-based systems, and neural network systems, although a wide range of other approaches have been explored. The different learning approaches mirror the different approaches to categorisation discussed in the previous section. Sebastiani (2002) shows a thorough comparison of the different learning approaches. All in all, there is significant variation between the different approaches: certainly there is no clear technique that has proven to be most effective under a wide range of circumstances. By and large, the machine learning approaches correspond to the category representations just described.

(a) *Defining features.*   Some of the oldest and best known machine learning techniques use defining features, especially in the form of decision trees. Good examples include ID3 (Quinlan 1983) and COBWEB (Fisher 1987). These use information-theoretic measures to build trees capable of classifying correctly all available examples.

(b) *Spatial boundaries.*   One of the most successful current approaches to text categorisation is that of support vector machines, or SVMs (Dumais 1998). SVMs typically transform the feature space before finding a linear boundary, so the actual boundary apparent in the original feature space may well be non-linear.

(c) *Prototypes.*   Rocchio's relevance feedback algorithm is a typical example of this approach, and can easily be used as a text categorisation technique (Joachims 1997). It works by building a single prototype in the document space, with TFIDF weighting, and using this to classify objects within the document space.

(d) *Probabilistic models.*   Probabilistic methods treat terms (or other features) as evidence relating to category membership, and then use evidence combination techniques (such as Bayes' rule) to combine this evidence to select the most probably category label. There are many variations on this theme (e.g., Anderson 1990). In practice, Bayes' rule assumes that the probabilities are independent (an assumption that is generally false for text categorisation) – yet it is robust enough to work reasonably well despite this limitation. Bayesian methods are especially widepsread in use for email filtering.

(e) *Exemplar models.*   Perhaps the best known of these is the k-nearest neighbour algorithm (kNN), which is perhaps the easiest model to implement as the learning part of it does not actually need to do very much. The classifier has to do all the work, finding the closest ($k = 1$) (or several closest, $k > 1$) examples, and using the most common category label. Small values of k are better at resolving boundary cases, but are more susceptible to noise in the data.

(f) *Hidden dependency models.*   There are many different approaches to this. Compression models (e.g., Teahan and Harper 2001) are a good example. They build a model of a category by compressing a set of example texts. When a new text is to be classified, it will compress more when added to examples from the same category, as compared with examples from a different category. In effect, they again use a similarity measure, but this measure is based on conditional Kolmogorov complexity. Kolmogorov complexity is a measure of the smallest size a piece of text can be compressed to. Conditional Kolmogorov complexity measures the relationship between two texts, and is a measure of the smallest size one piece of text can be compressed to, when given the second as a starting point. Compression models operate more or less according to this principle, using approximations to Kolmogorov complexity generated by advanced compression algorithms. However, there are other approaches to text categorisation which depend on more complex foldings of the document features, including: neural networks (Ruiz and Srinivasan 2002) and latent semantic indexing (Deerwester *et al.* 1990). Both of these exploit a kind of 'hidden layer' of association between features and documents, which can correspond to deeper (semantic) features.

There are other machine learning techniques which are important to text categorisation, but which do not fit neatly into this structure, or which combine several elements of it. Boosting is perhaps the most significant. Boosting is a way of building one good classifier from a number of weak ones; it builds a chain of classifiers, each of which can be trained as an independent system, so that documents which cannot be accurately classified by one system are then passed to the next in turn.

8.2.3 The feature selection component

One of the major differences between information retrieval systems and text categorisation systems is that feature extraction becomes much more significant. In information retrieval, queries tend to be short and highly varied, so essentially the whole information content of the document collection needs to be retained; this means information retrieval systems use feature weighting rather than feature extraction and filtering, although stemming and some stop-word filtering is still common. Feature extraction, by contrast, involves reducing the information content in the category representations to a sometimes very much smaller set of key features, which are particularly distinctive for the classes concerned.

There are many different approaches to feature extraction, but the most common approach is to use a statistical estimation of significance or information content, and to arbitrarily use those features which score highest. Common measures used include document frequency, information gain, and χ^2 (chi square). These usually correlate fairly well, and all offer good classification accuracy (Yang and Pedersen 1997).

Feature selection is a dimension reduction technique. This is not a coincidence: the reason for using feature selection is that a document is far bigger than a query, and not structured like a profile. Feature selection makes classification easier when using a large number of dimensions; of course, this is particularly problematic when using vector-space exemplar techniques such as kNN. Finding the closest match in 10 00 000 documents with 10 000 dimensions is far more computationally expensive than when you only need to worry about 500 dimensions. Other approaches to classification may use feature selection to make the category representations easier to construct (e.g., in the case of decision trees and rules), or may not need feature selection to the same extent, because it is 'built in' to the technique (e.g., in the case of SVM, probabilistic models, and Rocchio).

A few approaches work potentially with the entire feature set, and transform the document space to handle them. Mainly, this means latent semantic indexing (LSI) and support vector machines (SVM), which deserve a special place in a discussion of feature selection. Latent semantic indexing works by taking documents in one vector space and building a new space with a smaller set of dimensions. In this new space, dimensions no longer correspond one-to-one with terms, but in principle, they correspond to semantic concepts (Deerwester *et al.* 1990). These are derived through term co-occurrence. In practice, LSI derives two sets of associations with these dimensions, one associates each term with the reduced dimensions, the other associates each document with the reduced dimensions. The folding of the vector space means that this can quickly be turned into a matrix of document similarities. In effect, this reduces the number of features, so that those that correlate are collapsed into a single dimension. Essentially, feature selection is built into the process. Support vector machines have the same basic property, but work in the opposite way – they transform the document space into a larger number of dimensions; features which are not helpful at discriminating between categories become less important in this expanded space.

8.3   Evaluating Text Categorisation Systems

In practice, text categorisation is more like branches of machine learning than information retrieval as far as evaluation is concerned. That is, a training set is used to build the classification model, and then this is used with an unseen test set, and the accuracy on this test set can be used as a measure of the effectiveness of the classification technique. This is different to the normal metrics of information retrieval – in particular, precision and recall need to be measured in a slightly different way.

Any automated classification system is bound to classify some items wrongly. To evaluate the system's effectiveness, we need to assess the proportion of classifications that are correct as opposed to incorrect. These may, or may not, be equally serious. Again, spam is a good example: classifying spam as an email message is not necessarily as serious as classifying an ordinary email message as spam, because the outcome may involve inappropriate deletion of messages. There may, therefore, be different weightings attached to classification errors, although most of the time for evaluation purposes all errors are treated equally.

The normal approach is to consider each category separately as a binary classification (e.g., spam versus non-spam. When using a binary classification, the comparison in evaluation allows four different outcomes: true positives (TP – predicted and actual were both category matches), true negatives (TN – predicted and actual were both category non-matches), false positives (FP – predicted in the category, but incorrectly) and false negatives (FN – predicted not in the category, again incorrectly). The outcome from each member of the test set will fall into one of these categories, and by looking at the number of FPs and FNs, we can assess the effectiveness of a categorisation system.

Category-based precision and recall can then be calculated as follows:

$$precision = \frac{TP}{TP + FP}$$
$$recall = \frac{TP}{TP + FN}$$

With many algorithms, there is room for 'tuning' the result, perhaps by adjusting a threshold parameter. Under these circumstances, an algorithm can be turned into a 'trivial acceptor' (which marks everything as in the category, and therefore has good recall) or a 'trivial rejector' (which rejects everything, leaving precision undefined). For these reasons, a number of measures are used which help look at the overall quality of a categorisation system. The most common measures include:

- *Breakeven point.* This is the value where precision and recall are the same – in effect, where the proportion of false positives is the same as the proportion of false negatives. This is calculated by adjusting the threshold parameter to achieve this balance point, but some interpolation is often necessary to find this point.
- *11-point average precision.* This is calculated by adjusting the recall to 0.0, 0.1, and so on to 1.0, and then taking the mean interpolated precision at each point. This is very similar to common techniques for evaluating ranked retrieval systems.
- *Van Rijsbergen's F_β.* The weighted harmonic mean of precision and recall, this allows the parameter ß to be adjusted to achieve different tradeoffs between them. Most often, for evaluation purposes, ß = 1, which weights them equally. When there are no parameters that can be used to smoothly adjust precision and recall, F_β is the only option.

Unfortunately, the values obtained from these different measures cannot easily be compared. However, when the same measure is used consistently, with the same collection, different algorithms can be compared, and it is possible to establish an overall picture of the effectiveness of different techniques – at least within this context (Sebastiani 2002). For this reason, most evaluations of text categorisation have been conducted on partitions of the large sample of Reuters news stories, known as Reuters-21578, and briefly mentioned earlier. The starting point for most classification evaluations is to use a corpus and divide it into a training set and a test set. The training set can be used by a machine learning technique to establish models of the categories in some form, and the test set is then used to validate these models by comparing the predicted category (through the classification technique) against the actual category. Different researchers have used slightly different partitions between training and test sets, although the most common is probably based on the experiments of Apte *et al.* (1994). It consists of 9603 training documents and 3299 test documents, with 90 different categories represented (many of these, however, by only one or two documents).

This Reuters collection is drawn from newswire stories, especially in the fields of business and finance, and this, to some extent, has influenced the structure of the stories, as well as the categories into which they fall. This is because the story structure has a different purpose – the title in a Reuters story is there for editors, so they can accurately make a decision as to whether a particular story is relevant to their needs. In a typical newspaper, however, the headline is there to draw readers in and encourage them to read the rest of the article. This difference – both in audience and purpose – is reflected in the term usage in the collection. For example, in Reuters-21578, abbreviations are used extensively in the story title, and far less commonly in the story body, where the expansion of the abbreviation is more likely. Distinctive classification words are often used with less ambiguity in the title than in the body; for example 'sell' is significantly more common in the story title, 'sold', 'paid', and 'bought' are all significantly less common in the title than in the rest of the story. This subtlety shows why machine learning is important to text categorisation – each corpus comes with a social context which shapes its use, and which implicitly affects the word distribution, often in different parts of the document as well as through the document as a whole.

8.4 Genre: Text Structure and Purpose

This brings us back to the third of our main points, that the intended purpose of a document is reflected in its structure and layout, as well as in its use of language. Structure and layout have traditionally been used less for text categorisation, largely because they usually only make sense within a particular community. For those within a community, however, convergence on a set of standardised document structures is both natural and helpful (Yates and Orlikowski 1992). These structures reflect the purpose of the document, and are called genres.

8.4.1 An overview of genre

The term 'genre' is not always used in a consistent manner. Historically, it is a classification within a form of art; films come in different genres (e.g., comedy, western) as do books (e.g., romantic, biography) (Bakhtin 1986). In the world of text, common genres include business letters, memos, academic papers, blogs, and so on. The actual topic of the text is separate from the genre: blogs can be written on everything from US Republican politics to cooking. In both cases, there is an easily recognised structure (date, headline, text, comments), with some standardised use of terms ('permalink', 'tag').

Some examples may help to clarify this. Newspaper articles are a particularly good example of genre (Collins *et al.* 2001). A typical newspaper article is written as an 'inverted pyramid' – where the first sentence is a snapshot summary of the whole story. The first paragraph again summarises the story. Then the story is told in more detail, but the further the text goes, the more peripheral information is included. This actually dates from the days of lead type, where the story would literally be cut to fit the space available! The inverted pyramid genre meant that it could be cut almost anywhere and still be coherent as a story. The example in Figure 8.5 shows this in practice.

Although news stories follow an inverted pyramid genre, each genre may have a different structure. Genres emerge through social interaction in communities, and reflect the needs and aims of those communities, so there is no common set of structures. Academic papers, for example, typically have a very different genre – they have an abstract, introduction, conclusions, and references; in particular subfields, the middle sections may also be structured. Again, this format has a purpose: abstracts are there to help people decide whether or not to read the paper (and, originally, abstracting services

US Congress passes $787 billion stimulus package

The United States Congress passed a US$787 billion stimulus package late Friday night, in an effort to curb the recession and boost the faltering U.S. economy

The bill was passed by a vote of 60–38, barely reaching the minimum number of 60 votes needed to make the bill a law. Only three Republican senators voted for the measure. Shortly before the Senate vote, the U.S. House of Representatives approved of the stimulus by a margin of 246–183, with all 176 Republicans and seven Democrats voting against the bill.

The Senate vote was delayed until night so that Democratic Senator Sherrod Brown could fly back from Ohio, Where his mother had recently died. He cast the final and decisive sixtieth vote in favor of the stimulus bill.

President Barack Obama is expected to sign the bill into law in Denver, Colorado on Tuesday, February 17.

"This is a major milestone on our road to recovery, and I want to thank the members of Congress who came together in common purpose to make it happen," Obama said in his weekly address. "I will sign this legislation into law shortly, and we'll begin making the immediate investments necessary to put people back to work doing the work America needs done. This historic step won't be the end of what we do to turn our economy around, but the beginning."

Figure 8.5   A typical newspaper article, extract taken from article entitled 'U.S. Congress passes $787 billion stimulus package' from Wikinews, dated 16 February 2009, and retrieved on the 7 June 2009 (available at: http://en.wikinews.org/wiki/article?curid=121002). Contribution copyright Wikinews and used under the Creative Commons Attribution license, http://creativecommons.org/licenses/by/2.5/

enabled people to make this decision when actually obtaining a copy of the article was generally harder than it is today). The rest of the paper reflects the purpose of an article: to make an argument or case, concluding something from evidence or substantive points made in the middle of the article.

This is where genre reveals a divergence from topic-oriented models of text categorisation. These examples show that genre is often defined by structural rather than by linguistic features. There is little difference (linguistically) between the first sentence of a news story and the rest of it, but in practice, the first sentence is what people read first, and is significantly more important for classification purposes. For news articles, therefore, classification (although not retrieval) accuracy can often be improved by ignoring most of the text! Key words or phrases from the first sentence of the example in Figure 8.5 include 'Congress' and 'stimulus package'; not much more is needed to identify the most important aspects of the purpose of this document. Other information, such as the people involved, are far less important. Strangely enough – the news headline is often not as informative. They are written for a different purpose, to catch people's attention. Headlines such as 'Not baking but blogging: the new face of the WI' or 'Late drinking laws fall foul of residents' say something about the topic, but often much less than the article contents; they are often written using puns and linguistic tricks to engage people, rather than to actually communicate information.

Genre, then, is there for a purpose: to guide people about what to do with the text – it is an aid to decision-making, and as such, ought to be very close to the heart of the text categorisation problem. In the case of a news story, the purpose of the headline is to grab people's attention; in an academic paper, the abstract aids filtering. These are there to reduce cognitive load – there is no need for a person to read the whole text, the genre provides these filtering cues in its structure. In this view, genres behave as 'affordances'. Gibson's (1979) theory of affordances is an attempt to restructure the ways in which meaning and perception are related: it argues that, instead of perceiving objects (such as texts) and then adding meaning later, there are visual combinations of invariant properties of objects which cue a reader about how to act in relation to these objects. In the case of genre, these invariant properties are layout cues, predominantly, rather than linguistic ones.[4]

In the case of Reuters, the audience is slightly different. The Reuters collection does not really adhere to the typical news genre: its stories are not written for readers, but for journalists, editors, and

---

[4] This is an important assertion, and one which will be discussed in more depth later on in this chapter.

decision-makers: the title of a Reuters story is, therefore, much more a filtering tool to enable editors to judge whether or not this story is likely to be useful to them. The title of a Reuters story is very nearly as good for classification purposes as the whole of the rest of the story.

The Reuters categories allow people, especially in the finance and business sectors, to filter messages efficiently and focus down on those messages that are relevant to their particular needs. They therefore focus on, for example, particular currencies, particular commodities, particular (usually governmental or regulatory) organisations, as well as countries and people in key positions within those countries. The ideal text categorisation system, therefore, would automatically tag Reuters messages with codes so that anyone who may need to know about the messages can quickly find and track these messages.

8.4.2 Text categorisation and genre

Text categorisation has only begun to use genre in experimental work fairly recently (Dewdney *et al.* 2001; Finn and Kushmerick 2006). In this work, genre is intended to be orthogonal to topic, for example, Finn and Kushmerick (2006) say: 'By genre we loosely mean the style of text in the document. . . . This classification is based not on the topic of the document, but rather on the kind of text used.' Generally, this work has added new (genre-based) features to the feature extraction system, and explored its effects on classifying documents according to genre.

As to the results of this work, genre is still elusive: in the TREC HARD ('High Accuracy Retrieval from Documents') tracks, genres that were used included 'news', 'opinion/editorial', and 'other', within a collection of press articles. Muresan *et al.* (2006) reported that language modelling found detecting genre to be a difficult challenge. Using the Reuters collection, Dewdney *et al.* (2001) did find that genre could improve precision with little impact on recall. Finn and Kushmerick (2006) found that shallow text statistics could recognise genre classes almost as well as linguistic features, and improve overall results in one domain (identifying subjectivity) although the effect was less than successful in a second (distinguishing different types of review). Finn and Kushmerick's work could, arguably, have been improved by a sharper concept of genre: the fact *versus* opinion distinction is equivalent to the article *versus* editorial genres in press documents, and seems reasonable; but whether restaurant and movie reviews are properly distinct genres, or just examples of one genre, the review, is more open to question.

More recent work (e.g. Boese, 2005; Boese and Howe 2005; Santini 2006) has moved the work in a more definite direction, with convergence on a model of genre that is more explicitly orthogonal to that of topic. By and large, this work interprets structure (or form) and purpose as fundamental to genre, and where topic and content play a subordinate role rather than a leading role. As such, this modern research on genre in information retrieval brings out subjective and task-dependent issues that go significantly beyond earlier work that used genre primarily as a way of improving topic relevance.

8.4.3 The importance of layout

Since much of the work in the field has emphasised linguistic features of genre, which seem to be well within the current remit of text categorisation, we wanted to look in more detail at layout. In particular, we were interested in answers to the following two questions:

- Whether layout assisted people in classifying texts, and
- Whether layout features were independent of linguistic features.

We conducted a small experiment to investigate these questions. We selected a balanced email collection, containing conference calls for papers (which have a relatively well-defined genre structure) matched against other emails of the same length. We asked 8 users familiar with these types of messages to classify the documents. Each was presented a sample of 24 randomly selected email 'call for papers' messages, paired with 24 other email messages with a similar length and date of posting. In

Table 8.1   Results from the genre classification experiment

	Format preserved		Formatting removed	
	Speed	Accuracy	Speed	Accuracy
Letters unchanged	3.28s	97.9%	3.87s	93.8%
Letters 'X'	3.93s	76.0%	3.54s	58.9%

different conditions, we removed word information (changing letters to 'x's) and layout information (changing all whitespace to single spaces, and removing punctuation and other non-linguistic characters), and recorded both the speed and accuracy in timed-response decision task.

Our results were surprisingly dramatic. First of all, documents with the genre were classified significantly faster than those without (Mann–Whitney $U = 12900$, $n = 384$, p $\ll 0.001$), supporting the idea that the genre assists human classification. The more detailed response times are shown in Table 8.1. In fact, people could classify documents with reasonable accuracy, even with all linguistic information removed, relying on layout alone, although classification accuracy using only the linguistic information was high. However, the main effect was on speed – people were far faster when classifying documents with the layout information than without, and in terms of response time there was an interaction (2 × 2 ANOVA, $F = 4.43$, $p < .05$) between the effect of text and the effect of layout, suggesting that layout is influencing the process by which people classify texts.

The interaction is important – it is statistical evidence that linguistic and layout features are not independent of one another. This implies that it would not be sufficient to add layout features into the mix in a Bayesian or Bayesian-derived machine learning algorithm. It seems more probable that the layout features are providing invariant cues about which parts of the texts to attend to, both reducing the effort of and speeding up the classification process.

Of course, all this applies to people, not to information retrieval systems, which are not bound to follow the approaches used by people. Having said that, people are a lot more accurate than text categorisation algorithms in classifying texts, as well as being surprisingly quick at it. Taken with the timed-response results, this suggests that people are using layout perceptually to assist the classification process, probably by spotlighting certain parts of documents, and essentially using a process which combines evidence in pieces, rather than processing all the evidence simultaneously. This fits with research on natural categories, where classification can undergo strange reversals: people often classify things differently when time is very limited compared with when it is open.

Finally, until recently there has been a limiting factor on the use of genre: most documents do not really preserve layout. However, with the advent of the web, and especially now with XML, the layout is not only preserved, it is actually tagged, making it especially easy to use it to improve classification. Early experiments on INEX have shown definite promise (Clark and Watt 2007), although this work is continuing.

Genre is a complex field, crossing linguistics and information retrieval with the social sciences. However, many of the principles are common to other work in the field. Genre emphasises the context of use – what people are doing with these texts, how they are choosing to select some for reading and use, but not others.

8.5   Related Techniques: Information Filtering

Search engines adopt a particular conversational pattern with their users, taking the role of a librarian and mediating between the user and a collection. They can use techniques such as relevance feedback to help refine the search. This is exemplified by Belkin's (1980), 'anomalous state of knowledge' framework, which places search within a problem-solving framework.

Text categorisation does not fit this pattern, and text categorisation systems do not play the role of a librarian, at least, not in the pattern of mediating between the user and a collection. Information filtering is one very different pattern that they can adopt. In this case, the text categorisation system is effectively deciding whether or not the user should be informed about a particular issue (Belkin and Croft 1992). Information filtering is about building systems that reduce people's information load (Maes 1994) – in this way, information filtering systems are making decisions about whether people are likely to be interested.

The most frequent approach is to use what amounts to a probabilistic retrieval system, and then apply a threshold – if the likelihood exceeds this level then the system can act with an appropriate degree of confidence. Of course, the actual level will depend on various factors: the environment, the cost of failure, and the user's preferences. Under these circumstances, precision is likely to count for more than recall – in effect, the cost of a false positive is higher than that of a false negative.

In information filtering, there is no anomalous state of knowledge, nor is there a query. Typically, information filtering uses a 'profile' which represents a user's likely interests; in one sense, a profile is an expanded persistent query, and can be used more or less to replace a query to turn an information retrieval system, with a judiciously chosen threshold, into an information filtering one (Belkin and Croft 1992). However, a profile is not necessarily a query: Pazzani (2000), for example, adopts a kind of prototype representation; other representations include rules (Lewis *et al.* 1996), etc.

8.6   Applications of Text Categorisation

We have already seen one of the most obvious applications of text categorisation – spam filtering. There are several hundred different spam filtering packages in existence at the moment, most of which are based on variations of Bayesian machine learning. Email filtering is now also built into the more recent versions of popular email readers, such as Mozilla Thunderbird and Microsoft Outlook. The majority of email filtering systems use probabilistic Bayesian methods for classification. A few are open source, and are worth looking at for implementation techniques. However, there are many other applications. For example:

- *Knowledge management and sharing.*   In this case, the system uses the classification of a document to pass it to others who might be interested. Good examples include AnswerGarden (Ackerman and Malone 1990), InfoFinder (Krulwich and Burkey 1997), and KSE (Davies *et al.* 1998). This field is especially important in handling email and other relatively ephemeral knowledge (Moreale and Watt 2003).
- *Automated essay grading.*   Here, the task is to assign a grade to a text, principally by comparing it to a set of previously grade examples. Latent semantic analysis has been used for this (Landauer *et al.* 1998) with some success, as have systems such as e-rater, which use a wide range of linguistic features, many of which address higher level aspects of text use such as readability (Hearst 2000).
- *Recommender systems.*   Recommender systems (Resnick and Varian 1997) suggest documents that may be relevant to people, based on interest profiles. Typical recommender systems (such as PHOAKS: Terveen *et al.* 1997) include a classification system to help people select appropriate recommendations.
- *Keyword assignment.*   Although this is related to key phrase extraction, there may also be an aspect of classification. When a set of standardised categories are combined with key phrase extraction, the results are considerably improved (Hulth *et al.* 2001).
- *Authorship attribution.*   This is an important and emerging area in forensics. Usually, with authorship attribution, the number of candidates and texts is fairly small, so algorithms do not need to be of exceptionally high performance. Good techniques include those based on conditional Kolmogorov complexity (Malyutov 2005) and their related compression models (Teahan and Harper 2001).
- *Web classification.*   Some web search engines automatically classify websites or pages, to make it easier for people to browse rather than search by keyword. In many cases this is largely a manual (or semi-automatic) process.

8.7   Summary and the Future of Text Categorisation

Far from a tangent, text categorisation is a useful challenge within the field of information retrieval. Unlike the task of providing accurate 'hits' through a search engine, text categorisation can filter information, and help guide people to the information they need more quickly and easily. It plays an important role in browsing and filtering, and can be used to improve the precision of searching, by focusing it on the types of result that can be needed, without having too much of a negative effect on recall.

Within psychology at least, the possibilities of a purely similarity-based or a purely rule-based theory of categorisation have been more or less exhausted. The current problem is one of how to integrate them effectively. Intriguingly, Hahn and Chater (1997) have suggested that one possible solution may lie in the field of case-based reasoning, CBR. On one level, CBR classifies things through similarity to a set of previously seen exemplars, often using a variation on nearest-neighbour similarity matching kNN. However, this is far from the whole story: CBR also has a theory or rule-based aspect to it – features are inter-dependent, and one feature may be completed from others through a kind of weighted network, for example. This kind of structured approach to the category representation has many merits – yet it does begin to challenge the 'knowledge free' assumptions common to text categorisation (and often information retrieval too, for that matter).

Now we can start to reconceptualise text categorisation, or, rather, set it in a wider context. This is equivalent to Belkin's (1980) 'anomalous state of knowledge' in information retrieval, or newer theories of context, placing categorisation in a wider framework, in which human behaviour plays a very significant role. Unlike ASK, categorisation has a different purpose in assisting people to complete tasks, by reducing cognitive load, and this underpins many of the differences between text categorisation and other aspects of information retrieval. This framework is simple: category labels are there to enable people to predict how to use a particular document without having to treat each individual document separately. Text categories, therefore, are there to assist people in making the decisions they need to make about these texts, and for this reason, they vary widely according to the users and the texts they are working with.

As a research area in information retrieval, text categorisation does not have the instant applicability of straight retrieval algorithms, but it can help to augment them when placed in the right framework. Text categorisation, conceptualised as a tool to assist people in their daily tasks, can help people find the information that they need more quickly and easily. At the simplest, it can augment straight retrieval algorithms by improving precision with little impact on recall (Dewdney et al. 2001). At the other end of the range, it can be used to provide graphical interfaces that encourage browsing as well as searching. In both cases, text categorisation is an important complementary technique that has great potential to help people overcome the burdens of information overload in today's knowledge-intensive world.

Exercises

8.1 Consider a scenario where it is necessary to do some categorisation. This could be work related or personal, e.g. filing applications, photo album, etc. Describe in detail the purpose of the categorisation.

8.2 Consider your personal collection of books, videos, films or music, etc. Think about how you would classify these personal items by genre.

References

Ackerman, M. S. and Malone, T. W. (1990). Answer Garden: A Tool for Growing Organizational Memory. Paper presented at the *Conference on Supporting Group Work*, Cambridge, MA, pp. 31–39.

Anderson, J. R. (1990). *The Adaptive Character of Thought*. Hillsdale, New Jersey: Lawrence Erlbaum Associates.

Apte, C., Damerau, F. and Weiss, S. M. (1994). Towards Language Independent Automated Learning of Text Categorisation Models. Paper presented at the *17th Annual International ACM SIGIR Conference on Research and Development in Information Retrieval*, pp. 23–30.

Bakhtin, M. M. (1986). The Problem of Speech Genres. In *Speech Genres and other Late Essays*: University of Texas Press.

Belkin, N. J. (1980). Anomalous States of Knowledge as a basis for information retrieval. *Canadian Journal of Information Science*, **5**, 133–143.

Belkin, N. J. and Croft, W. B. (1992). Information filtering and information retrieval: two sides of the same coin? *Communications of the ACM*, **35**(12) 29–38.

Boese, E. S. (2005). *Stereotyping the Web: genre classification of web documents*. Unpublished MSc dissertation, Colorado State University.

Boese, E. S. and Howe, A. E. (2005). Effects of web document evolution on genre classification. In the *Proceedings of the 14th ACM International Conference on Information and Knowledge Management*, Bremen, Germany, pp. 632–639.

Clark, M. and Watt, S. N. K. (2007). Classifying XML documents by using genre features. In the *Proceedings of Database and Expert Systems Applications, DEXA* 2007, Regensburg, Germany, pp. 242–248.

Collins, T. D., Mulholland, P. and Watt, S. N. K. (2001). Using genre to support active participation in learning communities. Paper presented at the *European Conference on Computer Supported Collaborative Learning (Euro-CSCL'2001)*, Maastricht, The Netherlands, pp. 156–164.

Davies, J., Stewart, S. and Weeks, R. (1998). Knowledge Sharing Agents Over the World Wide Web. *British Telecom Technology Journal*, **16**(3) 104–109.

Deerwester, S., Dumais, S. T., Furnas, G. W., Landauer, T. K. Harshman, R. (1990). Indexing by Latent Semantic Analysis. *Journal of the American Society for Information Science*, **41**(6) 391–407.

Dewdney, N., Van-Ess-Dykema, C. and MacMillan, R. (2001). The Form is the Substance: Classification of Genres in Text. Paper presented at the *ACL 2001 Workshop on Human Language Technology and Knowledge Management*, Toulouse, France, 6–7 July 2001.

Domingue, J. B. and Motta, E. (2000). Planet-Onto: From News Publishing to Integrated Knowledge Management Support. *IEEE Intelligent Systems*, **15**(3) 26–32.

Dumais, S. (1998). Using SVMs for text categorization. *IEEE Intelligent Systems*, **13**(4) 21–23.

Finn, A. and Kushmerick, N. (2006). Learning to classify documents according to genre. *Journal of the American Society for Information Science and Technology*, **7**(5) 1506–1518.

Fisher, D. H. (1987). *Improving inference through conceptual clustering*. Paper presented at the *AAAI'87*, Seattle, Washington, pp. 461–465.

Freund, L., Toms, E. G. and Clarke, C. L. A. (2005). Modeling task–genre relationships for IR in the workplace. Paper presented at the *SIGIR'05*, Salvador. Brazil.

Gibson, J. J. (1979). *The Ecological Approach to Visual Perception*. Boston: Houghton Mifflin.

Hahn, U. and Chater, N. (1997). Concepts and similarity. In K. Lamberts and D. Shanks (eds), *Knowledge, Concepts and Categories* (pp. 43–92). Hove: Psychology Press/MIT Press.

Hearst, M. A., (ed.). (2000). Trends and controversies: the debate on automated essay grading. *IEEE Intelligent Systems*, **15**(5) 22–37.

Hulth, A., Karlgren, J., Jonsson, A., Boström, H. and Asker, L. (2001). Automatic Keyword Extraction Using Domain Knowledge. Paper presented at the *2nd International Conference on Computational Linguistics and Intelligent Text Processing (CICLing 2001)*, Mexico City, pp. 472–482.

Joachims, T. (1997). A probabilistic analysis of the Rocchio algorithm with TFIDF for text *categorization*. Paper presented at the *ICML-97, 14th International Conference on Machine Learning*, Nashville, pp. 143–151.

Krulwich, B. and Burkey, C. (1997). The InfoFinder Agent: Learning User Interests through Heuristic Phrase Extraction. *IEEE Expert*, **12**(5) 22–27.

Landauer, T. K., Foltz, P. W. and Laham, D. (1998). An Introduction to Latent Semantic Analysis. *Discourse Processes*, **25**, 259–284.

Lewis, D. D., Schapire, R. E., Callan, J. P. and Papka, R. (1996). Training algorithms for linear text classifiers. Paper presented at the *19th Annual International ACM SIGIR Conference on Research and Development in Information Retrieval*, Zurich, Switzerland, pp. 298–306.

Maes, P. (1994). Agents that Reduce Work and Information Overload. *Comunications of the ACM*, **37**(7) 31–40.

Malyutov, M. B. (2005). Authorship attribution of texts: a review. *Electronic Notes in Discrete Mathematics*, **21**(1) 353–357.

Masterton, S. J. and Watt, S. N. K. (2000). Oracles, bards, and village gossips, or, social roles and meta knowledge management. *Journal of Information Systems Frontiers*, **2**(3/4), pp. 299–315.

Moreale, E. and Watt, S. N. K. (2003). An agent-based approach to mailing list knowledge management. Paper presented at the *AAAI Spring Symposium on Agent-Mediated Knowledge Management (AMKM-03)*, Stanford, CA, pp. 118–129.

Muresan, G., Smith, C. L., Cole, M., Liu, L. and Belkin, N. J. (2006). Detecting Document Genre for Personalization in Information Retrieval. Paper presented at the *Hawaii International Conference on System Sciences (HICSS-39)*, Kauai, Hawaii, p. 50c.

Pazzani, M. J. (2000). Representation of Electronic Mail Filtering Profiles: A User Study. Paper presented at the *Intelligent User Interfaces (IUI'00)*, New Orleans, pp. 202–206.

Quinlan, J. R. (1983). Learning efficient classification procedures and their application to chess end games. In R. S. Michalski, J. G. Carbonell and T. M. Mitchell (eds), *Machine Learning: An Artificial Intelligence Approach* (pp. 436–382). Palo Alto: Tioga.

Resnick, P. and Varian, H. R. (1997). Recommender Systems. *Communications of the ACM*, **40**(3) 56–58.

Rosch, E. (1978). Principles of Categorisation. In E. Rosch and B. B. Lloyd (eds), *Cognition and Categorisation* (pp. 27–48). Hillsdale, New Jersey: Lawrence Erlbaum.

Rosch, E. and Mervis, C. (1975). Family Resemblances: Studies in the Internal Structure of Categories. *Cognitive Psychology*, **7** 573–605.

Ruiz, M. E. and Srinivasan, P. (2002). Hierarchical text categorization using neural networks. *Information Retrieval*, **5**(1) 87–118.

Santini, M. (2006). Towards a Zero-to-Multi-Genre Classification Scheme, *Journée ATALA 'Typologies de textes pour le traitement automatique'*, 9 December 2006, Paris, pp. 1–5.

Sebastiani, F. (2002). Machine Learning in Automated Text Categorization. *ACM Computing Surveys*, **34**(1) 1–47.

Sebastiani, F. (2005). Text categorization. In A. Zanasi (ed.), *Text Mining and its Applications to Intelligence, CRM and Knowledge Management* (pp. 109–129). Southampton, UK: WIT Press.

Sparck Jones, K., Robertson, S. and Walker, S. (2000). A probabilistic model of information retrieval: development and comparative experiments. *Information Processing and Management*, **36**(6) 809–840.

Teahan, W. J., & Harper, D. J. (2001). *Using compression-based language models for text categorization*. Paper presented at the *Workshop on Language Modeling and Information Retrieval*, Carnegie Mellon University.

Terveen, L., Hill, W., Amento, B., McDonald, D. and Creter, J. (1997). PHOAKS: A System for Sharing Recommendations. *Communications of the ACM*, **40**(3) 59–62.

Wittgenstein, L. (1953). *Philosophical Investigations*. Oxford: Basil Blackwell.

Yang, Y., & Pedersen, J. O. (1997). *A comparative study on feature selection in text categorization*. Paper presented at the *14th International Conference on Machine Learning, ICML-97*, Nashville, pp. 412–420.

Yates, J. and Orlikowski, W. J. (1992). Genres of Organizational Communication: A Structurational Approach to Studying Communication and Media. *Academy of Management Review*, **17**(2) 299–32.

9

Semantic Search

John Davies, Alistair Duke and Atanas Kiryakov

9.1 Introduction

This chapter will examine the use of semantic technology in information retrieval. We discuss the prospects for systems which combine semantics-aware information retrieval (IR) techniques to search information resources with the ability to browse and query against semantic annotations of those resources.

An example of semantic annotation is tagging (or marking up) a web page with the identifiers (e.g. URLs) of people, organizations, locations and other entities referred to. These identifiers can be considered as 'semantic features', since they constitute a feature-space where the resources are characterised by means of entities and concepts (instead of just tokens, lemmata, or stems as in traditional IR). This feature-space is usually of reduced dimensionality – there are a number of different strings which can represent a reference to one and the same concept; as for instance, 'UK', 'U.K.' and 'United Kingdom'. Another interesting quality of the semantic features space is that it is (or at least it can be) structured. One can use a database or a knowledge base (KB[1]) which contains structured information about the entities mentioned in the documents. An example would be a KB holding the facts that a person works for a company, which is registered in some city, which in turn is located in a particular country. This allows for retrieval approaches where probabilistic models are combined with structured queries. An example of this would be a search for documents referring to a person called John, who works for a non-government organisation in Bulgaria. A proper answer to such a request requires concrete factual knowledge, which may not be available in the documents where the person is mentioned. The latter means that for a classical IR system it is theoretically impossible to handle such query.

When semantic features are used as an alternative feature-space one gains the benefits typically associated with reduced dimension models (e.g. LSA[2]): better performance on noisy/redundant data and 'associative' retrieval. Typically, this leads to higher precision and lower recall; in the above example, not all documents containing the token 'john' will be extracted. Recall can be improved through query expansion based on the underlying KB.

---

[1] KB is a term with various interpretations; we use it as something broader than ontology; see Section 9.1.2.
[2] Latent Semantic Analysis (LSA) is presented in (Landauer and Dumais 1997); it is also discussed in Section 3.3 of Chapter 11 from the perspective of cross-language IR.

Information Retrieval: Searching in the 21st Century   edited by A. Göker & J. Davies

Another approach is to use the semantic features not as an alternative, but rather as an extension of the standard feature set used for full-text search. In such cases, semantic search can be characterised as 'low threshold, high ceiling' in the sense that where semantic annotations exist they are exploited for an improved information-seeking experience, but where they do not exist, a search capability is still available. Searching the full text of documents can ensure the high recall desirable in early stages of the information seeking process. In later stages of the search, when the user may typically be more interested in the precision of the retrieval results, it can be advantageous to put more emphasis on searching based on the semantic annotations.

The main objectives and the structure of this chapter can be summarized as follows:

- To outline some limitations of current search technology (Section 9.1.1);
- To introduce the notion of ontologies (Section 9.1.2);
- To introduce the Semantic Web and its relation to semantic search (Section 9.2);
- To discuss markup and annotation of text (Section 9.3);
- To discuss the role of semantic annotations (Section 9.4) and present a specific schema for semantic annotation of texts with respect to named entities (Section 9.5);
- To elaborate how search based on semantic annotations can be defined in IR terms (information need and satisfaction) and implemented (Section 9.6);
- To acquaint the reader with some of the most prominent semantic annotation and search systems (Section 9.7).

9.1.1   Limitations of current search technology

In general, when specifying a search, users enter a small number of terms in the query. The query describes the information need, and is commonly based on the words that people expect to occur in the types of document they seek. This gives rise to a fundamental problem, known as 'index term *synonymy*': not all documents will use the same words to refer to the same concept. Therefore, not all the documents that discuss the concept will be retrieved by a simple keyword-based search. Furthermore, query terms may of course have multiple meanings; this problem can be called 'query term *polysemy*'. As conventional search engines cannot interpret the sense of the user's search, the ambiguity of the query leads to the retrieval of irrelevant information.

Technically, the above two problems can be explained as follows: search engines that match query terms against a keyword-based index will fail to match relevant information when the keywords used in the query are different from those used in the index, despite having the same meaning. This problem can be overcome to some extent through thesaurus-based expansion of the query; this approach increases the level of document recall, but it may result in significant precision decay, i.e. the search engine returning too many results for the user to be able to process realistically.

Users can partly overcome query ambiguity by careful choice of additional query terms. However, there is evidence to suggest that many people may not be prepared to do this. For example, an analysis of the transaction logs of the Excite WWW search engine (Jansen *et al*. 2000) showed that web search engine queries contain on average 2.2 terms. Comparable user behaviour can also be observed on corporate intranets: an analysis of the queries submitted to the intranet search engine of BT[3] over a 4-month period between January and May 2004 showed an average query length of only 1.8 terms.

In addition to difficulties in handling synonymy and polysemy, conventional search engines are of course unaware of any other semantic links between terms (or, more precisely, the concepts which the terms represent). A major limitation of non-semantic IR approaches is that they cannot handle queries which either require knowledge and data which are not available in the documents; or require extraction, explicit structuring, and reasoning about some data. Consider for example, the following query:

'telecom company' Europe 'John Smith' director

---

[3] http://www.bt.com/

The information need appears to be for documents concerning a telecom company in Europe, a person called John Smith, and a board appointment. Note, however, that a document containing the following sentence would not be returned using conventional search techniques:

At its meeting on the 10th of May, the board of London-based O2 appointed John Smith as CTO.

In order to be able to return this document, the search engine would need to be aware of the following semantic relations:

- O2 is a mobile operator, which is a kind of telecom company;
- London is located in the UK, which is a part of Europe;
- A CTO is a kind of director.

These are precisely the kinds of relations which can be represented and reasoned over using semantic web technology.

9.1.2   Ontologies

Formal knowledge representation (KR) is about building models[4] of the world (of a particular state of affairs, situation, domain or problem), which allow for automatic reasoning and interpretation. Such formal models are called *ontologies*, whenever they (are intended to) represent a shared conceptualisation (e.g. a basic theory, a schema, or a classification). Ontologies can be used to provide formal semantics (i.e. machine-interpretable meaning) to any sort of information: databases, catalogues, documents, web pages, etc. Ontologies can be used as semantic frameworks: the association of information with ontologies makes such information much more amenable to machine processing and interpretation. This is because formal ontologies are represented in logical formalisms, such as OWL (Dean *et al*. 2004), which allow automatic inferencing over them and over datasets aligned to them. An important role of ontologies is to serve as schemata or 'intelligent' views over information resources[5]. Thus they can be used for indexing, querying, and reference purposes over non-ontological datasets and systems, such as databases, document and catalogue management systems. Because ontological languages are formal logical languages, ontologies allow inference of facts which are not explicitly stated from the explicit data. In this way, they can improve the interoperability and the efficiency of the usage of arbitrary datasets.

An ontology can be characterized as comprising a 4-tuple[6]:

$$O = < C, R, I, A >$$

where C is a set of **classes** representing *concepts* we wish to reason about in the given domain (invoices, payments, products, prices, . . .); R is a set of **relations** (also referred to as **properties** or **predicates**) holding between (instances of) those classes (Product *hasPrice* Price); I is a set of **instances**, where each instance can be an instance of one or more classes and can be linked to other instances or to **literal** values (strings, numbers, . . .) by relations (*product23* compatibleWith product348; *product23* hasPrice €170); A is a set of **axioms** (if a product has a price greater than €200, then shipping is free).

---

[4] The typical modelling paradigm is mathematical logic, but there are also other approaches, rooted in information and library science. KR is a very broad term; here we only refer to one of its main streams.

[5] Comments in the same spirit are provided in (Gruber 1992) also. This is also the role of ontologies on the semantic web.

[6] By tuple is meant an ordered list. A more formal and extensive mathematical definition of an ontology is given, for example, in (Ehrig *et al.* 2005). The characterisation offered here is suitable for the purposes of our discussion, however.

The ontologies can be classified as *lightweight* or *heavyweight* according to the complexity of the KR language used. Lightweight ontologies allow for more efficient and scalable reasoning, but do not possess the high predictive (or restrictive) power of the full-bodied concept definitions of heavyweight ontologies. The ontologies can be further differentiated according to the sort of conceptualisation that they formalise: *upper-level* ontologies model general knowledge, while *domain-* and *application-ontologies* represent knowledge about a specific domain (e.g. medicine or sport) or a type of applications (e.g. knowledge management systems). Basic definitions regarding ontologies can be found in (Gruber 1992; 1993) and (Guarino and Giaretta 1995; Guarino 1998).

Finally, ontologies can be distinguished according to the sort of semantics being modelled and their intended usage. The major categories from this perspective are:

- *Schema-ontologies*: ontologies which are close in purpose and nature to database and object-oriented schemata. They define classes of objects, their appropriate attributes and relationships to objects of other classes. A typical usage of such an ontology is that large sets of instances of the classes are defined and managed. Intuitively, a class in a schema ontology corresponds to a table in an RDBMS (relational database management system); a relation – to a column; an instance – to a row in the table for the corresponding class.
- *Topic-ontologies*: taxonomies which define hierarchies of topics, subjects, categories, or designators. These have a wide range of applications related to classification of different things (entities, information resources, files, web pages, etc.) The most popular examples are library classification systems and taxonomies, which are widely used in the KM field. Yahoo and DMoz[7] are popular large scale incarnations of this approach in the context of the web. A number of the most popular taxonomies are listed as encoding schemata in Dublin Core (DCMI 2005).
- *Lexical ontologies*: lexicons with formal semantics, which define lexical concepts[8], word-senses and terms. These can be considered as semantic thesauruses or dictionaries. The concepts defined in such ontologies are not instantiated, rather they are directly used as reference, e.g. for annotation of the corresponding terms in text. WordNet is the most popular general purpose (i.e. upper-level) lexical ontology; it is discussed in greater detail in Section 2 of Chapter 10.

This chapter is mostly concentrated on annotation, indexing and retrieval with respect to schema-ontologies and KBs built with respect to them. The usage of taxonomies for document categorization is discussed in Chapter 8. The usage of lexical ontologies for IR is discussed in Chapter 10.

PROTON (Terziev *et al.* 2004) is a light-weight upper-level schema-ontology developed in the scope of the SEKT project (Davies *et al.* 2005). It is used in the KIM system for semantic annotation, indexing and retrieval. We will also use it for ontology-related examples within this section. PROTON is encoded in OWL Lite and defines about 300 classes and 100 properties, providing good coverage of named entity types and concrete domains (i.e. modelling of concepts such as people, organizations, locations, numbers, dates, addresses, etc.) A snapshot of the PROTON class hierarchy is given in Figure 9.1.

9.1.3   Knowledge bases and semantic repositories

Knowledge base (KB) is a broader term than ontology. Similarly to an ontology, a KB is represented in a KR formalism, which allows automatic inference. It could include multiple axioms, definitions, rules, facts, statements, and any other primitives. In contrast to ontologies, however, KBs are not intended to represent a shared or consensual conceptualisation. Thus, ontologies are a specific sort of KB. Many KBs can be split into ontology and instance data parts, in a way analogous to the splitting of schemata and concrete data in databases. A broader discussion on the different terms related to ontology and semantics can be found in section 3 of (Kiryakov 2006).

---

[7] http://www.yahoo.com and http://www.dmoz.org respectively.

[8] We use 'lexical concept' here as some kind of a formal representation of the meaning of a word or a phrase. In Wordnet, for example, lexical concepts are modelled as synsets (synonym sets), while word-sense is the relation between a word and a synset.

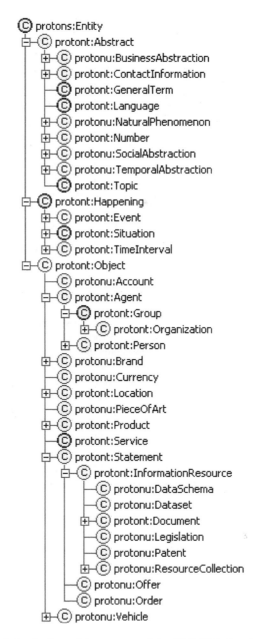

Figure 9.1 A view of the top part of the PROTON class hierarchy

Semantic repositories[9] allow for storage, querying, and management of structured data with respect to formal semantics; in other words, they provide KB management infrastructure. Semantic repositories can serve as a replacement for database management systems (DBMS), offering easier integration of diverse data and more analytical power. In a nutshell, a semantic repository can dynamically interpret

---

[9] 'Semantic repository' is not a well-established term. A more elaborate introduction can be found at http://www.ontotext.com/inference/semantic_repository.html.

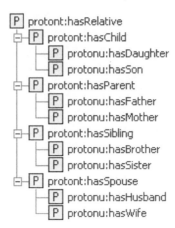

Figure 9.2   A hierarchy of family relationships

metadata schemata and ontologies, which determine the structure and the semantics of data and of queries against that data.

Compared with the approach taken in relational DBMSs, this allows for: (i) easier changes to and combinations of data schemata; and (ii) automated interpretation of the data. As an example, let us imagine a typical database populated with the information that John is a son of Mary. It will be able to 'answer' just a couple of questions: *Who are the son(s) of Mary?* and *Of whom is John the son?* Given simple family-relationships ontology (as the one in PROTON, see Figure 9.2), a semantic repository could handle a much bigger set of questions. It will be able infer the more general fact that John is a child of Mary (because hasSon is a sub-property of hasChild) and, even more generally, that Mary and John are relatives (which is true in both directions, because hasRelative is defined to be symmetric in the ontology). Further, if it is known that Mary is a woman, a semantic repository will infer that Mary is the mother of John, which is a more specific inverse relation. Although simple for a human to infer, the above facts would remain unknown to a typical DBMS and indeed to any other information system, for which the model of the world is limited to data-structures of strings and numbers with no automatically interpretable semantics.

9.2   Semantic Web

Semantic Web is about adding formal semantics to the content on the WWW (and private web-based systems such as company intranets). The current web contains mostly HTML, meant for human interpretation; the machine-processable metadata there is related to the formatting and the layout of the text, but not to its meaning. Berners-Lee (1999) introduced the notion of the Semantic Web thus: 'If HTML and the Web made all the online documents look like one huge **book**, RDF, schema, and inference languages will make all the data in the world look like one huge **database**'. The two major design rationales of the Semantic Web are outlined at the web site of W3C[10] as follows:

- Common formats for the interchange of data;
- Language for recording how the data relates to real-world objects.

As an Internet standardization authority, W3C promotes Semantic Web as one of its key initiatives.

---

[10] http://www.w3.org/2001/sw/

9.2.1   Semantic web and semantic search

One of the anticipated benefits of the Semantic Web is that information held on it will be accessible through semantic search engines and that such search engines will offer a more effective search capability than that offered by today's keyword-based search engines. Thus, the relationship between Semantic Web and semantic search is two-fold[11]:

- Semantic Web standards and technologies can be used to enable Semantic Search;
- Semantic Search can be the killer application of the Semantic Web.

One of the open issues for the Semantic Web is the establishment of effective information access methods.[12] For instance, classical IR has a single, basic, well-established and understood way of defining and satisfying the information need[13]: it is defined as a set of words (tokens) of interest and is satisfied with a list of documents relevant to those words. As regards the domain of relational databases, the information need is defined as an SQL (Structured Query Language) query and satisfied, typically, through a result table. An obvious question about the Semantic Web is:

How is the information need defined and satisfied within the Semantic Web?

The above question is stated incorrectly – like the current web, the Semantic Web cannot be expected to have a single access method, and therefore a single approach for the definition of the information need. The following questions thus seem more relevant:

- How does the Semantic Web extend the existing information access methods?
- What new information access methods become feasible?

Proposals which answer the above question should unite the following elements: proven user needs, a sound scientific theory, and a robust technology, which can implement efficient applications based on the theory. As outlined in Section 9.1.1, the models employed by the classical IR engines are far from perfect: the information need is poorly defined and imprecisely satisfied. However, search engines are popular because they meet a number of conditions that are critical for a wide acceptance of an information access method in the web context:

(i) To significantly improve the efficiency and effectiveness of users to access information on the web;

(ii) To ensure that no additional skills, effort, discipline, good will, or correctness are required from information providers;

(iii) To offer predictable behaviour and performance.

9.2.2   Basic semantic web standards: RDF(S) and OWL

A family of markup and KR standards were developed, under W3C-driven community processes, as a basis for the Semantic Web. RDF (Klyne and Carroll 2004) is a metadata representation language, which serves as a basic data-model for the Semantic Web. It allows resources to be described through relationships to other resources and literals. The resources are defined through unified resource identifiers (URIs) (, as in XML; e.g. URL). The notion of resource is virtually unrestricted; anything can

---

[11] Though note that the two are not interdependent: for instance, ontology-based semantic search can be used in knowledge management systems which do not use Semantic Web standards or technology.

[12] An information access method could be considered as a paradigm for: (i) definition of an information need (the question); and (ii) the way that need is satisfied (the answer/result type). This notion is related to the one of "models", from Section 1.2 of Chapter 1.

[13] Of course, this statement reflects an oversimplified view on what IR really does, but still this is the level of understanding that most users have, and it allows them to make use of the technology.

be considered as a resource and described in RDF: from a web page or a picture published on a web to concrete entities in the real world (e.g. people, organisations) or abstract notions (e.g. the number π and the musical genre Jazz). Literals (again as in XML) are any concrete data values e.g. strings, dates, numbers, etc. The main modelling block in RDF is the statement – a triple <Subject, Predicate, Object>, where:

- Subject is the resource, which is being described;
- Predicate is a resource, which determines the type of the relationship;
- Object is a resource or a literal, which represents the 'value' of the attribute.

A set of RDF triples can be seen as a graph, where resources and literals are nodes and each statement is represented by a labelled arc (the Predicate or relation), directed from the Subject to the Object. So-called blank nodes can also appear in the graph, representing unique anonymous resources, used as auxiliary nodes. A sample graph, which describes a web page, created by a person called Adam, can be seen in Figure 9.3.

Resources can belong to (formally, be *instances* of) classes – this can be expressed as a statement through the rdf:type system property as follows: <resource, rdf:type, class>. Two of the system classes in RDF are rdfs:Class and rdf:Property. The instances of rdf:Class are resources which represent classes, i.e. those resources which can have other resources as instances. The instances of rdf:Property are resources which can be used as predicates (relations) in triple statements.

The most popular format for encoding RDF is its XML syntax, (Becket 2004). However, RDF can also be encoded in a variety of other syntaxes. The main difference between XML and RDF is that the underlying model of XML is a tree of nested elements, which is rather different from the graph of resources and literals in RDF.

RDF Schema (RDFS), (Brickley and Guha 2000), is a schema language, which allows for definition of new classes and properties. OWL, (Dean *et al*. 2004), is an ontology language, which extends RDF(S)[14] with means for more comprehensive ontology definitions. OWL has three dialects: OWL-Lite, OWL-DL, and OWL-Full. Owl-Lite is the least expressive of these dialects but the most amenable to efficient reasoning. Conversely, OWL-Full provides maximal expressivity but is undecidable[15]. OWL-DL can be seen as a decidable sub-language inspired by the so-called description logics.

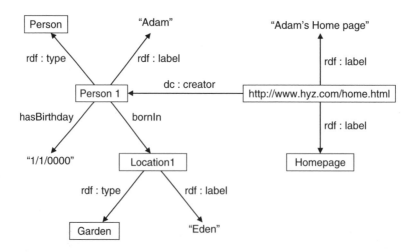

Figure 9.3   RDF graph describing Adam and his home page

---

[14] RDF(S) is a short name for the combination of RDF and RDFS.

[15] An undecidable logical language is one for which it is a theoretical impossibility to build a reasoner which can prove all the valid inferences from any theory expressed in that language.

These dialects are nested such that every OWL-Lite ontology is a legal OWL-DL ontology and every OWL-DL ontology is a legal OWL-Full ontology.

Below we briefly present all the modelling primitives used in the PROTON ontology, comprising most of the RDFS constructs and few of the simplest ones from OWL:

- All resources, including classes and properties, may have titles (literals[16], linked through property rdf:label) and descriptions or glosses (literals linked through rdf:comment);
- Classes can be defined as sub-classes, i.e. specializations, of other classes (via rdf:subClassOf). This means that all instances of the class are also instances of its super class. For example, in PROTON City is a sub-class of Location.
- In OWL, properties are distinguished into object- and data-properties (instances respectively of owl:ObjectProperty and owl:DataProperty). The object-properties are binary relationships, relating entities to other entities. The data-properties can be considered as attributes – they relate entities to literals.
- Domains and ranges of properties can be defined. A domain (rdfs:domain) specifies the classes of entities to which this property is applicable. A range (rdfs:range) specifies the classes of entities (for object-properties) or data-types of the literal values (in case of data-properties), which can serve as objects in statements predicated by this property. For instance, the property hasSister might typically have the class Person as its domain and Woman as its range. Whenever multiple classes are provided as domain or range for a single property, the intersection of those classes is used.
- Properties can be defined as sub-properties, i.e. specializations of other properties (via rdf:subPropertyOf). Imagine that there are two properties, p1 and p2, for which <p1,subPropertyOf,p2>. The formal meaning of this statement is that for all pairs for which p1 takes place, i.e. <x,p1,y>, p2 also takes place, i.e. <x,p2,y> is also true. The hierarchy of family relationships discussed above (see Figure 9.2) provides a number of intuitive examples of sub-properties.
- Properties can be defined as a symmetric (via owl:SymmtericProperty) and transitive (via owl:TransitiveProperty) ones. If p1 is a symmetric property then whenever <x,p1,y> is true, <y,p1,x> is also true. If p2 is a transitive property and <x,p2,y> and <y,p2,z> are true, it can be concluded that <x,p2,z> is also true. hasRelative is an example of a property which is both symmetric and transitive.
- Object-properties can be defined to be inverse to each other (via owl:inverseOf). This means that if <p1,inverseOf,p2> then, whenever <x,p1,y> holds, <y,p2,x> can be inferred and vice versa. An obvious example is <hasChild, owl:inverseOf, hasParent>.

9.3   Metadata and Annotations

Metadata is a term of wide and sometimes controversial or misleading usage. From its etymology, metadata is 'data about data'. Thus, metadata is a role that certain (pieces of) data could play with respect to other data. Such an example could be a particular specification of the author of a document, provided independently from the content of the document, say, according to a standard like Dublin Core (DC), (DCMI 2005). RDF has been introduced as a simple language that is to be used for the assignment of semantic descriptions to information resources on the web. Therefore an RDF description of a web page represents metadata. However, an RDF description of a person, independent from any particular documents (e.g., as a part of an RDF(S)-encoded dataset), is not metadata – this is data about a person, not about other data. In this case, RDF(S) is used as a KR language. Finally, the RDFS definition of the class Person, will typically be part of an ontology, which can be used to structure datasets and metadata, but which is again not a piece of metadata itself.

A term which is often used as a synonym for metadata, particularly in the natural language processing (NLP) community, is *annotation*. In this section, we discuss annotation of documents in

---

[16] Recall that literals are values such as strings or numbers.

general, while the next section presents a discussion of 'semantic annotation' in the Semantic Web context.

Annotations on text documents can be distinguished into two groups according to their scope:

- *Document-level annotations*, which refer to the whole document. Such examples are the DC elements (Title, Subject, Creator, etc.);
- *Character-level annotations*, which refer to a fragment of a document, determined by start and end characters. An example might be a comment attached to a particular part of a document. Character-level annotations are usually meant when the term 'annotation' is used for text documents without further clarification.

It is worth mentioning that *hyperlinks* can be considered as a specific sort of character-level annotation, when the metadata is, essentially, a reference to another document or part of document.

Further, annotations can also be distinguished with respect to the way in which they are attached to the text. The basic choices here are:

- *Embedded markup*, when the annotations are incorporated within the document. In this case the metadata is bundled together with the data. Examples are markup languages such as HTML, where the annotations are specified through pairs of start and end tags, e.g. abc<tag>de</tag>gf. When document-level annotations have to be represented this way, they are sometimes attached in a special section at the start or end of the document – one example is the <head> section in the HTML files. An example of character-level embedded annotations is a footnote.
- *Standoff references*, when the annotations are maintained separately from the document to which they refer. In the case of character-level annotations, the reference should also specify the specific part of the document. One approach for this is based on position, e.g. offset and length; another possibility is the usage of some sort of anchoring and linking mechanism. An example of specification based on standoff annotations is TIPSTER (Grishman 1997), a US government-funded effort to advance the state of the art in text handling technologies.

The different types of annotation are shown diagrammatically in Figure 9.4. In his thesis on architectures for language engineering Cunningham (2000), provides an overview of various annotation models and discusses their advantages and disadvantages in the context of text processing systems and applications. Similar analysis, but in the context of open hypermedia systems (OHS), can be found in (van Ossenbruggen *et al*. 2002). Here we will only briefly mention few of the main characteristics of the embedded and the standoff models:

- Embedded markup is not applicable in cases when the author of the metadata has no write-permission to the document;
- Standoff annotations may become inconsistent in the event of change to the document to which they refer;

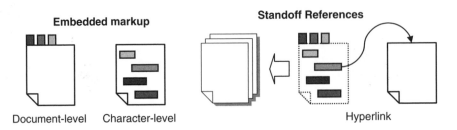

Figure 9.4 Types of Annotations

- Access to embedded annotations requires processing (e.g. parsing) of the documents. Thus, such annotations are not appropriate for applications where random (non-sequential) access to the annotations is important. Conversely, standoff annotations can be maintained and queried efficiently in structured form (e.g. in a database);
- Embedded annotations are simpler to encode, read and manage when the volume of the markup is relatively small. However, they are inappropriate when the volume of the markup becomes comparable to or bigger than that of the text itself;
- Tagging mechanisms based on embedded annotations have difficulties in handling overlapping (as opposed to nested) annotations;
- Embedded annotations should always be distributed together with the document, which can cause IPR issues, unnecessary redundancy or conflict when multiple sets of annotations are available for one and the same document.

9.4   Semantic Annotations: the Fibres of the Semantic Web

If we abstract the current web away from the transport, content type, and content formatting aspects, it could be regarded as a set of documents with some limited metadata, attached to them (document-level annotations about title, keywords, etc.), and with hyperlinks between the documents (see the left-hand side part of Figure 9.5).

What does the Semantic Web add to this picture? Essentially, *semantic* metadata of different kinds, both on the document- and the character-level. Figure 9.5 compares links on the current web to those on the semantic web. Typically, the Semantic Web has a greater number of more meaningful annotations, as compared with the current WWW. Many of those annotations represent links to external knowledge, which constitute a new sort of connectivity that is not presented on Figure 9.5, but is extensively discussed in this section (Figures 9.6 and 9.7).

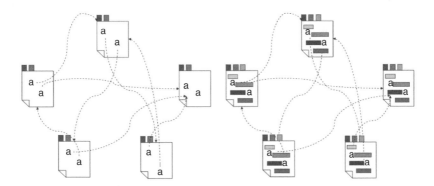

Figure 9.5   The Current WWW (left) and the Semantic Web (right)

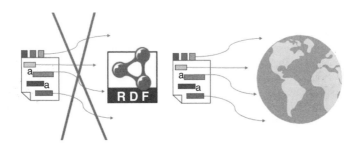

Figure 9.6   Metadata about the world, not about RDF

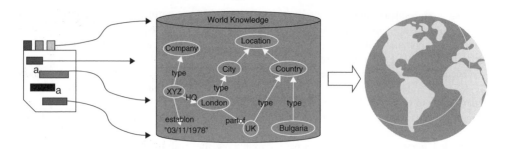

Figure 9.7   Metadata Referring to World Knowledge

In order to uncover the added value of the Semantic Web, it is crucial to elaborate a little further regarding the nature of semantic metadata. Suppose, we add a tag <2134> to some portion of a document as follows '... Abc <2134>xyz...'. Is this metadata useful? Can we call it Semantic? Without further assumptions, the answers are negative. In order to have metadata useful in a Semantic Web context, it should mean something, i.e. the symbols (or expressions or references) that constitute it should allow further interpretation. Interpretation in this context means allowing the assigning of some additional information to the symbols. It is important to realize that interpretation is only possible with respect to something; to some domain, model, context, (possible) world. This is the domain that (the interpretations of) the symbols are 'about'. Obviously, annotations in RDF(S), OWL, or some other language refer to a model of the world. Annotations can be expressed in RDF(S), but they are not *about* RDF(S), as depicted in Figure 9.6.

Further, the metadata can hardly refer to (or be interpreted directly with respect to) the world. Such references cannot be formal and unambiguous. What the semantic metadata can be expected to refer to directly is a knowledge base (KB), a formal model of (some aspect of) the world, as depicted in Figure 9.7. Such a KB specifies some world knowledge which serves as a semantic link from the metadata to the world. Note, that in the Semantic Web context such a KB can be as scattered and heterogeneous as the metadata is. Guha and McCool (2003), consider this KB itself to be the Semantic Web: 'the Semantic Web is not a web of documents, but a web of relations between resources, denoting real world objects'. In our view the Semantic Web is the combination of the KB and the semantic annotations referring to it (not just the KB).

For automatic processing, the interpretations of metadata should be performed automatically by machines in a strict and predictable fashion. This requires a formal definition of how the metadata should be interpreted and, because of this, a formal definition of the context. Assuming that one and the same context can be modelled in different ways, allowing different (and potentially ambiguous) interpretations, what has to be specified is the conceptualisation – as defined in (Gruber 1993): 'a conceptualisation is an abstract, simplified view of the world that we wish to represent for some purpose'. This is where ontologies are used to act as logical theories for the 'formal specification of a conceptualisation' (again in Gruber 1993, see also Section 9.1.2.)

Although the above discussion is informally presented, we consider it rather important for the realisation of the Semantic Web. It is the intuition of the authors that the KR and modelling issues related to the development or generation of useful semantic annotations require more specific attention. RDF(S) and OWL are designed to serve well for data modelling in as diverse and heterogeneous environments as possible. Thus, they provide very little modelling guidance and constraints. For instance, an RDF(S) annotation of an HTML page can include at the same time a definition of the class Person (which is a piece of ontological knowledge), the description of a specific person Mr X (which should normally be world knowledge, part of a KB) and the fact that this person is an author of the web page (which is the only piece of actual metadata describing the web page). We believe that the development of real world Semantic Web applications require some concrete knowledge modelling commitments to be made and the corresponding design and representation principles to be set out. Sections 9.5 and 9.6 below present some specific modelling patterns and applications based on them.

Linguistic approaches for automatic generation of semantic annotations are discussed in Chapter 10 of this volume.

9.5   Semantic Annotation of Named Entities

Semantic annotation of named entities (SANE) is a specific metadata generation process, aiming to enable new information access methods and to extend some of the existing ones. The annotation schema, discussed here, is based on the intuition that named entity (NE, see Section 9.5.1) references constitute an important part of the semantics of the documents in which they occur. Moreover, via the use of KBs of external background knowledge, those entities can be related to formal descriptions of themselves and related entities and thus provide more semantics and connectivity to the web.

In a nutshell, SANE is character-level annotation of mentions of entities in the text with references to their semantic descriptions (as presented in Figure 9.8). This sort of metadata provides both class-level and instance-level information about the entities. Such semantic annotations enable many new types of applications: highlighting, indexing and retrieval, categorization, generation of more advanced metadata, smooth traversal between unstructured text and available relevant structured knowledge. Semantic annotation is applicable for any sort of text – web pages, non-web documents, text fields in databases, etc. Further knowledge acquisition can be performed on the basis of the extraction of more complex dependencies – analysis of relationships between entities, event and situation descriptions, etc.

To use SANE in information retrieval, two basic tasks need to be addressed:

1. Identify and mark references to named entities in textual (parts of) documents and link these references to descriptions of the entities in a KB[17];
2. Index and retrieve documents with respect to the entities they refer to.

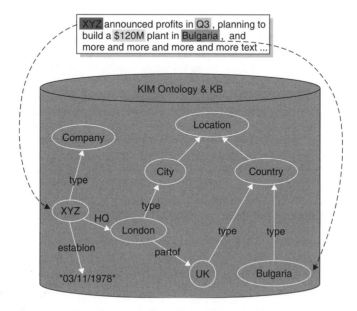

Figure 9.8   Semantic Annotation

---

[17] There can also be references to various ontologies. On one hand, there could be a direct reference from the annotation to the class of the entity, as it is defined in an ontology. On the other, some instances can be part of ontologies.

The first task resembles at the same time a basic press-clipping exercise, a typical IE[18] task, and hyperlinking. The resulting annotations then provide the semantic data for document enrichment and presentation, which can be further used to enhance information retrieval (the second task) as discussed in Section 9.6.

9.5.1 Named entities

In the Natural Language Processing (NLP) field, and particularly the Information Extraction (IE) tradition, **named entities** (NE) are considered: *people, organizations, locations*, and others, referred to by name, (Chinchor and Robinson 1998). In a wider interpretation, these can also include scalar values (*numbers, dates, amounts of money), addresses*, etc.

NEs should be handled in a different, special way because of their different nature and semantics compared with general words (terms, phrases, etc.). While NEs denote particulars (individuals or instances), the general terms can denote universals (concepts, classes, relations, attributes). Even a basic level of formal semantic definition of general word senses involves modelling of lexical semantics and common sense[19]. On the other hand, useful descriptions of named entities can be modelled on the basis of much simpler and more specific 'factual' world knowledge.

9.5.2 Semantic annotation model and representation

In this section we discuss the structure and the representation of SANE, including the necessary knowledge and metadata. The basic prerequisites for the representation of semantic annotations are:

- An ontology, defining the entity classes and allowing unambiguous references to those;
- Entity identifiers, which allow these to be distinguished and linked to their semantic descriptions;
- A knowledge base (KB) with entity descriptions.

Entity descriptions actually make up the non-ontological part of formal knowledge in the semantic repository. The entity descriptions represent a KB, a body of instance knowledge or data. Such KB can either be available as pre-populated background knowledge and/or be extended through information extraction from the documents.

As with other sorts of annotations, a major question about the representation is: 'To embed or not to embed?' There are a number of arguments, giving evidence that semantic annotations are best decoupled from the content they refer to. One key reason for this is the ambition to allow for dynamic, user-specific, semantic annotations – conversely, embedded annotations become a part of the content and may not change according to the interest of the user or to the context of usage. Further, complex embedded annotations would have a negative impact on the volume of the content and could complicate its maintenance – e.g. imagine that a page with three layers of overlapping semantic annotations needs to be updated without compromising their consistency.

Given that semantic annotations should preferably be kept separate from the content to which they refer, the next question is whether or not (or to what extent) the annotations should be integrated with the ontology and the KB. It is the case that such an integration seems profitable – it would be easier to keep the annotation in sync with the class and entity descriptions. However, there are at least three important considerations to be made in this regard:

---

[18] Information extraction (IE) is a relatively young discipline within Natural Language Processing (NLP), which conducts partial analysis of text in order to extract specific information, (Cunningham 1999). IE and named entities are discussed in greater detail in Section 3 of Chapter 10.

[19] WordNet is the most popular large scale lexical database, providing partial descriptions of the word senses in the English language. It can be considered also as a lexical ontology or a KB. See the discussion around WordNet and its usage for semantic annotation and IR in Section 2 of Chapter 10.

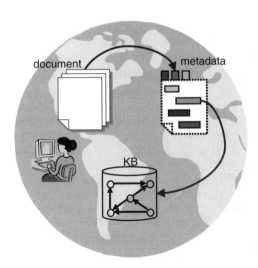

Figure 9.9   Distributed Heterogeneous Knowledge

- *Number and the complexity of the annotations*:   these differ from those of the entity descriptions – the annotations are simpler, but more numerous than the entity descriptions. Even considering middle-sized corpora of documents, the number of annotations typically reaches tens of millions. Suppose that 10 M annotations are stored in an RDF(S) store together with 1 M entity descriptions. Suppose also that on average annotations and entity descriptions are represented with 10 statements each. The difference, regarding the inference approaches and the hardware that is capable of efficient reasoning and access to a 10 M-statement semantic repository and to a 110 M-statement repository, is considerable.
- *Separation of concerns*:   if the world knowledge (ontology and instance data) and the document-related metadata are kept independent, this would mean that for one and the same document, different extraction, processing, or authoring methods will be able to deliver alternative metadata, referring to one and the same knowledge store.
- *Distributivity*:   importantly, it should be possible that the ownership and the responsibility for the metadata and the knowledge are distributed. In this way, different parties can develop and separately maintain the content, the metadata, and the knowledge.

On the basis of these arguments, we propose a general model which allows for decoupled representation and management of the documents, the metadata (annotations), and the formal knowledge (ontologies and instance data), as illustrated in Figure 9.9

9.6   Semantic Indexing and Retrieval

As already mentioned, one of the key tasks which can be performed on top of semantic annotations is indexing and retrieval of documents with respect to the corresponding semantic features. This is a modification of the classical IR task – documents are retrieved on the basis of relevance to concepts instead of words. However the basic model is quite similar – a document is characterised by the bag of tokens[20] which constitute its content, disregarding its structure. While the basic IR approach considers

---

[20] Sometimes 'token' is used with a specialized meaning in the NLP field – in essence, tokens are the elements of the text, as they are separated by white spaces and punctuation (the delimiters are also considered tokens). (Kiryakov and Simov (1999) introduce the term 'atomic text entities' (ATE) as a general notion of token in IR context, to avoid ambiguity with the NLP usage of 'token'.

the word lemmata (base forms) or stems as tokens, there have been considerable efforts for the last decade related to indexing with respect to two sorts of higher-level semantic features, namely:

- Word-senses, lexical concepts, references to controlled vocabularies;
- Named entities (including numbers, dates, etc.).

Both types of indexing can serve as a basis for cross-language IR (see Chapter 11). Lexical concepts in one language can be related to such in another language or to some sort of interlingua – this was one of the main objectives for the development of the EuroWordNet[21] lexical ontology. On the other hand, once properly recognised, the named entities references are language independent (both the mentions of 'London' and 'Llundain' will be tagged with one and the same entity identifier).

9.6.1 Indexing with respect to lexical concepts

Many of the words in natural languages are polysemous, i.e. they can have more than one meaning. For instance, the word 'bank' can denote both a financial institution and a river bank. In WordNet (and other lexical ontologies) this linguistic phenomena is handled through association of the word with different lexical concepts[22] representing their different meanings. One of the main problems in the course of semantic annotation with respect to lexical ontologies is word-sense disambiguation (WSD) – the selection of the correct lexical concept, which represents the meaning of the word in the specific context. Once WSD is performed and documents are annotated with respect to lexical concepts, indexing and retrieval with respect to them solves the problem with query term polysemy and index term synonymy mentioned in Section 9.1.1. WSD and the usage of lexical resources for IR are discussed in Chapter 10. Here we will comment only on the principle advantages that such indexing can provide, given a lexical ontology with super-concept/sub-concept relationships, as well as possible indexing and retrieval techniques, as discussed in Kiryakov and Simov (1999). Suppose the following IR setup with hierarchically structured feature space:

- The documents are indexed with respect to (occurrence in the documents of references to) lexical concepts.
- The lexical concepts are properly related to the corresponding more general concepts (hypernyms) and less general concepts (hyponyms).
- The query is specified through lexical concepts (either because the user directly selected them, or because a 'bag of keywords' query has been semantically annotated).

Let us define *hyponym-matching* as a retrieval operator which matches more general concepts in the query with more specific ones in the document. Using such an operator a query containing the concept for 'bird' will match documents referring to 'duck' or 'eagle'; this follows the intuition that if a specific species of bird is mentioned, then this is also a reference to bird. On the contrary, a document which refers to 'bird' will not be hyponym-matched to a query including 'hen'; intuitively, there is a no guarantee that a document about birds has something to say about 'hens'.

First, let us note that hyponym-matching offers clear benefits for the users as compared with the mainstream IR techniques (e.g. vector-space model using word lemmata as features). In a typical search engine, documents mentioning 'duck', but not mentioning explicitly 'bird', will not be returned as a result for a query for 'bird' – the lack of hyponym-matching leads to poor recall. To fix this 'manually' the user should include in the query all possible species of birds and their synonyms, which would be an unreliable and inefficient exercise. Further, the words from the expanded query will match words in the text, which may be used in a different meaning there – for example, the query expansion will

---

[21] http://www.illc.uva.nl/EuroWordNet/

[22] The lexical concepts in WordNet are called synsets (from synonym set, as already mentioned).

contain 'duck' which in a specific text could have been used for cloth[23]. The effect is a reduction in precision.

An interesting question is how hyponym-matching can be integrated into an existing system, e.g. a vector-space-based probabilistic model. Let us see first how indexing and retrieval can be performed with respect to lexical concepts (disregarding hyponymy). Suppose that before being indexed the documents are pre-processed as follows: (i) they are semantically annotated with lexical concepts; and (ii) each occurrence of a word (or a multi-word token) is replaced by the identifier of the corresponding concept. The queries can be pre-processed in the same manner. In a simplified world, this is an easy way to make an existing IR engine implement semantic search – the vector-based similarity between queries and documents should be a good model for relevance, as when documents are indexed with respect to word lemmata.

One straightforward solution for extending this model with hyponym-matching is query expansion. Each of the concepts in the query can be replaced with the set of itself plus all of its hyponyms (sub-concepts). This approach is simple and can prove sufficient in many contexts, but one should be aware of its disadvantages:

- A concept with a bigger set of hyponyms will gain bigger weight in the relevance calculation as opposed to such with no or just a few hyponyms. This problem can partly be solved if the IR engine supports disjunction (i.e. OR operator) – however, this is not a natural feature for the engines based on probabilistic models.
- The set of all sub-concepts of all the concepts in the query, could grow to thousands of elements, which can cause problems with the performance of the IR engine.

An alternative approach (let us name it *hypernyms-indexing*) is to modify the indexing strategy, so, that the documents are indexed with respect to the hypernyms of the concepts they refer to. In such case, a document mentioning the concept 'eagle' will also appear in the reverse index for its super-concept (e.g. 'bird') and the super-concept of the super-concept (e.g. 'animal') and so forth following the subsumption chain to the top concept. In cases when there is no control of the engine's indexing strategy, this effect can be achieved if each word gets annotated not only with the specific lexical concept, corresponding to its meanings, but also with the super concepts. Then in the document pre-processing phase, the identifiers of the super-concepts will also appear in the document index.

Query expansion is no longer necessary, when hypernym-indexing is involved, because the relevance of the document to the more general concepts has been reflected in the indices. A query for 'bird' will retrieve a document which only mentions 'eagle', without any need for query modification. The problems of this approach can be summarised as follows:

- The overall size of the indices will grow, due to the fact that each token in the document appears in multiple indices. The growth can be estimated as a factor close to the average depth of the of the hypernymy hierarchy;
- The indices for the most general concepts will get huge and cause efficiency problems.

These problems can be addressed to some extent if limited query- or index-expansion is performed. For instance, one can put a constraint on the number of levels of the hypernyms to be considered or to the total number of hyponyms to be used for expansion. In all cases, it should be clear that the adaptation of a probabilistic IR model (tuned for a flat feature set) to deliver good performance for a hierarchically structured feature set is far from trivial. Experiments show that the adaptation of a "general-purpose" lexical ontology for this task is problematic, (Voorhees 1998). An interesting implementation is reported in (Mahesh *et al*. 1999): a large-scale lexical ontology, built for the specific IR task, is used for the IR engine built into one of the major RDBMS systems. The evaluation of the system on some of the tasks of the TREC competition, prove clear performance benefits for this approach.

---

[23] According to WordNet 2.1 (http://wordnet.princeton.edu/perl/webwn) one of the meanings of 'duck' is: a heavy cotton fabric of plain weave; used for clothing and tents.

9.6.2   Indexing with respect to named entities

A recent large-scale human interaction study on a personal content IR system of Microsoft, (Dumais *et al*. 2003), demonstrates that, at least in some cases, named entities are central to user needs:

> The most common query types in our logs were People/Places/Things, Computers/Internet, and Health/Science. In the People/Places/Things category, names were especially prevalent. Their importance is highlighted by the fact that 25% of the queries involved people's names, which suggests that people are a powerful memory cue for personal content. In contrast, general informational queries are less prevalent.

As the volume of web content grows rapidly, the demand for more advanced retrieval methods increases accordingly. Based on semantic annotation of named entities (SANE), efficient indexing and retrieval techniques can be developed, involving an explicit handling of the named entity references.

In a nutshell, SANE could be used to index both 'NY' and 'N.Y.' as occurrences of the specific entity 'New York', as though a unique identifier for that entity occurred in the text in place of the different syntactic variations of the strings used to denote it. Since many present systems do not involve entity recognition, they will index on 'NY' (for the former), and 'N' and 'Y' (for the latter), which demonstrates well some of the problems with the keyword-based search engines. It should be noted that there have been previous attempts to conflate synonymous terms to the same unique index token (e.g. Robertson *et al*. 1991), though these have tended to lack the natural language processing technology deployed more recently. A survey of recent approaches to semantic annotation and human language technology can be found in Bontcheva *et al*. (2006).

Given metadata-based indexing of content, advanced semantic querying becomes feasible. A query against a repository of semantically annotated documents can be specified in terms of restrictions on the entity's type, name, attribute values, and relations to other entities. For instance, a query can request all documents that refer to Person-s that hold some Position-s within an Organisation, and which also restricts the names of the entities or some of their attributes (e.g. a person's gender). Further, semantic annotations can be used to match specific references in the text to more general queries. For instance, a query such as 'Redwood Shores company' could match documents mentioning specific companies such as ORACLE and Symbian, which are located in this town.

Hybrid query modes, such as the one mentioned above, can provide unmatched analytical levels through a combination of:

- Database-like structured queries, extended with the reasoning capabilities of the semantic repositories;
- IR-like probabilistic models.

Although the above sketched enhancements look promising, further research and experimentation are required to determine to what extent and in which way(s) they can improve existing IR systems. It is hard, in a general context, to predict how semantic indexing will combine with the symbolic and the statistical methods currently in use. Large-scale experimental data and evaluation efforts (similar to TREC) are required for this purpose.

9.6.3   Retrieval as spreading activation over semantic network

Rocha *et al*. (2004) describe a search architecture that applies a combination of spread activation and conventional information retrieval to a domain-specific semantic model in order to find concepts relevant to a keyword-based query. The essence of spread activation, as applied in conventional textual searching, is that a document may be returned by a query, even if it contains none of the query keywords. This happens if the document is linked to by many other documents which do contain the keywords.

In the case described here, the user expresses the query as a set of keywords. This query is forwarded to a conventional search engine which assigns a score to each document in its index in the usual way. In addition to a conventional index, the system contains a domain-specific KB, which includes instance nodes pointing to web resources. As usual in RDF, each instance is described in terms of links labelled with properties in compliance with the ontology. The basic assumption is that weightings that express the strength of each instance relation can be derived[24]. Thus the resulting network has, for each relationship, a semantic label and an associated numerical weight. The intuition behind this approach is that better search results can be obtained by exploiting not only the relationships within the ontology, but also the strength of those relationships.

Searching proceeds in two phases: as mentioned, a traditional approach is first used to derive a set of documents from a keyword-based query. As discussed, these documents are annotated with instances nodes – this set of nodes is supplied as an initial set to the spread activation algorithm, using the numeric ranking from the traditional retrieval algorithm as the initial activation value for each node. The set of nodes obtained at the end of the propagation are then presented as the search results. Two case studies are reported, showing how in some cases the combination of the traditional and spread activation techniques performs better than either on its own.

9.7   Semantic Search Tools

This section presents several tools which adopt different approaches for semantic search. These represent concrete implementations which demonstrate applications of semantic search and semantic annotation:

(i) QuizRDF was a relatively early semantic search tool, supporting search over document-level annotations expressed in RDF(S);
(ii) TAP augments conventional search results with relevant information aggregated from distributed semantic information sources;
(iii) KIM is a platform providing infrastructure and services for automatic semantic annotation, indexing, and retrieval of unstructured and semi-structured content;
(iv) Squirrel builds on KIM and a number of other component technologies to provide an advanced semantic search engine.

9.7.1   Searching through document-level RDF annotations – QuizRDF

Even the simplest semantic Web standard, RDF(S), allows for content annotation in ways much more flexible than those of the syntactic annotation formalisms, such as XML:

- RDF(S) is descriptive not prescriptive – XML dictates the format of individual documents; whereas RDF(S) allows the description of any content and the RDF(S) annotations need not be embedded within the content itself.
- More than one RDF(S) ontology, schema, or dataset can be combined to describe the same content; there could be alternative annotations for one and the same document, provided from different sources for different purposes.
- RDF(S) has a well-defined semantics regarding, for example, the sub-class relation; it allows the definition of a set of relations between resources as described in Sections 9.2.2 and 9.4.

The QuizRDF search engine (Davies *et al*. 2003) combines free-text search with a capability to exploit RDF annotation querying. QuizRDF combines search and browsing capabilities into a single

---

[24] A number of different approaches to this derivation are taken and the authors state that no single weight derivation formula is optimal for all application areas.

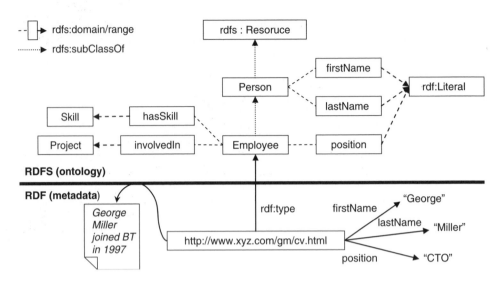

Figure 9.10   Semantic Annotation Datamodel in QuizRDF

tool and allows document-level RDF annotations to be exploited (searched over) where they exist, but will still function as a conventional search engine in the absence of those annotations

A simplified example of semantic annotation in QuizRDF is given in figure 9.10. The data model employed for semantic annotation and the indexing scheme of QuizRDF can be summarized as follows:

- The ontology is represented as an RDF schema. According to the classification in Section 9.1.2, QuizRDF assumes a schema-ontology, e.g. there are classes (such as Person) with their appropriate attributes (properties with literal values, e.g. lastName) and relations to other classes (e.g. hasSkill), which are being instantiated.
- Each document is linked to one or more subjects via the rdf:type property. The RDF resource (identifying the document) is considered to be an instance of the class representing the subject. This way, the CV of a person and the person herself are modelled with a single RDF resource[25].
- A document can be described with properties appropriate for the class of its subject. For example. a document associated with subject Employee can be given attributes firstName, lastName (inherited from the Person super-class) and position.
- Full-text indexing is applied not only to the content of the documents, but also the literal values of any relations. This way a keyword search for 'CTO' will return Miller's CV document, although this string may not appear in the document itself[26].

This data model would have been clearer if the documents and the objects they are about were defined as separate RDF resources, e.g. <cv,isAbout,emp1>, <emp1,type,Employee>, <emp1,firstName,'George'>. On the other hand, this model has all the advantages of being more simple and efficient.

The user enters a query into QuizRDF as they would a conventional search engine. A list of documents is returned, ranked according to the resource's relevance to the user's query using a

---

[25] Note that the modelling approach described in identifying a web page (CV) as an instance of class Employee could be seen as epistemologically naïve: a web page is a document, not a person. In a later version of QuizRDF, a more sophisticated approach is taken and is discussed in Davies *et al.* (2003).

[26] The information about the position of George Miller could have been acquired as background knowledge added to the KB from some database or extracted from another document. The usage of background knowledge is discussed in the next section.

traditional vector space approach (Salton *et al*. 1975). The subjects of the documents returned in response to a query are ascertained and displayed along with the traditional ranked list. By selecting one of the displayed subjects, the user can filter the retrieval list to include only those documents associated with it. QuizRDF also displays the properties and classes related to the selected class. Each class displayed has an associated hyperlink that allows the user to browse the RDF(S) ontology: clicking on the class name refreshes the display to show that class properties and related classes[27] and again filters the results list to show only URIs of the related class. Where properties have literal types (e.g. string) as their range, QuizRDF enables users to query against these properties. An example of search of the documents about the employee George Miller is given in Figure 9.11; Note that, based on the ontology from Figure 9.10, one might imagine there would be a fourth field labelled 'position' – which relations are displayed for instances of a given class is a configurable parameter in QuizRDF and in this case, the relation 'position' has not been selected for display. This can be important in an ontology where classes have many relations, some perhaps inherited from other classes). Another example, would be a search for documents associated with class Painting, which has a property hasTechnique with range of type 'string'. If a set of documents has been returned of class Painting, a user could enter 'oil on canvas' as a value for hasTechnique and the document list would be filtered to show only those documents which have the value 'oil on canvas' for this property.

Thus QuizRDF has two retrieval channels: a keyword query against the text, and a much more focussed query against specific RDF properties, as well as supporting ontology browsing. Searching the full text of the documents ensures the desired high recall in the initial stages of the information seeking process. In the later stages of the search, more emphasis can be put on searching the RDF annotations (property values) to improve the precision of the search.

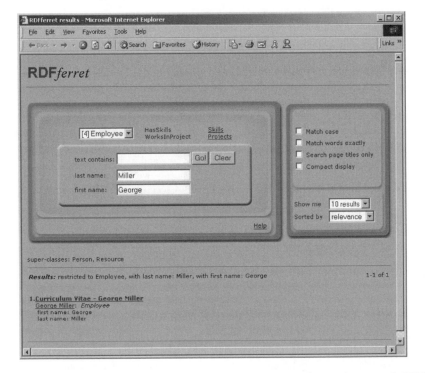

Figure 9.11 QuizRDF Search Interface (screen shot from an early version named RDFferret) (Reproduced by Permission of © 2009 British Telecommuncations Plc)

---

[27] A related class in this context is a class which is the domain or range of a property of the original class.

Experiments with QuizRDF on an RDF-annotated web site showed improvements in performance as compared to a conventional keyword-based search facility (Iosif *et al.* 2003).

9.7.2 Exploiting massive background knowledge – TAP

As discussed above, conventional search engines have no model of how the concepts denoted by query terms may be linked semantically. When searching for a paper published by a particular author, for example, it may be helpful to retrieve additional information that relates to that author, such as other publications, curriculum vitae, contact details, etc. A number of search engines are now emerging that use techniques to apply ontology-based domain-specific knowledge to the indexing, similarity evaluation, results augmentation and query enrichment processes.

TAP, (Guha and McCool 2003), is Semantic Web architecture, which allows RDF(S)-compliant consolidation and querying of structured information. Guha *et al.* (2003) describe a couple of Semantic Web-based search engines: ABS – activity-based search and W3C Semantic Search. In both cases TAP is employed to improve traditional search results (obtained from Google, http://www.google.com) when seeking information in relation to people, places, events, news items, etc. TAP is used for two tasks:

- Result augmentation: the list of documents returned by the IR system is complemented by text and links generated from the available background knowledge.
- Query term disambiguation: the user is given the opportunity to choose the concrete entity she is searching for, then the system attempts to filter the results of the IR system to those referring only to this entity. An approach using several statistics for this purpose is sketched in (Guha *et al.* 2003) without details on the implementation.

The semantic search application, which runs as a client of TAP, sends a user-supplied query to a conventional search engine. Results returned from the conventional search engine are augmented with relevant information aggregated from distributed data sources that form a knowledge base (the information is extracted from relevant content on targeted websites and stored as machine-readable RDF annotations). The information contained in the knowledge base is independent of and additional to the results returned from the conventional search engine. A search for a musician's name, for example, would augment the list of matching results from the conventional search engine with information such as current tour dates, discography, biography, etc. Figure 9.12 shows a typical search result from ABS. Special attention is paid to the selection of a dataset to show and its presentation.

Relatively simple heuristics are used to find the concepts which are relevant to the query. No considerable processing of the query terms is performed – according to Guha *et al.* (2003), concepts are considered relevant if they have a label (name) that contains one of the query terms.

The couple of semantic search engines mentioned above do not perform any pre-processing or indexing of the documents – they are based on an existing search engine. Dill *et al.* (2004) presents a system called SemTag, which performs automatic semantic annotation of texts with respect to large scale knowledge bases available though TAP, solving a task similar to the one presented in the next section.

9.7.3 Character-level annotations and massive world knowledge – KIM

The KIM platform (Popov *et al.* 2003), provides infrastructure and services for automatic semantic annotation, indexing, and retrieval of unstructured and semi-structured content.

As a baseline, KIM performs character-level semantic annotation of named entities, as described in Section 9.5. The automation of this task is possible through information extraction technology, which the General Architecture for Text Engineering (GATE) (http://www.gate.ac.uk). KIM analyzes texts and recognizes references to entities (such as persons, organizations, locations, dates). Then it tries to match the reference with a known entity that has a unique URI and description. In cases when

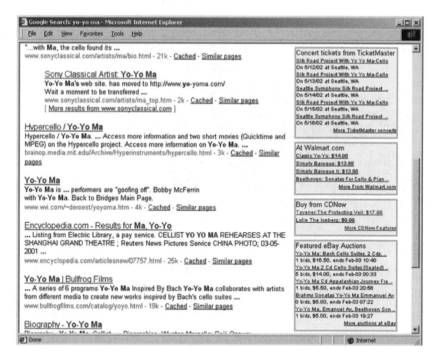

Figure 9.12   Semantic Search with TAP

there is no match, a new URI and description are generated automatically – this is the situation when ontology population takes place. Finally, the reference in the document is annotated with the URI of the entity, as presented in Figure 9.8.

KIM is equipped with an Internet Explorer plug-in, which uses these annotations for highlighting and hyperlinking, as presented in Figure 9.13. The mentions are coloured in accordance with the class (type) of the entity; hyperlinks provide access to popup forms presenting their descriptions in the KB. As the later include also references to other related entities, the user can further traverse the KB. This way the plug-in allows smooth transition from the text to the KB and exploration of the available structured knowledge.

In order to enable the easy bootstrapping of applications, KIM is based on the PROTON ontology (Terziev *et al*. 2004), which consists of about 250 classes and 100 properties. Furthermore, a knowledge base (KIM's World KB, WKB), pre-populated with about 200 000 entity descriptions, is bundled with KIM. Its role is to provide as a background knowledge (resembling a human's common culture) a quasi-exhaustive coverage of the entities of general importance – those, which are considered well-known and thus not explicitly introduced in the documents, which makes it hard to get their descriptions automatically extracted. KIM uses the OWLIM high-performance semantic repository (http://www.ontotext.com/owlim) to manage the WKB together with the extracted instance data.

A unique entity ID is inserted in the text at the places where the entity is referred. The application of entity co-reference resolution means that the system would regard the strings 'Gordon Brown,' 'Mr Brown,' 'the Prime Minister' as referring to the same entity in the KB. Then the texts are passed for indexing to a standard full-text search engine; in its basic configuration KIM uses for this purpose the Lucene engine (http://lucene.apache.org/java/docs/).

This allows KIM to offer the semantic queries which combine structured queries, reasoning, and full-text search. The most generic search interface (named Entity Pattern Search), allows the specification of queries about any type of entity, relations between such entities and required attribute values (e.g. 'find' all documents referring to a Person that hasPosition 'CEO' within a Company, locatedIn

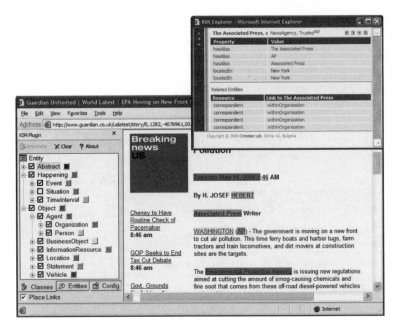

Figure 9.13   Semantic Browsing and Navigation in KIM   (Reproduced by Permission of Ontotext AD)

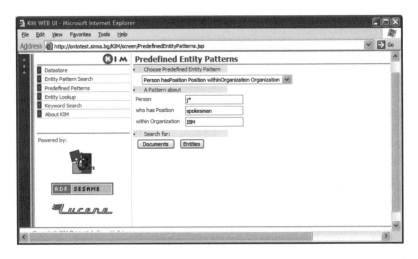

Figure 9.14   Semantic Querying in KIM   (Reproduced by Permission of Ontotext AD)

a Country with name 'UK'). To answer the query, KIM applies the semantic restrictions over the entities in the KB. The resulting set of entities is matched against the semantic index and the referring documents are retrieved with relevance ranking according to these entities.

KIM provides also a simplified search interface for several predefined patterns. In Figure 9.14 a semantic query is specified, concerning a person whose name begins with J, and who is a spokesman for IBM.

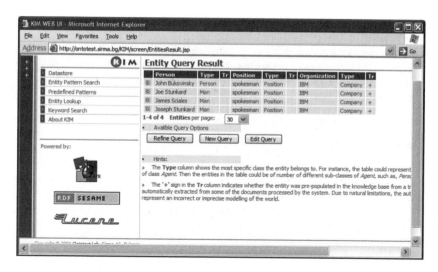

Figure 9.15 Semantic Query Results (Reproduced by Permission of Ontotext AD)

Figure 9.15 shows that four entities have been found in the documents indexed. It is then possible to browse a list of documents containing the specified entities and KIM renders the documents, with entities from the query highlighted (in this example IBM and the identified spokesperson).

In other work, Bernstein *et al*. (2005) describe a controlled language approach whereby a subset of English is entered by the user as a query and is then mapped into a semantic query via a discourse representation structure. Vallet *et al*. (2005) propose an ontology-based information retrieval model using a semantic indexing scheme based on annotation weighting techniques.

9.7.4   Squirrel

Squirrel (Duke *et al*. 2007) provides combined keyword-based and semantic searching. The intention is to provide a balance between the speed and ease of use of simple free text search and the power of semantic search. In addition, the ontological approach provides the user with a rich browsing experience. Squirrel builds on and integrates a number of semantic technology components:

(i) PROTON, the lightweight ontology and world knowledge base discussed above is used, and user profiles reflecting end-user personal interests are modelled in PROTON.

(ii) Lucene[28] is used for full-text indexing.

(iii) The KAON2 (Motik and Studer 2005) ontology management and inference engine provides an API for the management of OWL-DL and an inference engine for answering conjunctive queries expressed using the SPARQL[29] syntax. KAON2 also supports the Description Logic-safe subset of the Semantic Web Rule Language[30] (SWRL). This allows knowledge to be presented against concepts that goes beyond that provided by the structure of the ontology. For example, one of the attributes displayed in the document presentation is 'Organisation'. This is not an attribute of a document in the PROTON ontology; however, affiliation is an attribute of the Author concept and has the range 'Organisation'. As a result, a rule was introduced into the ontology to infer that the organisation responsible for a document is the affiliation of its lead author.

(iv) OntoSum (Bontcheva 2005), a Natural Language Generation (NLG) tool, takes structured data in a knowledge base as input and produces natural language text, tailored to the presentational

---

[28] http://lucene.apache.org/

[29] http://www.w3.org/TR/rdf-sparql-query/

[30] http://www.w3.org/Submission/SWRL/

context and the target reader. In the context of the Semantic Web and knowledge management, NLG is required to provide automated documentation of ontologies and knowledge bases and to present structured information in a user-friendly way. When a Squirrel search is carried out and certain people, organisations or other entities occur in the result set, the user is presented with a natural language summary of information regarding those entities. In addition, document results can be enhanced with brief descriptions of key entities that occur in the text.

(v) KIM (see above) is used for massive semantic annotation.

9.7.4.1 Initial search

Users are permitted to enter terms into a text box to commence their search. This initially simplistic approach was chosen based on the fact that users are likely to be comfortable with it due to experience with traditional search engines. If they wish, users can specify which type of resource (from a configurable list) they are looking for, e.g. publications, web articles, people, organisations, etc. although the default is to search in all of these.

The first task Squirrel carries out after the user submits a search is to call the Lucene index and then use KAON2 to look up further details about the results, be they textual resources or ontological entities. In addition to instance data, the labels of ontological classes are also indexed. This allows users to discover classes and then discover the corresponding instances and the documents associated with them without knowing the names of any of the instances, e.g. a search for 'Airline Industry' would match the 'Airline' class in PROTON. Selecting this would then allow users to browse to instances of the class where they can then navigate to the documents where those instances are mentioned. This is an important feature since with no prior knowledge of the domain it would be impossible to find these documents using a traditional search engine.

Textual content items can by separated by their type, e.g. Web Article, Conference Paper, Book, etc. Squirrel is then able to build the meta-result page based upon the textual content items and ontological instances that have been returned.

9.7.4.2 Meta-result

The meta-result page is intended to allow users to quickly focus their search as required and to disambiguate their query if appropriate. The page presents the different types of result that have been found and how many of each type. In order not to introduce unnecessary overhead on the user, the meta-result page also lists a default set of results, allowing the user to immediately see those results deemed most relevant to their query by purely statistical means.

The meta-result for the 'home health care' query is shown in Figure 9.16 under the subheading 'Matches for your query'. The first items in the list are the document classes. Following this is a set of matching topics from the topic ontology. In each case, the number in brackets is the number of documents attributed to each class or topic. Following the topics, a list of other matching entities is shown. The first five matching entities are also shown, allowing the user to click the link to go straight to the entity display page for these. Alternatively they can choose to view the complete list of items. Squirrel can be configured to separate out particular entity types (as is the case with topics and organisations as shown in Figure 9.16) and display them on their own line in the meta-result.

9.7.4.3 Refining by topic

Alongside the results, the user is presented with the topics associated with the documents in the result set. Not all topics are shown here since in the worst case where each document has many distinct topics the list of topics presented to the user would be unwieldy. Instead an algorithm takes the list of topics that are associated with the collection of documents in the result set, and generates a new list of topics representing the narrowest 'common ancestors' of the documents' topics.

Having selected a topic and viewed the subset of documents from the result set, the user can switch to an entity view for the topic. The user can also reach this view by selecting a topic from

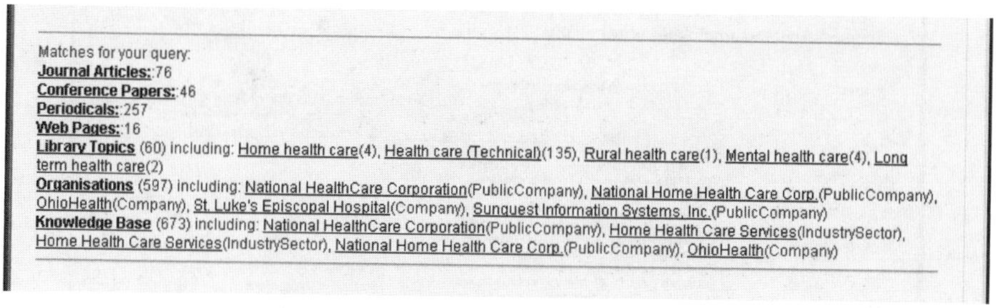

Matches for your query:
Journal Articles: 76
Conference Papers: 46
Periodicals: 257
Web Pages: 16
Library Topics (60) including: <u>Home health care</u>(4), <u>Health care (Technical)</u>(135), <u>Rural health care</u>(1), <u>Mental health care</u>(4), <u>Long term health care</u>(2)
Organisations (597) including: <u>National HealthCare Corporation</u>(PublicCompany), <u>National Home Health Care Corp.</u>(PublicCompany), <u>OhioHealth</u>(Company), <u>St. Luke's Episcopal Hospital</u>(Company), <u>Sunquest Information Systems, Inc.</u>(PublicCompany)
Knowledge Base (673) including: <u>National HealthCare Corporation</u>(PublicCompany), <u>Home Health Care Services</u>(IndustrySector), <u>Home Health Care Services</u>(IndustrySector), <u>National Home Health Care Corp.</u>(PublicCompany), <u>OhioHealth</u>(Company)

Figure 9.16   Meta-result

the meta-result section. Instead of showing the documents of the topic, the metadata for the topic is shown, which includes the broader, narrower and related topics. This allows the user to browse around the topic ontology. Each topic is shown with the number of documents it contains in brackets after it. Two links to document results are also shown. The first takes the user to a list of documents that have been attributed to the topic itself. The second takes the user to a list of documents that include all subtopics of the current topic. The layout of entity views in Squirrel are defined by templates. This allows the administrator to determine what metadata is shown and what is not. The use of these templates is discussed further in Section 9.7.4.6 where a 'Company' entity view is described.

9.7.4.4   Attribute-based refinement

Any document result list has a link which opens a 'refiner window'. This allows the user to refine the results based upon the associated metadata. The metadata shown and the manner in which it is displayed are configurable through the use of entity-type-specific templates that are configured by an administrator or knowledge engineer. Documents can be refined by the user based upon their authors, date of publication, etc. The approach adopted has been to allow the user to enter free text into the attribute boxes and to re-run the query with the additional constraints. An alternative would be to list possible values in the result set. However, the potential size of this list is large and is difficult to present to the user in a manageable way. The downside to the free-text approach is that the user can reduce the result set to zero by introducing an unsatisfiable constraint – which is obviously undesirable. Squirrel attempts to address this by quickly showing the user the size of the result set once constraints have been set. The user can then modify them before asking Squirrel to build the result page.

9.7.4.5   Document view

The user selects a document from the reduced result set, which takes them to a view of the document itself. This shows the metadata and text associated with the document and also a link to the source page if appropriate – as is the case with web pages. Since web pages are published externally with specific formatting data, the text of the page is extracted at index-time. Any semantic markup that is applied at this stage can then be shown on the extracted text at query-time. However, the user should always be given the option to navigate to the page in its original format. A screenshot of the document view is shown in Figure 9.17.

The document view also shows the whole document abstract, marked up with entity annotations. 'Mousing-over' these entities provides the user with further information about the entity extracted from the ontology. Clicking on the entity itself takes the user to the entity view.

9.7.4.6   Entity view

The entity view for 'Sun Microsystems' is shown in Figure 9.18. It includes a summary generated by OntoSum. The summary displays information related not only to the entity itself, but also information about related entities such as people who hold job roles with the company. This avoids users having

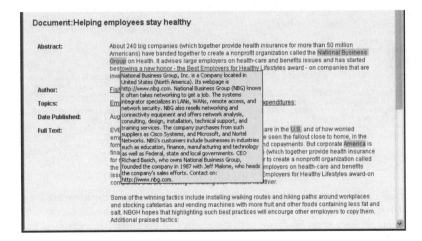

Figure 9.17   Document view

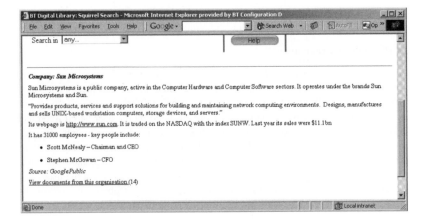

Figure 9.18   Company entity view

to browse around the various entities in the ontology that hold relevant information about the entity in question. The relationship between people and companies is made through a third concept called JobPosition. Users would have to browse through this concept in order to find the name of the person in question.

9.7.4.7   Consolidated results

Users can choose to view results as a consolidated summary of the most relevant parts of documents rather than a discrete list of results. This view is appropriate when a user is seeking to gain a wider view of the available material rather than looking for a specific document. The view allows them to read or scan the material without having to navigate to multiple results.

Figure 9.19 shows a screenshot of a summary for a query for 'Hurricane Katrina'. For each subdocument in the summary the user is able to view the title and source of the parent document, the topics into which the subdocument text has been classified or navigate to the full text of the document. A topic tree is built which includes a checkbox for each topic. These allow the user to refine the summary content by checking or unchecking the appropriate checkboxes.

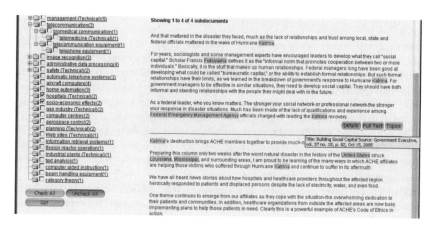

Figure 9.19   Consolidated results

9.7.4.8   Evaluation

Squirrel has been subjected to a three-stage user-centred evaluation process with users of a large digital library. Results are promising regarding the perceived information quality (PIQ) of search results obtained by the subjects. From 20 subjects, using a 7 point scale the average (PIQ) using the existing library system was 3.99 compared with an average of 4.47 using Squirrel – a 12% increase. The evaluation also showed that users rate the application positively and believe that it has attractive properties. Further details can be found in Thurlow and Warren (2008).

9.7.5   Other approaches

In this section, we briefly mention some other systems which have been described in the published literature: note that this list – along with the systems described above – is intended to be indicative, rather than exhaustive.

9.7.5.1   Searching for semantic web resources

We have seen in the earlier sections a variety of approaches to searching XML, RDF and OWL annotated information resources. The Swoogle search engine (Ding *et al*. 2004) is tackling a related but different problem: it is primarily concerned with finding ontologies and related instance data.

Finding ontologies is seen as important to avoid the creation of new ontologies where serviceable ones already exist, thus, it is hoped, leading to the emergence of widely used canonical (or reference) ontologies. Swoogle supports querying for ontologies containing specified terms. This can be refined to find ontologies where such terms occur as classes or properties, or to find ontologies that are in some sense about the specified term (as determined by Swoogle's ontology retrieval engine). The ontologies thus found are ranked according to Swoogle's OntologyRank algorithm which attempts to measure the degree to which a given ontology is used.

In order to offer such search facilities, Swoogle builds an index of semantic web documents (defined as web-accessible documents written in a Semantic Web language). A specialised crawler has been built using a range of heuristics to identify and index semantic web documents.

The creators of Swoogle are building an ontology dictionary based on the ontologies discovered by Swoogle[31].

---

[31] http://swoogle.umbc.edu/

9.7.5.2 Semantic browsing and navigation

Web browsing complements searching as an important aspect of information-seeking behaviour. Browsing can be enhanced by the exploitation of semantic annotations and below we describe systems which offer a semantic approach to information browsing (in some cases combined with searching).

Magpie (Domingue *et al*. 2004) is an internet browser plug-in which assists users in the analysis of webpages. Magpie adds an ontology based semantic layer onto web pages on-the-fly as they are browsed. The system automatically highlights key items of interest (in a way similar to KIM, see Figure 9.13), and for each highlighted term it provides a set of 'services' (e.g. contact details, current projects, related people) when you right-click on the item. This relies, of course, on the availability of a domain ontology appropriate to the page being browsed – similarly to TAP, Magpie annotates on the basis of match of a label from the KB.

CS AKTiveSpace (Glaser *et al*. 2004) is a Semantic Web application which provides a way to browse information about the UK Computer Science Research domain, by exploiting information from a variety of sources, including funding agencies and individual researchers. The application exploits a wide range of semantically heterogeneous and distributed content. AKTiveSpace retrieves information related to almost 2000 active computer science researchers and over 24 000 research projects, with information being contained within thousands of published papers, located in various university websites. This content is gathered on a continuous basis using a variety of methods, including harvesting publicly available data from institutional websites, bulk translation from existing databases, as well as other data sources. The content is mediated through an ontology, and stored as RDF triples; the indexed information comprises around 10 000 000 RDF triples in total.

CS AKTive Space supports the exploration of patterns and implications inherent in the content, using a variety of visualisations and multi-dimensional representations to give unified access to information gathered from a range of heterogeneous sources.

Alonso (2006) presents an impressive proof-of-concept prototype of a KM solution, using various features of ORACLE 10gR2 to implement semantic metadata-based search and browse. The system combines RDF support, full-text indexing, and clustering functionality. The system's user interface implements several navigation modes, based on visualisation components available as libraries from third parties.

Siderean's Seamark Navigator[32] allows for faceted search based on document-level metadata represented in RDF: subject, author, publisher, date. Internally, the system combines an RDF repository (which may or may not incorporate reasoning) and a full-text search engine. The search user interface allows for interactive focussing: for each facet the user is presented with the most popular values. For each value (e.g. a specific author) the system presents the number of documents matching this value – this provides information to the user about the 'selectivity' of the specific values. The user can choose a value of a facet and it is added to the filter, following which the values and the selectivity figures presented are altered to consider only the documents matching the current filter.

9.8 Summary

In this chapter, we began by discussing some of the limitations of current search technology and proceeded to discuss ontologies and Semantic Web technology and how they could be used to annotate document collections semantically, particularly with regard to named entities occurring within documents. We then explained how semantic annotations could be used in systems to overcome some of the problems we earlier identified with respect to current search technology. Finally, a number of systems using semantic search techniques were described. In general, these systems allow a greater precision in search and can also link entities occurring in document collections to ontologies, effectively providing a 'background knowledge base' to augment search results. Furthermore, the formal nature of the ontological KBs means that it is possible to reason over them (taking a simple example, if searching for an organisation in France, an organisation in Paris might be returned even if it is

---

[32] http://www.siderean.com/products_suite.aspx)

nowhere in the document collection linked explicitly to France). Of course, the cost of generation and maintenance of such KBs and of semantic annotation of documents need to be weighed against the potential benefits of such an approach for each potential domain of application. It is worth noting that the automatic or semi-automatic generation of ontologies and semantic annotations is a focus for much current research (see, for example, Davies *et al*. 2009), but space has not permitted a detailed discussion of the prospects for such technology in this chapter.

Basic definitions regarding ontologies can be found in Gruber (1992; 1993) Guarino and Giaretta (1995) and Guarino (1998).

Those wishing to find out more about semantic technology are referred to (Davies et al. 2006).

Acknowledgements. Paul Warren is thanked for his comments on a draft of this chapter.

Exercises

9.1 Search engine users have been found to enter only a small number of query terms. Outline two difficulties arising from this fact associated with current search engine technology. Explain briefly how the semantic search approach aims to overcome these.

9.2 Given the ontology shown in Figure 9.8, explain how a query containing the terms *London* and XYZ could match a document containing the terms *UK* and *Company*.

References

Alonso, O. 2006. Building semantic-based applications using Oracle. Developer's Track at WWW2006, Edinburgh, May 2006. http://www2006.org/programme/item.php?id = d16

Beckett, D. 2004. RDF/XML Syntax Specification (Revised). http://www.w3.org/TR/2004/REC-rdf-syntax-grammar-20040210/

Berners-Lee, T. 1999. *Weaving the Web*. Orion Books, London, UK.

Bernstein, A., Kaufmann, E., Goehring, A. and Kiefer, C. 2005. Querying Ontologies: A Controlled English Interface for End-users, *Proceedings of the 4th International Semantic Web Conference*, ISWC2005, Galway, Ireland, November 2005, Springer-Verlag, 2005.

Bontcheva, K. 2005. Generating Tailored Textual Summaries from Ontologies, *Proceedings of the 2nd European Semantic Web Conference*, Crete, Greece, June 2005, pp. 531–545.

Bontcheva, K., Cunningham, H., Kiryakov, A. and Tablan, V. 2006. Semantic Annotation and Human Language Technology, in Davies *et al.* (2006), pp. 29–49.

Brickley, D., Guha, R. V, eds. 2000. Resource Description Framework (RDF) Schemas, W3C http://www.w3.org/TR/2000/CR-rdf-schema-20000327/

Chinchor, N. and Robinson, P. 1998. MUC-7 Named Entity Task Definition (version 3.5). In *Proceedings of the MUC-7*.

Cunningham, H. 2000. *Software Architecture for Language Engineering*. PhD Thesis, University of Sheffield.

Cunningham, H. 1999. *Information Extraction: a User Guide* (revised version). Department of Computer Science, University of Sheffield, May, 1999.

Davies, J., Fensel, D. and van Harmelen, F. 2003. *Towards the Semantic Web*, Wiley, UK, 2003.

Davies, J., Weeks, R. and Krohn, U. 2003. QuizRDF: Search technology for the semantic web. In (Davies, Fensel, van Harmelen 2003).

Davies, J., Studer, R., Sure, Y. and Warren, P. 2005. Next Generation Knowledge Management, *BT Technology Journal*, **23**(3) 175–190.

Davies, J., Studer, R. and Warren, P. *Semantic Web Technology: Trends and Research*, Wiley, UK, 2006.

Davies, J., Mladenic, D. and Grobelnik, M. 2009. *Semantic Web: Integrating Ontology Management, Knowledge Discovery and Human Language Technologies*, Springer, 2009.

DCMI Usage Board. 2005. *DCMI Metadata Terms*. http://dublincore.org/documents/2005/06/13/dcmi-terms/

Dean, M and Schreiber, G. 2004. (Eds); Bechhofer, S., van Harmelen, F., Hendler, J., Horrocks, I., McGuinness, D. L., Patel-Schneider, P. F. and Stein, L. A. 2004. OWL Web Ontology Language Reference. W3C Recommendation 10 February 2004. http://www.w3.org/TR/owl-ref/

Dill, S., Eiron, N., Gibson, D., Gruhl, D., Guha, R., Jhingran, A., Kanungo, T., McCurley, K. S., Rajagopalan, S., Tomkins, A., Tomlin, J. A. and Zienberer, J. Y. 2003. A Case for Automated Large Scale Semantic Annotation. *Journal of Web Semantics*, **1**(1), 2004, 115–132.

Ding, L., Finin, T., Joshi, A., Pan, R., Cost, R. S., Peng, Y., Reddivari, P., Doshi, V. and Sachs. J. 2004. Swoogle: A Search and Metadata Engine for the Semantic Web, *Conference on Information and Knowledge Management CIKM04*, Washington DC, USA, November 2004, pp. 652–659.

Domingue, J., Dzbor, M. and Motta, E. 2004. Collaborative Semantic Web Browsing with Magpie in The Semantic Web: Research and Applications Davies, J., Bussler, C., Fensel, D. and Studer, R. (ed). Proceedings of ESWS, 2004. LNCS 3053, Springer-Verlag, pp. 388–401.

Duke, A., Glover, T. and Davies, J. 2004. Squirrel: an advanced semantic search and browse facility. In *The Semantic Web: Research and Applications* Franconi, E., Kifer, M. and May, W. (eds), *Proc. 4th European Semantic Web Conference*, 2004. LNCS 4519, Springer, Berlin, pp. 341–455.

Dumais, S., Cutrell E., Cadiz J., Jancke G., Sarin R. and Robbins D. 2003. Stuff I've Seen: A system for *personal information retrieval and re-use*. In *Proceedings of SIGIR'03*, 2003. Toronto, ACM Press, pp. 72–79.

Ehrig, M., Haase, P., Hefke, M. and Stojanovic, N. 2005. Similarity for ontologies – a *comprehensive framework*. *Proceedings of the 13th European Conference on Information Systems, May 2005*.

Glaser, H., Alani, H., Carr, L., Chapman, S., Ciravegna, F., Dingli, A., Gibbins, N., Harris, S., Schraefel, M. C. and Shadbolt, N. 2004. CS AKTiveSpace: Building a Semantic Web Application in *The Semantic Web: Research and Applications* Davies, J., Bussler, C., Fensel, D. and Studer, R. (eds). *Proceedings of ESWS, 2004*, pp. 388–401. LNCS 3053, Springer-Verlag.

Grishman, R. 1997. *TIPSTER Architecture Design Document Version 2.3*. Technical report, DARPA, 1997. http://www.itl.nist.gov/iaui/894.02/related_projects/tipster/

Gruber, T. R. 1992. A translation approach to portable ontologies. *Knowledge Acquisition*, **5**(2) 199–220, 1993. http://ksl-web.stanford.edu/KSL_Abstracts/KSL-92-71.html

Gruber, T. R. 1993. *Toward principles for the design of ontologies used for knowledge sharing*. In N. Guarino and R. Poli, (eds), International Workshop on Formal Ontology, Padova, Italy, 1993. http://ksl-web.stanford.edu/KSL_Abstracts/KSL-93-04.html

Guarino, N.; Giaretta, P. 1995. Ontologies and Knowledge Bases: Towards a Terminological Clarification. In N. Mars (ed.) *Towards Very Large Knowledge Bases: Knowledge Building and Knowledge Sharing*. IOS Press, Amsterdam: pp. 25–32

Guarino, N. 1998. Formal Ontology in Information Systems. In N. Guarino (ed.) *Formal Ontology in Information Systems. Proceedings of FOIS'98*, Trento, Italy, June 6–8, 1998. IOS Press, Amsterdam, pp. 3–15.

Guha, R.and McCool, R. 2003. Tap: A semantic web platform. *Computer Networks*, **42** 557–577, 2003.

Guha, R., McCool, R. and Miller, E. 2003. *Semantic Search*. WWW2003, May 20–24, 2003, pp. 700–709 Budapest, Hungary.

Iosif, V., Mika, P., Larsson, R. and Akkermans, H. 2003. Field Experimenting with Semantic Web Tools in a Virtual Organisation. In (Davies, Fensel, van Harmelen 2003), pp. 219–244.

Jansen, B. J., Spink, A. and Saracevic, T. 2000. Real life, real users, and real needs: A study and analysis of user queries on the web. *Information Processing and Management* **36**(2) 207–227.

Kiryakov, A. 2006. Ontologies for Knowledge Management. Chapter 7 in: Davies, J; Studer, R; Warren, P. (eds.). *Semantic Web Technologies: Trends and Research in Ontology-based Systems*. Wiley, UK, 2006, pp. 115–138.

Kiryakov A. and Simov K. Iv. 1999. Ontologically Supported Semantic Matching. In *Proceedings of NODALIDA'99: Nordic Conference on Computer Linguistics*, Trondheim, Dec. 9–10, 1999, pp. 9–10.

Klyne, G. and Carroll, J. J. 2004. *Resource Description Framework (RDF): Concepts and Abstract Syntax*. W3C recommendation 10 February 2004. http://www.w3.org/TR/rdf-concepts/

Landauer T. and Dumais S. 1997. A solution to Plato's problem: the Latent Semantic Analysis theory of acquisition, induction and representation of knowledge. *Psychological Review* **104**(2) 211–240.

Mahesh K., Kud J. and Dixon P. 1999. *Oracle at TREC8: A Lexical Approach*. In *Proceedings of the 8th Text Retrieval Conference (TREC-8)*, pp. 207–216.

Motik, B. and Studer, R. 2005. KAON2 – A Scalable Reasoning Tool for the Semantic Web. In: *Proceedings of the 2nd European Semantic Web Conference (ESWC'05)*, Poster session: http://kaon2.semanticweb.org (2005)

Popov, B., Kiryakov, A., Kirilov, A., Manov, D., Ognyanoff, D. and Goranov, M. 2003. *KIM – Semantic Annotation Platform*. In *Proceedings of 2nd International Semantic Web Conference (ISWC2003)*, 20–23 October 2003, Florida, USA. LNAI Vol. 2870, pp. 834–849, Springer-Verlag Berlin Heidelberg 2003

Robertson, S., M. Hancock-Beaulieu, A. Gull, M. Lau, K. Sparck-Jones, P. Willett, E. Keen, *Okapi at TREC*, http://research.microsoft.com/users/robertson/papers/trec_pdfs/okapi_trec1.txt, accessed on 28/2/08.

Rocha, C., Schwabe, D. and de Aragao, M. P. 2004. A hybrid approach for searching in the semantic web. *WWW 2004*, May 17–22, 2004, New York, USA, pp. 374–383.

Salton, G., Wong, A. and Yang, C.S. 1975. A vector space model for automatic indexing. *Comms ACM* **18**(11) 613–620.

Terziev, I., Kiryakov, A. and Manov, D. 2004. *D1.8.1. Base upper-level ontology (BULO) Guidance*, report EU-IST Integrated Project (IP) IST-2003-506826 SEKT), 2004. http://proton.semanticweb.org/D1_8_1.pdf

Thurlow, I. and Warren, P. 2008. Deploying and Evaluating Semantic Technologies in a Digital Library, in Davies *et al*. (2008), pp. 237–257.

Vallet, D., Fernandez, M. and Castells, P. 2005. An Ontology-based Information Retrieval Model, *Proceedings of the 2nd European Semantic Web Conference, ESWC2005*, Heraklion, Crete, May/June 2005, Springer-Verlag, Berlin. Editors: Gómez-Pérez, A. and Euzenat, J. LNCS 3532/2005. Springer Berlin/Heidelberg

van Ossenbruggen, J., Hardman L. and Rutledge L. 2002. Hypermedia and the Semantic Web: A Research Agenda. *Journal of Digital Information* **3**(1) 7–37.

Voorhees, E. 1998. Using WordNet for Text Retrieval. In *WordNet: an electronic lexical database*. Fellbaum, C. (ed.), MIT Press, pp. 285–303.

10

The Role of Natural Language Processing in Information Retrieval: Searching for Meaning and Structure

Tony Russell-Rose and Mark Stevenson

10.1   Introduction

Information retrieval can be thought of as providing tools that allow a user to satisfy an information need. The best known method for achieving this is through document retrieval where the retrieval engine searches through a set of documents to identify those which meet that need. One limitation of this paradigm is that it requires users to carry out some extra work to identify the information they require: they must read through the documents to find the information they were looking for. Users often seek quite specific pieces of information, for example 'How old was Mozart when he died?' The answer to this question may be contained in a document describing the life of Mozart, but this would also contain a large amount of irrelevant content which must be examined before the information need can be met. It can be straightforward to identify particular pieces of information in documents which contain structured data, but it is more challenging to automatically interpret those written in natural language.

Standard approaches to text retrieval such as Boolean, vector space and probabilistic models, rely on *index terms* to describe the content documents. Index terms are most easily obtained from the documents themselves and often consist of simply a list of the words in the document. This approach is sometimes referred to as the 'bag of words' model. Well-established techniques such as the removal of stop words and term weighting are usually applied to make the best use of the index terms (Robertson and Spark Jones 1997). However, the usefulness of the 'bag of words' model is not ideal for processing natural language documents. In particular the model fails to take account of several characteristics of natural language:

1. *Words can have several possible meanings.* For example, 'bat', can mean 'sports equipment' (as in 'cricket bat') or 'nocturnal mammal' (as in 'fruit bat'), etc. Consequently, if we are provided

Information Retrieval: Searching in the 21st Century   edited by A. Göker & J. Davies
© 2009 John Wiley & Sons, Ltd

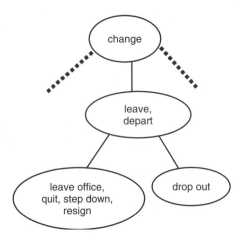

Figure 10.1 Fragment of the WordNet hierarchy showing hypernym relations between a sample of synsets

with no additional information, it is impossible to tell whether a document which contains the word 'bat' as an index term refers to the domain of sports or animals.

2. *The meaning of terms is affected by the order in which they appear in a document.* A popular example of this effect is that 'Venetian blinds' are not the same as 'blind Venetians', despite the fact that they contain the same terms.

3. *Ideas can be expressed in multiple ways.* Natural languages allow the same statement to be made in many different ways and this is enabled by devices such as synonymy, paraphrase and metaphor. For example, consider the sentences 'Starbucks coffee is the best' and 'The place I like most when I need to feed my caffeine addiction is the company from Seattle with branches everywhere'. While they do not share any (significant) words in common, they say much the same thing.

4. *The topic of a document is not determined simply by the terms it contains.* Linguistic constructions such as negation can be used to indicate that a piece of text does not contain any information about a particular area while at the same time mentioning it. For example the sentence 'A cricket bat is not a nocturnal mammal, or any other type of animal!' contains the terms 'bat', 'nocturnal mammal' and 'animal', but does not contain information about the 'nocturnal animal' meaning of 'bat'. However, a bag of words model is likely to identify this as a document containing information about those animals.

Natural language processing (NLP) techniques aim to intelligently analyse documents and capture their meaning. This has led many to suggest that these techniques could be productively applied to information retrieval problems. The remainder of this chapter is organised as follows. Section 10.2 describes some of the techniques and resources commonly used within NLP which have been applied to IR. Section 10.3 discusses two information access applications (text mining and question answering) closely associated with NLP. Section 10.4 discusses the relation between IR and NLP.

The learning objectives of this chapter are to:

- Introduce various areas of natural language processing technology that are of potential relevance to those interested in information retrieval;
- Describe how these technologies have been applied to information retrieval and evaluate the usefulness of this contribution;
- Critically assess future prospects for the application of natural language processing technology to information retrieval.

10.2   Natural Language Processing Techniques

10.2.1   Named entity recognition

Named entity recognition (NER) is the process by which key concepts such as the names of people, places and organizations can be identified within a document. This is best illustrated using an example, such as the following sentence from a news story, in which several entities are mentioned:

> 'Mexico has been trying to stage a recovery since the beginning of this year and it's always been getting ahead of itself in terms of fundamentals,' said Matthew Hickman of Lehman Brothers in New York.

The key entities we might expect to extract from this sentence would be:

Persons	Organisations	Cities	Countries
Matthew Hickman	Lehman Brothers	New York	Mexico

The accurate identification of such entities confers a number of benefits. Firstly, by identifying such concepts in a document, it is possible to index documents at a more fine-grained level, which in turn allows more precise searching. The word '*Bush*', for example, can refer to either a plant or the President of the USA – so by differentiating between the different usages we can maximize the likelihood that information seekers will receive only those documents that are relevant to their query.

Secondly, by identifying names within a document we can provide links to other related resources. Several content providers already use this technique to highlight the names of the key people mentioned in a news story, and provide links to their biographies. Similarly, it is possible to create more precise filters based on those names, allowing personalised delivery of content to individual users. For example, an SMS service might supply all the latest news about 'New York' as opposed to 'new' events in the city of 'York'.

Thirdly, the output of NER adds valuable structure to a document, and thus facilitates subsequent post-processing transformations, such as machine translation (e.g. translating English documents into Spanish) or even text-to-speech output (e.g. providing spoken versions of stock quotes for access over the telephone network).

Research and development into NER has increased significantly in recent years, within both academic institutions and commercial organisations. Much of this activity was originally precipitated by a number of US government-sponsored evaluation exercises, in which a variety of approaches to the problem have been compared. Two such examples are the series of Message Understanding Conferences (MUCs) for information extraction from newswire text (Hirschman 1998), and the Text Retrieval Conferences (TREC)[1] for information retrieval (Voorhees and Buckland 2002). These evaluations are organised as open challenges where research groups can submit their own systems to perform fixed tasks on known data, and many would argue their existence has made a significant contribution to the progress witnessed over the last decade on basic tasks such as named entity recognition. Because the tasks and data are fixed in advance, it makes it much easier to make direct comparisons between alternative techniques and systems, and thus for research groups to learn from each other. As a result, NER accuracy has increased significantly, to as high as 90%+ for English newswire text (Hirschman *et al.* 2002), and consequently it is now being applied in commercial environments to a wide range of content.

Most NER systems are designed to identify certain predefined types of entity, such as the names of people, places and organisations. Some commercially available systems also include the ability to

---

[1] http://trec.nist.gov

identify additional items, such as postcodes, telephone numbers and so on. In addition, some systems have been built to identify more esoteric items such as the names of proteins (see Section 10.3.1).

The most accurate NER systems are typically the product of ongoing academic research (as opposed to commercially available products). It is common for such systems to employ some or all of the following techniques (Yeh *et al.* 2005):

1. Markov modelling
2. Support vector machines
3. Manually generated rules

The first two techniques are machine learning approaches, in which the NER system is 'trained' using manually annotated examples from which it 'learns' the patterns corresponding to a named entity. As is common with machine learning approaches, the availability of high-quality annotated data is crucial: without an abundance of this, performance will be limited. However, creating large quantities of hand-annotated data is time consuming and expensive.

It is often the case that the choice of features can be as important as the overall approach used. Some NER systems focus on features internal to the text itself (such as word bigrams, POS tags, character substrings, word shapes, NE bigrams, etc.) whilst others utilise external evidence such as gazetteers, web querying techniques (Dingare *et al.* 2004), other corpora, etc. In addition, some NER systems also employ pre- or post-processing stages or even voting systems to improve accuracy (Zhou *et al.* 2004).

10.2.2   *Information extraction*

Information Extraction is an NLP technology which aims to identify well-defined pieces of information from documents. Examples of the types of information which might be extracted include the movements of company executives, victims of terrorist attacks, information about mergers and acquisitions and interactions between genes and proteins in scientific articles. When the relevant information has been identified it is then stored in a highly structured format known as a template.

For example, the Message Understanding Conferences (MUC) (Grishman and Sundheim 1996) were an international evaluation exercise in which information extraction systems were tested against one another using common documents and scoring systems. In the sixth MUC conference participants were asked to identify facts about the movements of executives between companies. The following paragraph describes an event in which 'John J. Dooner Jr' becomes chairman of the company 'McCann-Erickson':

> Now, Mr James is preparing to sail into the sunset, and Mr Dooner is poised to rev up the engines to guide Interpublic Group's McCann-Erickson into the 21st century. Yesterday, McCann made official what had been widely anticipated: Mr James, 57 years old, is stepping down as chief executive officer on July 1 and will retire as chairman at the end of the year. He will be succeeded by Mr Dooner, 45.

This fact is encoded in the following template structure:

```
<SUCCESSION_EVENT-2> :=
    SUCCESSION_ORG:
        <ORGANIZATION-1>
    POST: "chairman"
    IN_AND_OUT: <IN_AND_OUT-4>
    VACANCY_REASON: DEPART_WORKFORCE

<IN_AND_OUT-4> :=
    IO_PERSON: <PERSON-1>
```

```
    NEW_STATUS: IN
    ON_THE_JOB: NO
    OTHER_ORG: <ORGANIZATION-1>
    REL_OTHER_ORG: SAME_ORG

<ORGANIZATION-1> :=
    ORG_NAME: "McCann-Erickson"
    ORG_ALIAS: "McCann"
    ORG_TYPE:  COMPANY

<PERSON-1> :=
    PER_NAME: "John J. Dooner Jr."
    PER_ALIAS: "John Dooner" "Dooner"
```

MUC templates employ a complex object-oriented structure (Krupke and Rau 1992) which is organised as follows: each SUCCESSION_EVENT contains the name of the POST, organisation (SUCCESSION_ORG) and references to at least one IN_AND_OUT sub-template, each of which records an event in which a person starts or leaves a job. Several of the fields, including POST, PERSON and ORGANIZATION, may contain aliases which are alternative descriptions of the field filler and are listed when the relevant entity was described using different terms in the text. For example, the organisation in the above template has two descriptions: 'McCann-Erickson' and 'McCann'.

These template structures contain a large amount of information that enables complex queries to be carried out against the extracted data. For example, a user could ask for a list of all the events where a finance officer left a company to take up the position of CEO in another company. Current document retrieval technologies could not identify information as specific as this within text. Once the templates have been filled for a set of documents they can be made use of in two ways: firstly, they can be used to populate a database containing only the facts of interest in each document, and secondly, they can be linked to the document as metadata for use in a document retrieval process.

However, information extraction is a difficult task. One of the reasons for this is that the description of an event is often spread across several sentences or paragraphs of a text. For example, the following pair of sentences contains information about management succession events:

> Pace American Group Inc. said it notified two top executives it intends to dismiss them because an internal investigation found evidence of 'self-dealing' and 'undisclosed financial relationships'.
> The executives are Don H. Pace, cofounder, president and chief executive officer; and Greg S. Kaplan, senior vice president and chief financial officer.

The name of the organisation and fact that the executives are leaving are contained in the first sentence, and the names of the executives and their posts are listed in the second sentence. However, the second sentence does not mention the fact that the executives are leaving these posts. The succession event can therefore only be fully understood through a combination of the information contained in both sentences. Combining the required information across sentences is not a simple task since it is necessary to identify phrases which refer to the same entities, i.e. 'two top executives' and 'the executives' in the above example. Additional difficulties occur because the same entity may be referred to by a different linguistic unit. For example, 'International Business Machines Ltd' may be referred to by an abbreviation (IBM), nickname (Big Blue) or anaphoric expression such as 'it' or 'the company'. Problems such as this are one of the reasons that it has proved difficult to create reliable information extraction systems. Scores for the template filling task in the MUC evaluations were generally in the 50–60% F-measure range. This level of accuracy may not be sufficient for many applications.

An intermediate step which is somewhere between named entity extraction and full template filling is relation extraction. This task is simpler than full template filling since: (i) it only aims to identify (normally binary) relations between named entities; and (ii) only relations which occur within a single sentence are considered. These simplifications have meant that the task is more achievable than full template filling.

Brin (1998) developed a relation extraction system for the web. The motivation behind this system is the fact that information about the relation between items, for example actors and the movies they star in or authors and their books, are often spread across multiple web pages, It would be convenient if this information could be identified automatically and collected together. Brin's system, dual iterative pattern relation expansion (DIPRE), was provided with a small number of examples of pairs of related items. DIPRE would then search the Internet for web pages containing both items and create a pattern consisting of a set of regular expressions which matched each pair. These patterns were then reapplied to identify other instances of the relation and the process repeated in an iterative manner. In this way the redundancy of the Internet (i.e. the fact that the same information may be available from multiple sources) could be exploited to gather instances of a relation with only a few manually provided examples. The approach was extended by Agichtein and Gravano (2000) who developed improved methods for generating the patterns which identified pairs of items. One problem with the relation extraction task is that relations which are described across more than one sentence are often ignored. Stevenson (2004) showed that only around 40% of events in the extraction task used for the sixth MUC evaluation could be fully identified by only searching for events which were described within a single sentence.

10.2.3 WordNet

WordNet (Fellbaum 1998) is a semantic database for English originally designed to model the psycholinguistic construction of the human mental lexicon, but quickly proved to be an extremely popular resource in language processing since it has been made freely available[2]. The basic building block of WordNet is the *synset* (as in 'synonym set') – a group of words with closely related meanings. Words with more than one meaning belong to multiple synsets. For example, the noun 'car' has at least four different meanings (senses), including a 'railway car', 'elevator car', 'automobile', 'cable car', etc. The synset for this concept thus consists of the members: 'car', 'motorcar', 'machine', 'auto' and 'automobile'.

WordNet synsets are connected through a number of relations, the most important of which is *hypernymy*, the 'is a kind of' relationship. For example, the hypernym of the synset containing the term 'car' as previously mentioned consists of the terms 'motor vehicle' and 'automotive vehicle'. This relation organises the synsets hierarchically, with separate hierarchies for nouns, verbs, adjectives and adverbs.

The structure of WordNet makes it an attractive resource for use in information retrieval. Synsets can be used to identify the alternative ways in which a concept can be described. Voorhees (1993) attempted to exploit this structure for text retrieval. The correct synset for ambiguous terms in the query were identified by searching for the largest subportion of the WordNet hierarchy which contained a particular meaning for the term, but none of the other possible meanings. Voorhees found that retrieval performance was consistently worse when WordNet was used for retrieval this way. She did not quantitatively evaluate the performance of the disambiguation algorithm, but attributed the poor performance of the retrieval to the mistakes it made.

A multilingual version, EuroWordNet (Vossen 1998), has also been developed specifically for cross-lingual information retrieval (see Chapter 11 on cross-lingual IR). This extends the original WordNet to include coverage of Dutch, Italian, Spanish, French, German, Czech and Estonian. Through a number of initiatives it has been extended to cover Basque, Portuguese and Swedish[3]. The various language-specific WordNets are linked through a language-independent inter-lingual index (ILI). In practise the ILI is closely based on a version of the original English WordNet. A set of equivalence relations are used to identify concepts which are equivalent across different languages (Vossen 1998). The multilingual aspect of EuroWordNet means that it is possible to semantically tag text in one language and use the resource to identify information about possible translations.

---

[2] http://wordnet.princeton.edu
[3] http://www.globalwordnet.org

10.2.4   Word sense disambiguation

The fact that words can have multiple meanings, or senses, is known as *polysemy*. In addition to making information retrieval more challenging, polysemy is a problem for language processing in general. NLP has developed techniques to automatically identify the meanings of words as they appear in text, a process known as word sense disambiguation (WSD).

One of the main sources of ambiguity is syntactic category: words often have different meanings, depending upon whether they are being used as a noun, verb, adjective etc. For example, two of the possible meanings of 'light' are 'not heavy' ('magnesium is a light metal') and 'illumination' ('The light in the kitchen is quite dim'). The first meaning can only be used when 'light' is being used as an adjective and the second when it is used as a noun. Consequently, knowledge of a word's part of speech can be a useful guide to identifying its meaning.

NLP provides technology for identifying the grammatical category of words in the form of part of speech (PoS) taggers. This technology attaches a tag to each word in a text which indicates whether the word is being used as a noun, verb, adjective, determiner, etc. The set of tags used by a particular PoS tagger may provide more detail for a particular grammatical category. For example, the Penn Treebank tagset (Marcus *et al.* 1993), which is commonly used by PoS taggers, has four tags for noun: NN (singular common noun or mass noun), NNP (singular proper noun), NNS (plural common noun) and NNPS (plural proper noun). Part of speech tagging has been researched since the 1950s (Voutilainen 2003) and a number of tagging systems are now easily available. These systems use a number of different strategies, including handwritten rules (Karlsson 1990), hidden Markov models (Cutting *et al.* 1992) and n-grams learned from text annotated with part of speech information (Church 1998). Wilks and Stevenson (1998) showed that word sense ambiguity could be carried out effectively using a part of speech tagger when only broad differences between different word meanings were being considered. For example, this technique could be used to differentiate between the 'turn aircraft' ('The plane banked sharply') and 'financial institution' senses of 'bank', but couldn't distinguish between more closely related meanings such as 'a car crashed into the bank' and 'the bank gave me a good rate on my mortgage'.

It seems a reasonable assumption that the added information in part of speech tags will be beneficial for information retrieval. Krovetz (1997) experimented with the use of part of speech tagging for information retrieval and wished to 'determine how well part of speech differences correlate with different word meanings, and to what extent the use of meanings determined by these differences will affect the performance of a retrieval system' (Krovetz 1997, p. 76). A standard document collection and query set was tagged using Church's part of speech tagger (Church 1998) and the documents were indexed by their word and part of speech (e.g. 'light/noun') rather than using the standard IR method of indexing by word alone. He found that the performance of his retrieval system decreased with the added information, but further error analysis showed that the decrease was caused by poor performance of the tagger.

If part of speech disambiguation does not provide enough information to improve IR performance then it might be worth experimenting with whether WSD, which attempts to provide more detailed information about word meanings, has any benefit. There has been a long history of research into WSD in NLP during which a wide variety of techniques were investigated. Early approaches such as Wilks (1975) and Hirst (1987) applied techniques from artificial intelligence that relied on hand- crafted knowledge sources containing detailed semantic information. However, these approaches were limited by the fact that the information they relied on had to be manually created. Large-scale machine-readable lexical resources, such as WordNet, started to become available in the 1980s and were quickly used for WSD. They have the advantage of providing both a list of possible senses for each word and information which can be used to identify the correct sense, such as the hypernym hierarchy in WordNet. A commonly used approach to WSD is based on dictionary definitions derived from these resources (Lesk 1986). Using this technique words are disambiguated using their definitions, and senses are chosen which share the most words in their definitions. Lesk estimated that this approach correctly disambiguates 50–70% of words. This approach is similar to the bag of words model which is pervasive in IR and which represents the semantics of a document using the words it contains; the meaning of a word sense is represented using the words in its definition. Another approach to WSD makes use of information derived from large collections of text (known as *corpora*). Yarowsky (1992) developed a method to identify the most appropriate sense from *Roget's Thesaurus* (Chapman 1977) using statistical models learned from unannotated text. This approach was tested on 12 polysemous words and achieved 92% correct disambiguation.

There has been some disagreement about the usefulness of WSD for IR. Some have argued that the benefits which might be gained from disambiguation are limited. Krovetz and Croft (1992; see also Krovetz 1997) manually disambiguated a standard test corpus and found that a perfect WSD engine would improve retrieval performance by only 2%. Sanderson (1994) performed similar experiments in which ambiguity was artificially introduced to a test collection by automatically creating 'pseudowords'. These are formed by concatenating unrelated pairs of words, for example 'banana' and 'kalashnikov' would become 'banana/kalashnikov', and replacing each occurrence of the individual word in a corpus with their concatenation. Since the original meaning is retained, these ambiguities could be easily resolved. Sanderson simulated various levels of WSD accuracy by automatically resolving the ambiguity, but occasionally making mistakes. He found that a 20–30% error rate in disambiguation led to text retrieval which was at the same level, or possibly worse, than if the ambiguity was left unresolved. Sanderson also found that queries containing fewer than five terms were less sensitive to mistakes in the disambiguation. He concluded that WSD was only useful if it was very accurate or the queries were short. However, in these experiments pseudowords were generated by choosing random pairs of words which Sanderson (2000) acknowledged was unlikely to generate the same kind of ambiguities which are observed in text. Stokoe (2005) carried out experiments using a more accurate model of ambiguity and which suggested that the minimum accuracy necessary to improve retrieval effectiveness may be lower than Sanderson's experiments imply. In particular Stokoe was interested in exploring the difference between polysemy (when a word's senses are related) and homonymy (when they are not). He used WordNet to generate pseudowords that model polysemy and a similar method to the one used by Sanderson to generate those modelling homonymy. Stokoe found that when additional homonymy was introduced to the collection, disambiguation accuracy of at least 76% was required to improve retrieval effectiveness. However, when additional polysemy was introduced, a lower disambiguation accuracy of 55% was found to be beneficial.

Others have demonstrated that WSD can be used to improve IR performance. Schutze and Pedersen (1995) showed that disambiguation can substantially improve text retrieval performance; demonstrating an improvement between 7 and 14% on average. Rather than using a lexical resource, such as WordNet, to define the possible word meanings they analysed a large database of documents to infer a set of word senses. The context of each word to be disambiguated was analysed and then clustered according to their similarity. Jing and Tzoukermann (1999) have also reported improvements of 8.6% in retrieval performance. Their disambiguation algorithm computes the word sense similarity in the local context of the query, the similarity of the lexical-occurrence information in the corpus, and the morphological relations between words.

Cross-language information retrieval is a form of text retrieval in which the user submits the query in one language while the documents being retrieved are written in a different language. This interesting problem combines elements of machine translation with information retrieval. Stevenson and Clough (2004) showed that performance on this task could be improved using WSD to identify a suitable meaning from EuroWordNet.

10.2.5 Evaluation

The most common approach to evaluation within NLP is to compare machine output against human judgement. For example, part of speech tagging is normally evaluated by manually annotating a document and then applying a part of speech tagger to the same text. Performance is calculated by comparing the two sets of annotations. However, this process is often complicated by two factors: first, problems of agreeing on the correct annotation, and second, difficulties in comparing annotations. The first problem is caused by the fact that people often disagree on the correct annotation for a particular piece of text – for example, they may believe that different part of speech tags apply. Steps are normally taken to maximise agreement between annotators, but if a sufficient level cannot be reached then meaningful evaluation may not be possible. Difficulties in the comparison of annotations can arise from the complexity of information that NLP technologies attempt to add to text. Information extraction, which often involves the completion of complex template structures, is a good example, but the problem is still present when the information is less complex, since it can be difficult to determine how partial matches should be scored. For example, consider the sentence 'Bill Gates is CEO of Microsoft'. If a system suggests 'Gates' is the name of a person (rather than 'Bill Gates') what credit, if any, should this, partially correct answer be assigned? Proposing 'Gates' as the name of a person, while not ideal, is certainly better than suggesting 'Microsoft' as a person.

There are large differences between the typical performance of NLP technologies. Part of speech tagging and named entity recognition can normally be carried out quite reliably (above 90% accuracy) while accuracy for the information extraction system is typically below 70%. NLP applications (such as those discussed in Section 10.3) often combine together several component technologies, for example part of speech tagging and named entity recognition. There are often dependencies between these components; for example the named entity recogniser may rely on output from the part of speech tagger. Inaccurate output from one component can propagate through to subsequent processes that rely on them.

10.3   Applications of Natural Language Processing in Information Retrieval

There are some applications within IR which seem to benefit from the use of NLP techniques and which can, at some level, be said to rely on them to be effective. In this section we examine two of these in some detail: text mining and question answering.

10.3.1   Text mining

Most readers will be familiar with the term *data mining*, in which the objective is to identify (or *mine*) new and interesting relationships among the concepts in a relational database. However, in many environments, the source data is in the form of unstructured natural language texts rather than structured databases tables. Discovering useful knowledge from such sources has become known as *text mining*. For example, Swanson and Smalheiser (1997) searched for pairs of terms which co-occurred within sentences and then used this information to identify connections between terms which were not mentioned together. For example, if terms A and B commonly occur together within a sentence and terms B and C are often seen together, a possible connection between terms A and C may be identified. This approach was used to identify a previously unknown link between migraine headaches and magnesium deficiency.

Text mining is becoming an increasingly important application of NLP in a variety of domains, including media/publishing, counter terrorism and competitive intelligence. But the greatest initial potential may be found within life sciences – a domain characterized by an almost exponentially increasing publication rate. From modest beginnings in the 1950s, Medline (the US National Library of Medicine's (NLM®) database of biomedical citations and abstracts) now contains over 14 million entries, with around 560 000 added in 2003 alone (Rebholz-Schuhmann *et al.* 2005).

Not surprisingly therefore, the market potential for text mining and related technologies is significant, with some analysts claiming that the life science text mining market could be worth as much as £200 m in 2014[4]. However, other analysts adopt a more cautious view, noting that considerable skills are still required to use text mining tools and products effectively, and until such time as they become usable by non-specialists the market potential for such technologies will remain modest[5].

---

[4] http://www.itilifesciences.com/defaultpage131cd0.aspx?pageID=734&rlID=398

[5] http://www.cfo.com/article.cfm/5347924?f=search

In many ways, text mining represents the combination of a number of NLP techniques within an IR/IE framework, as many of the techniques alluded to above must be combined to provide a complete solution. But text mining and IE are not synonymous – in fact, text mining can operate on a number of levels, with varying complexity. Some examples from the life sciences domain would be (Krallinger *et al.* 2005):

- Tagging biological entities. In biomedical literature, the identification of entities such as genes, proteins, diseases and chemical compounds is a vital step in facilitating accurate document retrieval.
- Information Extraction. Once the relevant entities have been identified (as above), then IE techniques can be applied to identify biologically meaningful structural relations within the text (such as interactions between proteins).
- Knowledge Discovery. Having extracted patterns or facts from the text, this information can then be combined with other knowledge sources (such as external ontologies, etc.) to *infer* or discover new knowledge, which could lead to new scientific discoveries or new insights to direct future research.

However, the current state of the art for biomedical text mining still falls somewhat short of the accuracy levels needed by biologists to adopt them for anything but the simplest of tasks. There may be a number of commercial organisations offering various text mining products, but few if any of these can really be said to constitute a complete 'knowledge discovery' solution. Consequently, many of these companies choose to work closely with selected academic institutions to harness the best research and create the framework for commercialising that know how as and when the market for text mining products matures.

One of the keys to further progress may be the adoption of international competitive evaluations like MUC (Chinchor 1997) and TREC[6]. In recognition of this, the biomedical text mining community has been busy in recent years organising its own evaluation competitions. The first of these was BioCreative (Hirschman *et al.* 2005), which ran in 2003 and consisted of 3 tasks:

1. an entity recognition task (the extraction of gene names from Medline abstracts);
2. a term normalisation task (the generation of lists of unique gene identifiers for genes mentioned in abstracts); and
3. a functional annotation task (the identification of text passages in full-text articles that provide evidence of GO[7] annotations about a particular protein).

The second evaluation was NLPBA (Kim *et al.* 2004), which was organised as part of COLING 2004 (one of the major NLP conferences). NLPBA focussed on the task of named entity recognition but used five classes of entity ('*protein*', '*DNA*', '*RNA*', '*cell line*' and '*cell type*'). The data was based on the GENIA corpus, a collection of MEDLINE abstracts annotated with around 35 named entity classes.

Since then, significant further effort has been applied to the problem of biomedical text mining, involving numerous attempts to modify existing NER systems to work with biomedical content. However, few of these have achieved satisfactory performance, due largely to the special characteristics of biomedical content, i.e. highly descriptive, extended naming conventions, extensive use of conjunction and disjunction, high occurrence of neologisms and abbreviations (Zhou *et al.* 2005).

10.3.2   Question answering

The standard IR paradigm, in which the user is provided with a ranked list of documents thought to contain the information they are looking for, requires the user to search through the documents.

---

[6] http://trec.nist.gov
[7] http://www.geneontology.org

Another approach to meeting the user's information need in a more focused way is to provide specific answers to specific questions. Question answering research has a long history (Green *et al.* 1961; Woods 1973) and was introduced as a task in the Text Retrieval Conference (TREC) IR evaluations in 1999 (Voorhees 1999). Within the TREC framework question answering can be thought of as a specialisation of the standard document retrieval problem in which the queries are expressed as natural language questions and the system is expected to identify the portion of the document in which the answer can be found. Like document retrieval, question answering is often carried out against a well-defined collection of documents which may, or may not, meet the user's information need.

Questions can be posed in a number of different ways which automatic question answering systems should aim to process. Examples of possible question formats include:

- **Yes/no questions** 'Is George W. Bush the current president of the USA?' 'Is the Sea of Tranquillity deep?' **'Who' questions** 'Who was the British Prime Minister before Margaret Thatcher?' 'When was the Battle of Hastings?'
- **List questions** 'Which football teams have won the Champions League this decade?' 'Which roads lead to Rome?'
- **Instruction-based questions** 'How do I cook lasagne?' 'What is the best way to build a bridge?'
- **Explanation questions** 'Why did World War I start?' 'How does a computer process floating point numbers?'
- **Commands** 'Tell me the height of the Eiffel Tower.' 'Name all the Kings of England.'

The standard method for tackling question answering is to approach the problem in three separate stages: (1) question analysis; (2) document retrieval; and (3) answer extraction. The aim of the first stage, question analysis, is to predict the type of answer expected (for example, the expected answer for 'When was Mozart born?' is a date) and creating a query which can then be passed to an IR system. Since the questions are expressed in natural language it is not surprising that NLP techniques have been found helpful for this stage. For example, Greenwood *et al.* (2002) parse the question to produce a logical representation from which the expected answer type can be extracted. The problem of generating a query from the question has not exploited NLP techniques to the same extent although some approaches, for example synonym expansion (Harabagiu *et al.* 2000; Hovy *et al.* 2000), have been explored. The second stage of processing passes the query to an IR engine with the aim of retrieving documents which contain the answer to the question. The final stage of processing, answer extraction, aims to extract the answer(s) from these documents, using the expected answer type information determined in the first stage. Various approaches have been applied to this problem and make use of a variety of NLP techniques. Straightforward methods include simple regular expressions which match the text (Ravichandran and Hovy 2002; Soubbotin and Soubbotin 2001). More complex methods make use of deep linguistic processing. For example, Harabagiu *et al.* (2001) and Scott and Gaizauskas (2000) carry out detailed syntactic and semantic analysis of the retrieved documents which is then queried to identify specific answers to the question. Named entity recognition (see Section 10.2.1) is often used to identify the items in the text which are of the same type as the expected answer and are therefore possible answers to the question.

The nature of the question answering task, which includes the interpretation of natural language questions and identification of specific information within documents, makes the standard document retrieval approach inadequate: NLP techniques are required to carry out some extra processing. In this way question answering is similar to cross-language retrieval (see Section 10.2.3 and both tasks seem to benefit from the use of NLP techniques.

10.4   Discussion

NLP and IR are both language processing technologies, and it might seem reasonable to expect these areas to be mutually beneficial. But this is not always the case, and many would argue that the number of cases in which NLP has improved IR performance are relatively few. However, there are others who believe that the prospects for applying NLP within IR are somewhat more promising and that the lack of 'success stories' to date is more a reflection of the IR community's overemphasis on precision and recall rather than other measures of performance or value. For example, Elworthy *et al.* (2001) demonstrate the application of NLP technology to the retrieval of captioned images, and in so doing also provide a novel method of results presentation and visualisation that could not have been achieved without the use of NLP techniques. It is difficult to quantify the value of such functionality by measuring retrieval accuracy alone. Consequently, there is often debate over precisely what benefit there might be from using NLP technologies in IR at all (a good example of which is the discussion over the relation between WSD and IR discussed earlier).

The majority of recent practical developments in IR have been driven by web search engines. The main recent developments in this area have been gained from ranking algorithms which analyse the link structure of web pages (Brin and Page 1998) rather than deeper analysis of their contents. The sheer size of the web makes the use of computationally intensive NLP techniques impractical since the amount of processing required would quickly become prohibitive. However, commercial web search engines have made use of some NLP technologies, such as providing the option to automatically translate web pages and mechanisms which attempt to automatically correct spelling errors in queries by suggesting more likely alternatives (this has been reported to be one of the most popular features in a major commercial search engine). There is also some evidence that major search engines make use of disambiguation techniques to make results of searches more usable.

There are also information access applications for which some form of NLP is necessary. Examples include those mentioned here: text mining, question answering and cross-language information retrieval. The simple bag of words model, which may be sufficient for document retrieval, is not rich enough to perform all of the processing required for these applications. However, it is often a crucial component – for example, question answering systems often make use of standard approaches to document retrieval to identify passages that may contain the answer to a question.

The Internet is increasingly being used by non-English speakers. Estimates[8] suggest that around 70% of Internet users do not speak English as their native language. The largest of these groups, Chinese speakers, currently account for 15% of all Internet users. Since 2000 there have been substantial increases in the number of Internet users for certain language groups, such as 1576% for Arabic, 474% for Chinese and 422% for French. Internet usage (i.e. the proportion of the population with access to Internet) is still relatively low for some language groups, including Chinese and Arabic, indicating that there is significant potential for further increases in the number of Internet users from these language groups. NLP techniques are likely to be beneficial in this multilingual environment with applications such as cross-language information retrieval and machine translation being used to access information. In addition, the writing systems used for some language groups on the increase (including Chinese and Arabic) are different from the alphabetic system used for English. Retrieval of texts written in these languages raises new challenges that techniques from NLP may help solve.

10.5 Summary

This chapter has reviewed a number of NLP techniques and their application within IR. In general, NLP techniques attempt to provide a deeper analysis of text than the approaches which are currently used in IR and, as we have seen, there are a number of the reasons why this additional information may be expected to be of benefit to IR. In exploring this issue we have examined various NLP technologies and reviewed the ways in which they may be productively applied to IR problems.

While there is some debate over the contribution of NLP techniques to various IR problems it is clear that there are certain information access applications for which the use of NLP is necessary. Two such examples – text mining and question answering – have been discussed in detail. We have also seen that these applications offer many interesting research questions, in addition to significant commercial prospects as the underlying technology matures. In closing, we have examined some

---

[8] http://www.internetworldstats.com/stats7.htm

alternative methods by which the value of NLP to IR can be measured, and reviewed some of the future prospects for this discipline.

Further discussion of the application of NLP to IR can be found in the following books:

- Manning, C. and Schutze, H. *Foundations of Statistical Language Processing*, MIT Press, 1999.
- Jurafsky, D. and Martin, J. *Speech and Language Processing*, Prentice-Hall, 2000.

Both provide an excellent introduction to modern approaches to NLP, and also include introductory material on information retrieval.

The following article provides an excellent introduction to information retrieval, written specifically for an NLP audience, and also provides a discussion of the contribution of NLP to Information Retrieval:

- Tzoukermann, E., Klavans, J. and Strzalkowski, T. Information Retrieval In Mitkov, R. (ed.) *The Oxford Handbook of Computational Linguistics*, 2005, pp. 529–544.

Exercises

10.1 Polysemy and Word Sense Disambiguation (Section 10.2.4). Think of some polysemous words (examples include 'ball', 'bank', 'port', but there are many others). Use these as queries in a search engine and comment on what you notice about the results.

10.2 Question Answering (Section 10.3.2). Several commonly used search engines allow users to formulate their query as a natural language question rather than lists of search terms. Choose one of these and try asking a few questions (for example 'What is the height of the Eiffel Tower in metres?'). What do you notice about the answers? Now try to answer the same questions with a normal web search (by entering a set of query terms instead of a natural language question). Are the results of this search any better or worse? Would it have been easier for you to find the answer to your query by asking a question or by searching in the usual way? Try a range of different questions and see whether this makes any difference to the results.

References

Agichtein, E. and Gravano, L. (2000) Snowball: Extracting relations from large plain-text collections. In *Proceedings of the 5th ACM International Conference on Digital Libraries (ACM DL '00)*, pp. 85–94.

Brin, S. (1998) Extracting Patterns and Relations from the World Wide Web. In *WebDB Workshop at 6th International Conference on Extending Database Technology, EDBT'98*, 1998.

Brin, S. and Page, L. (1998) The Anatomy of a Large-Scale Hypertextual Web Search Engine. *Computer Networks and ISDN Systems* **30**(1–7) 107–117.

Chapman, R. (1977) *Roget's International Thesaurus*. Harper and Row, New York, NY.

Chinchor, N. (1977) *MUC-7 Named Entity Task Definition Dry Run Version, Version 3.5, 17 September 1997*. Available via ftp or telnet at *online.muc.saic.com*, in the file *NE/training/guidelines/ NE.task.def.3.5.ps*.

Church, K. (1998) A Stochastic Parts Program and Noun Phrase Tagger for Unrestricted Text. *Proceedings of the 2nd Conference on Applied Natural Language Processing*, Austin TX, pp. 136–143.

Cutting, D., Kupiec, J., Pedersen, J. and Sibum, P. (1992) A Practical Part-of-Speech Tagger. *Proceedings of the 3rd Conference on Applied Natural Language Processing*, pp. 133–140.

Dingare, S., Finkel, J.R., Manning, C., Nissim, M. and Alex, B. (2004) Exploring the boundaries: Gene and protein identification in biomedical text. In *Proceedings of the BioCreative Workshop*.

Elworthy, D., Rose, T. G., Clare, A. and Kotcheff, A. (2001) A Natural Language System for the Retrieval of Captioned Images. *Journal of Natural Language Engineering*, Cambridge University Press, 2001, vol. 7, pp. 117–142.

Fellbaum, C. (1998) *WordNet: An Electronic Lexical Database and some of its Applications* MIT Press, Cambridge, MA.

Green, B., Wolf, A., Chomsky, C. and Laughery, K. (1961) BASEBALL: An Automatic Question Answerer. *Proceedings of the Western Joint Computer Conference*, **19** 219–224.

Greenwood, M., Roberts, I. and Gaizauskas, R. (2002) The University of Sheffield TREC 2002 Q&A System *Proceedings of the 11th Text Retrieval Conference*.

Grishman, R. and Sundheim, B. (1996) Message Understanding Conference – 6: A Brief History. *Proceedings of the 16th International Conference on Computational Linguistics (COLING-96)* Copenhagen, Denmark, pp. 466–470.

Harabagiu, S., Moldovan, D., Pasca, M. Surdeanu, M., Bunescu, R., Girju, R., Rus, V. and Morarescu, P. (2000) FALCON: Boosting Knowledge for Answer Engines. *Proceedings of the 9th Text Retrieval Conference*.

Hirschman, L. (1998) The evolution of evaluation: Lessons from the message understanding conferences. *Computer Speech and Language* **12** 281–305.

Hirschman, L., Park, J.C., Tsujii, J., Wong, L. and Wu, C.H. (2002) Accomplishments and challenges in literature data mining for biology.. *Bioinformatics* **18** 1553–1561.

Hirschman, L., M. Colosimo, *et al.* (2005) Overview of BioCreAtIvE task 1B: normalized gene lists.. *BMC Bioinformatics* **6**(Suppl 1) S11.

Hirst, G. (1987) *Semantic Interpretation and the Resolution of Ambiguity* Cambridge University Press, Cambridge, England.

Hovy, E. Gerber, L., Hermjakob, U., Junk, M. and Lin, C. (2000) Question Answering in Webclopedia. *Proceedings of the 9th Text Retrieval Conference*, pp. 655–673.

Jing, H. and Tzoukermann, E. (1999) Information retrieval based on context distance and morphology. *Proceedings of the 22nd Annual International ACM SIGIR Conference on Research and Development in Information Retrieval (SIGIR'99)*, pp. 90–96.

Karlsson, F. (1990) Constraint Grammar as a Framework for Parsing English Running Text. *Proceedings of the 13th International Conference on Computational Linguistics*, pp. 168–173.

Kim, Jin-Dong, Tomoko Ohta, Yoshimasa Tsuruoka, Yuka Tateisi and Nigel Collier. (2004). Introduction to the Bio-Entity Recognition Task at JNLPBA. *Proceedings of the International Workshop on Natural Language Processing in Biomedicine and its Applications (JNLPBA-04)*, pp. 70–75.

Krallinger, M. Alonso-Allende Erhardt, R. and Valencia, A. (2005) Text mining approaches in molecular biology and biomedicine. *Drug Discovery Today*, **10**(6) 439–45.

Krovetz, R. (1997) Homonymy and Polysemy in Information Retrieval. *Proceedings of the 35th Meeting of the Association for Computational Linguistics and the 8th Meeting of the European Chapter of the Association for Computational Linguistics (ACL/EACL-97)*.

Krovetz, R. and Croft, B. (1992) Lexical ambiguity and information retrieval. *ACM Transactions on Information Systems* **10**(2) 115–141, pp. 72–79.

Krupke, G. and Rau, L. (1992) GE Adjunct Text Report: Object-oriented Design and Scoring for MUC-4. *Proceedings of the 4th Message Understanding Conference(MUC-4)*. McLean, Virginia, pp. 78–84.

Lesk, M. (1986) Automatic sense disambiguation using machine readable dictionaries: how to tell a pine cone from an ice cream cone. In *Proceedings of ACM SIGDOC Conference*, Toronto, Canada, pp. 24–26.

Marcus, M., Santorini, B. and Marcinkiewicz, M. (1993) Building a Large Annotated Corpus of English: The Penn Tree Bank. *Computational Linguistics* **19**(2) 313–330.

Ravichandran, D. and Hovy, E. (2002) Learning Surface Text Patterns for a Question Answering System. *Proceedings of the 40th Annual Meeting of the Association for Computational Linguistics*, pp. 41–47.

Rebholz-Schuhmann, D., Kirsch, H. and Couto, F. (2005) Facts from text – is text mining ready to deliver?. *PLoS Biol* **3**(2).

Robertson, S. and Spark Jones, K. (1997) Simple Proven Approaches to Text Retrieval. *Technical Report TR356*, Cambridge University Computer Laboratory.

Sanderson, M. (1994) Word sense disambiguation and information retrieval. *Proceedings of the 17th ACM SIGIR Conference*, pp. 142–151.

Sanderson, M. (2000) Retrieving with Good Sense. *Information Retrieval* **2**(1) 49–69.

Schutze, H. and Pedersen, J. (1995) Information Retrieval Based on Word Senses. In *Symposium on Document Analysis and Information Retrieval (SDAIR)*, Las Vegas, NV, pp. 161–175.

Scott, S. and Gaizauskas, R. (2000) University of Sheffield TREC-9 Q & A System. *Proceedings of the 9th Text Retrieval Conference*.

Soubbotin, M. and Soubbotin, S. (2001) Patterns of Potential Answer Expressions as Clues to the Right Answers, *Proceedings of the 10th Text Retrieval Conference*, pp. 175–182.

Stevenson, M. (2004) Information Extraction from Single and Multiple Sentences. *Proceedings of the 20th International Conference on Computational Linguistics (COLING-2004)*, Geneva, Switzerland, pp. 875–881.

Stevenson, M. and Clough, P. (2004) EuroWordNet as a Resource for Cross-language Information Retrieval. In *Proceedings of the 4th International Conference on Language Resources and Evaluation*, Lisbon, Portugal, pp. 777–780.

Stokoe, C. (2005) Differentiating Homonymy and Polysemy in Information Retrieval. *Proceedings of the Human Language Technology Conference and Conference on Empirical Methods in Natural Language Processing (HLT/EMNLP)*, pp. 403–410.

Swanson, D. and Smalheiser, N. (1997) An interactive system for finding complementary literatures: a stimulus to scientific discovery. *Artificial Intelligence*, **91** 183–203.

Voorhees, E. (1993) Using WordNet to disambiguate word senses for text retrieval. In *Proceedings of the 16th Annual International ACM SIGIR conference on Research and Development in Information Retrieval* Pittsburgh, PA pp. 171–180.

Voorhees, E. (1999) The TREC8 Question Answering Track Report. *Proceedings of the 8th Text Retrieval Conference*, pp. 77–82.

Voorhees, E.M. and Buckland, L.P. (ed.) (2002) *The Eleventh Text REtrieval Conference (TREC 2002)*. NIST Special Publication 500-XXX, Gaithersburg, Maryland, USA. http://trec.nist.gov/pubs/trec11/t11proceedings.html

Vossen, P. (1998) Introduction to EuroWordNet. In *Computers and the Humanities* **32**(2–3) 73–89, Special Issue on EuroWordNet.

Voutilainen, A. (2003) Part of speech Tagging. In *The Oxford Handbook of Computational Linguistics* Mitkov, R. (ed.) pp. 219–232 Oxford University Press.

Wilks, Y. (1975) A Preferential, Pattern Seeking, Semantics for Natural Language Inference. *Artificial Intelligence* **6** 83–102.

Wilks, Y. and Stevenson, M. (1998) The Grammar of Sense: Using part-of-speech tags as a first step in semantic disambiguation. *Natural Language Engineering* **4**(3) 135–143.

Woods, W. (1973) Progress in Natural Language Understanding – An Application to Lunar Geology. *AFIPS Conference Proceedings*, vol. 42, pp. 441–450.

Yarowsky, D. (1992) Word-Sense Disambiguation Using Statistical Models of Roget's Categories Trained on Large Corpora. *Proceedings of the 14th International Conference on Computational Linguistics (COLING-92)*, Nantes, France, pp. 454–460.

Yeh, A.S., Morgan, A., Colosimo, M. and Hirschman, L. (2005) BioCreAtIvE task 1A: gene mention finding evaluation.. *BMC Bioinformatics*, **6**(Suppl 1) S2.

Zhou, G.D., Shen, D., Zhang, J., Su, J. and Tan, S.H. (2005) Recognition of protein/gene names from text using an ensemble of classifiers. *BMC Bioinformatics* **6**(Suppl 1).

Zhou, G. *et al*. (2004) Recognizing names in biomedical texts: a machine learning approach. *Bioinformatics*, **20**(7).

11

Cross-Language Information Retrieval

Daqing He and Jianqiang Wang

11.1 Introduction

Cross-Language information retrieval (CLIR) addresses the problem of finding information in one language (e.g. Spanish) in response to queries expressed in another (e.g. English). With the rapid development of web technologies and the explosive growth of electronic information in many languages, it becomes increasingly important to be able to support searching for relevant information in spite of language barriers (Oard and Diekema 1998). From 2000 to 2005, for example, Internet usage increased 126% throughout the world. In particular, it has tripled in the Middle East, Latin America, and Africa, and continues to grow rapidly in Asia and Europe[1]. By the end of 2005, languages used on the Internet were increasingly diverse, with the use of the ten most popular languages on the Internet accounting for only 79.6% of total Internet language use. Although English continues to be the most widely-used language on the Internet, it only accounts for 30.6% of total usage. This means that, increasingly, people's need to access information will cross language boundaries, and without the support of CLIR systems, huge amounts of unique information will be accessible only through a language other than English. At the same time, a great number of Web users will be forced to use English rather than their native languages to search for English materials. Each of these scenarios would hinder the dissemination of human knowledge across the world.

Although Web search engines have become an important part of the current Internet infrastructure, they only support searching for information with queries and documents in the same language. For the past decade, the majority of CLIR activities have been limited to exploring research questions and techniques in laboratory settings. However, with the recent significant progress in machine translation technologies, it will become possible to deploy operational search systems with CLIR capability in the near future.

This chapter aims at providing an introduction to the state of art about CLIR for graduate students, researchers, and information specialists in corporations. In the literature, CLIR is sometimes called 'translingual information retrieval'. It also has a closely related topic area called 'multilingual information retrieval' (MLIR). MLIR concerns the tasks involved in satisfying a query with information

---

[1] http://www.InternetWorldStats.com

from documents in multiple languages which, to a certain degree, can be seen as an aggregation of a set of CLIR tasks (Oard *et al*. 1999). Although there are unique problems in MLIR, MLIR and CLIR share many of the same fundamental challenges and techniques. Therefore, in this chapter we focus our discussion on CLIR, examining the major challenges and related techniques in designing and developing a useful CLIR system. The set of questions that we will explore within this chapter are as follows. Why – and how – does translation play such an important role in CLIR systems? How are the appropriate linguistic units for translating queries or documents determined? What resources can be used in translation? Why are their qualities so important to the CLIR process? Why are translation ambiguities important issues in CLIR, and what major methods have been used to address them? How can searchers examine the translations of the queries and select potential relevant documents written in a language that they do not understand? Finally, if people claim that the challenges in designing CLIR systems have been solved, why is there no commercial CLIR system widely available on the market?

The remainder of this chapter is organized as follows. In Section 11.2, we identify three major challenges in designing and developing techniques for matching queries and documents across languages. This sets the stage for more detailed discussion of important research studies and related techniques for meeting these challenges in Sections 11.3–11.5. In Section 11.6, we discuss CLIR in a more complex and realistic context, in which the evolving characteristics of user information needs and the interactive nature of information retrieval are considered. Section 11.7 provides an overview of the research on evaluating CLIR systems. We conclude the chapter with some remarks on major research directions for CLIR in Section 11.8.

11.2   Major Approaches and Challenges in CLIR

Modern information retrieval (IR) systems often rely on term representations of the searcher's information need to perform the retrieval task. Term mismatch has been recognised as a critical issue in information retrieval (Croft 1995). To succeed in the retrieval task, the searcher must ultimately discover the query terms used by the authors of the documents that are sought. For example, the searcher who wants articles about 'automobiles' might discover the term 'vehicle' as a potential query term after noticing that several relevant documents use the latter instead of the former. If the searcher uses a CLIR system, s/he would need to determine query terms that would lead, after translation, to the retrieval of relevant documents in the other language. There are two major drawbacks to using translation-based CLIR. First, it must be done indirectly by examining translations, rather than the original documents. Second, there are numerous opportunities for errors in translation between the two languages. This means that searchers using CLIR techniques need more support to match the same effectiveness of using monolingual IR (He *et al*. 2003a).

However, ambiguities exist in translations. The exact meaning of some words can only be revealed in context. For example, the word 'bank' has several different meanings depending upon the rest of the sentence: 'I went to the bank and asked to borrow money' or 'we paddled around the island to the town, climbed the river bank to find a group of people playing banjos and singing'. This poses problems for CLIR systems as different meanings often involve different translations. Many techniques have been developed to resolve this so-called translation ambiguity problem (e.g. Ballesteros and Croft 1998; Pirkola 1998), which can be found in all major approaches to develop effective CLIR systems.

CLIR is traditionally viewed as the intersection of machine translation (MT) and information retrieval (IR) (Grefenstette 1998). Translation can be done either manually or automatically. Although often highly accurate in translation quality, manual translation is extremely slow and very expensive when a large number of documents are to be translated. Therefore, commonly applied translation methods in CLIR involve some form of automatic translation. Depending on what is to be translated, translation-based CLIR has three major approaches to translation problems: query translation, document translation, and interlingual techniques (Oard and Diekema 1998).

The query translation approach has been the most popular among the three. It involves translating input queries into the document language. Query translation is usually efficient and flexible and can be performed on the fly after the queries are entered. However, its limitations include the lack of context information for handling translation ambiguities in query terms.

The document translation approach takes the opposite route by translating documents into the same language as the query. Because documents are usually much longer than queries, they provide more context information for resolving translation ambiguities. However, this approach also has some limitations (Oard 1998). As the translation usually involves a large number of documents, it can only be performed offline. In addition, if the system accepts queries in more than one language, it requires that the documents be translated into more than one language, greatly increasing the translation workload. Finally, if the translation materials or translation target have changed, the translation process must be repeated.

The interlingual technologies attempt to map both queries and documents to a language independent representation. Multilingual thesauri were used in the early development of CLIR research to create such representations. However, the thesauri are expensive to construct, and the automatic mapping of terms in queries and documents to the thesauri is itself an open research question (Hlava *et al.* 1997; Loukachevitch and Dobrov 2002). Yet, multilingual thesauri have their merits for handling CLIR among multiple languages. This is why Euro Wordnet was developed for major European languages (Vossen 1998), and has been used in CLIR studies (Gonzalo *et al.* 1999). Automated interlingual techniques, such as latent semantic indexing (Littman *et al.* 1998), the generalized vector space model (Carbonell *et al.* 1997), and kernel canonical correlation analysis (Vinokourov *et al.* 2002), have been considered as the possible alternatives in the interlingual technologies. However, they all rely heavily on having a high- quality on-domain parallel collection with a large quantity of documents in the right languages. This requirement is probably not an issue to resource rich languages such as English in the news domain, but it is a significant problem for resource poor languages such as Cebuano. Recently, Wang and Oard (2006) developed a 'cross-language meaning matching' model that estimates the probability of each pair of query-document words sharing the same meaning based on bidirectional translation and synonymy knowledge. The probabilities are then used to estimate the term frequencies and document frequencies of query terms.

Sometimes CLIR can be achieved without involving any translation. Cognate matching is an example of such a technique in which letter sequences with similar sounds are created to help identify terms in two languages that share similar pronunciation (Buckley *et al.* 2000; Gey and Chen 1997; He *et al.* 2003a). Although these techniques can be applied independently, it is more common for them to be integrated with other CLIR approaches (e.g., He *et al.* 2003a).

There are three major challenges in translation-based CLIR: what to translate, how to obtain translation knowledge, and how to apply the translation knowledge (Oard and Diekema 1998). First, the correct linguistic unit to be used in the translation process has to be determined and then, the corresponding translation resources and language processing tools have to be located or developed. Finally, depending on the availability of translation resources and tools, the strategies for using the resources and tools have to be identified. We will talk about how CLIR research has met these three challenges in detail in Sections 11.3–11.5.

11.3   Identifying Translation Units

Translation units can be word stems, words, character n-grams, or phrases. The selection of translation units depends on what type of translation resource is available. Generally, there are three types of translation resources: bilingual dictionaries, bilingual corpora, and machine translation (MT) systems (Wang 2005). Bilingual dictionaries and corpora focus on word translation and sometimes, phrase translation. Machine translation systems sometimes use many linguistic features, and are usually capable of translating individual words, phrases, or sentences.

Several natural language processing (NLP) techniques have been applied to obtaining and preprocessing translation units. They include tokenisation, stemming, phrase identification, and stop-word removal.

11.3.1 Tokenisation

Tokenisation is the process of recognising words. It often includes isolating words from each other (i.e., word segmentation) and from punctuation marks. Word segmentation can be relatively simple for languages such as English, based on the white space between words. However, word segmentation for languages such as Chinese requires sophisticated algorithms because these languages do not have explicit word boundaries. A sentence can be segmented in different ways even though only one segment conveys the intended meaning of the sentence. Segmentation can also refer to splitting compound noun phrases in German and Finnish, which is also called 'decompounding'. Three sources of evidence can help in automatic word segmentation: lexical information such as lexicons containing known words in that language, algorithmic knowledge such as a heuristic preference for the longest strings, and statistical evidence acquired from a representative collection of text in similar domains (Chen 2002; Monz 2000).

Each source of evidence has its limitations. Lexicon-based segmentation cannot handle words not included in the lexicon. A greedy left-to-right search for the longest matching substring is a simple and effective heuristic approach, but it fails when a shorter substring is more appropriate. Statistical segmentation approaches can help overcome these problems and can further refine the selection among alternative segmentations. The accuracy of statistical segmentation, however, depends on how closely the corpus used represents the text to be segmented. Therefore, practical segmentation schemes exploit multiple sources of evidence (Huang *et al.* 2003). Tokenisation also includes processes for standardizing words. These processes involve recognising and standardising abbreviations and acronyms, correcting word splits due to hyphenation, transforming words to lower case, and removing accents if necessary.

11.3.2 Stemming

Stemming identifies stems shared by morphological variants. In CLIR systems, stemming can be performed either before or after query/document translation. For example, in the query translation approach, post-translation stemming attempts to match the translations of query words with the document words that share the same stems (hence, hopefully sharing the same meanings). The main purpose of pre-translation stemming is to help translate words not covered by the translation resource. That is, when direct matching between query words and words in translation resources fails, matching their stems may allow translation to proceed. Stemming algorithms can be rule-based, as in the Porter stemmer (Porter 1980), or induced statistically by discovering suffixes through examining lexicographically similar words contained in a large corpus (Chen and Gey 2002; Oard *et al.* 2000). The rule-based stemmers are accurate, but more expensive to build.

Whichever type of stemmer is used, stemming 'depth' could be critical since too 'light' stemming will have little effect while too 'aggressive' stemming may pick up words that do not share the same meanings.

11.3.3 Phrase identification

A phrase is a syntactic structure that consists of more than one word and its meanings may not be inferred correctly from component words. Phrases are often less ambiguous than words, therefore, the application of phrase translation in CLIR systems can consistently lead to a higher level of retrieval effectiveness (Ballesteros and Croft 1998). However, the translations of phrases are less likely to be available in a dictionary than are individual words, which means that phrase translation is more likely to fail. Therefore, there is a tradeoff between obtaining phrases and maintaining the coverage. A practical approach is to consider both phrases and individual words as units, so that the translation is applied to phrases first; if that fails, individual words would be translated (He *et al.* 2003a).

Phrases can be identified by phrase dictionaries or may be learned from corpora. There are three types of corpus-based approaches to phrase identification: statistical recognition, part of speech (POS) tagging, and syntactic parsing. Statistical recognition exploits term co-occurrence information (Gey and Chen 1997). For example, to identify two-word phrases, a corpus can first be segmented into overlapping word bigrams (i.e, chunks containing two words), which in turn are ranked in decreasing order by some combination of their term frequency and inverse document frequency. Those receiving high ranks are recognised as phrases. Another way to recognise phrases is to automatically assign POS tags to each word in a text using a probabilistic or rule-based POS tagger. Phrases then can be recognised by POS patterns of consecutive words (Ballesteros and Croft 1998). In addition, syntactic parsers can also be used to perform phrase identification (Gao *et al.* 2000). Syntactic parsing of input text creates syntactic structures that can be regarded as phrases.

11.3.4 Stop-words

Finally, as in monolingual IR, function words (e.g. of, or) and words appearing in many documents (e.g. and, the) are of little importance for IR and should be removed. These function words and frequent words are called stop-words. For CLIR systems based on word-for-word translation, the main reason for removing stop-words prior to performing translation is to reduce translation noise. For example, if the translation knowledge is derived statistically from corpora (see more details in Section 11.4), a stop-word can be incorrectly translated into many different words. Some of them are rare words, which could significantly increase the retrieval of non-relevant documents (Levow and Oard 2000).

11.4 Obtaining Translation Knowledge

In CLIR, translation of queries/documents aims at helping to match what the information searcher means with what the document authors meant. The knowledge applicable in a term-to-term translation often includes all possible translations for the term, and estimates the probability that the term translates to each of these alternatives.

The acquisition of translation knowledge is achieved through two steps: acquiring translation resources and extracting translation knowledge from the translation resource. We focus on two types of translation resources here, bilingual dictionaries and bilingual corpora, since they are both important and widely used translation resources.

11.4.1 Obtaining bilingual dictionaries and corpora

A common way to acquire bilingual dictionaries is to search on the Web. Many language resources are available for free on the Web. The 'Freedict' website is an example[2]. However, dictionaries found in this way often have limited vocabulary coverage, which results in a less effective CLIR systems. Furthermore, such resources are available in only a few languages that have many web-based resources.

Another potential translation resource involves printed bilingual dictionaries. Printed dictionaries are designed for humans to read and often have elaborate structures and excessive information. However, certain patterns can still emerge from regular and repeated structures existing across different lexical entries. This makes it possible to develop techniques that combine machine learning and optical character recognition (OCR) to extract the translation knowledge from printed dictionaries (Ma *et al.* 2003). However, translation resources obtained this way often have suboptimal quality due to the erroneous parsing in dictionary structures and OCR output, which reduces the retrieval effectiveness of CLIR systems (Darwish and Oard 2002). Nevertheless, given the vast volume of printed bilingual dictionaries, research in this area has attracted more and more interest (Kolak *et al.* 2003).

The third and most promising translation resources for CLIR systems are parallel corpora. A parallel corpus consists of pairs of documents that are translations of each other in two languages. There are many human-generated parallel corpora from a variety of sources. For example, the United Nations (UN) has many rules and regulations published in several languages. The European Union publishes

---

[2] http://www.freedict.com.

official documents in several European languages. Documents from the Canadian Parliament are available in both English and French. In Hong Kong, the news and laws are available in Chinese and English. For languages having a large quantity of electronic resources on the web (called high-density languages) and those having only a few (low-density languages), the Bible can be easily constructed to be a parallel corpus (Resnik *et al.* 1999). In addition, parallel corpora can be constructed automatically from the web. Work by Resnik (1998) and Nie *et al.* (1999) used simple, but effective, techniques to find parallel web pages (that is, web pages that are translations of each other). In these studies, simple heuristics such as anchor text and HTML structure were used to select candidate web sites.

11.4.2   Extracting translation knowledge

In addition to the translations of a term, parallel corpora can also provide translation probabilities as part of the translation knowledge. This knowledge can be obtained from parallel corpora using the word alignment technique, which is also an important technique in statistical machine translation (MT). Different from rule-based MT, which requires human-written linguistic rules (syntax, grammars, etc.) for the source language and the target language, statistical MT is a data-driven technique that seeks to automatically 'learn' translation knowledge from bilingual corpora. Sentences in a parallel corpus are aligned pair-wise based on their location, length, etc. (Gale and Church 1991). If two words co-occur frequently in parallel sentence pairs, they are likely to be translations of one another. Brown *et al.* (1993) and Och (2003) give detailed descriptions of statistical MT. A statistical MT toolkit called GIZA++ has been developed and is widely used in CLIR research for obtaining translation knowledge (Och & Ney 2000).

The cross-language latent semantic indexing (CL-LSI) by Littman *et al.* (1998) is another technique for extracting translation knowledge using bilingual corpora. CL-LSI has been discussed as an interlingual method for CLIR tasks. It uses a parallel training corpus aligned at the document level to create a high-dimensional term-vector space, with translation of each document being combined with that document. Singular value decomposition techniques are then used to map the relatively sparse term-vector space to a dense semantic space where each word is represented by a short vector of real numbers giving its position in the reduced semantic space. In this space, words that are closely related to one another will position closely together. Each document in the collection can then be mapped into the space by using the weighted sum of its constituent words. A query can be represented by a vector in the same way, whether it is in the same language as the documents or in another language. In this way, CLIR tasks can be carried out using vector space retrieval techniques developed for monolingual retrieval.

11.4.3   Dealing with out-of-vocabulary terms

When there is no translation for a term in the available resources, the term cannot be translated, which means that it will not contribute to query-document matching. Such terms are often known as out-of-vocabulary (OOV) terms. Because OOV terms are often proper names, technical terms, abbreviations, and acronyms, it is important for CLIR systems to be able to translate them in some way (McNamee and Mayfield 2002; Demner-Fushman and Oard 2003, Wu *et al.*, 2008).

The major techniques for solving OOV problems include transliteration, backoff translation, and pre-translation query expansion. Transliteration can be used with orthographic mapping or phonetic mapping. For languages sharing similar alphabets, orthographic rules specify how certain substrings in one language are spelled in another language. OOV terms then can be transliterated using these rules (Buckley *et al.* 2000; Pirkola *et al.* 2003. For languages that do not share an alphabet (such as Chinese and English), orthographic mapping rarely works. In this case, phonetic mapping can be considered. OOV terms in the source language are first converted into their phonetic representations, which can then be mapped into another language using phonetic mapping rules between the two languages. Finally, phonetic representations in the target language are converted into character sequences. Phonetic mapping rules between two languages can be derived using statistical approaches (Knight and

Graehl 1997). Qu *et al.* (2003) have demonstrated that transliteration results in consistent improvement of CLIR effectiveness.

Backoff translation technique is built upon stemming (Oard *et al.* 2000). The technique consists of four stages: (1) match the surface form of an input term to surface forms of source language terms in the translation dictionary; (2) if this fails, match the stem of the input term to surface forms of source language terms in the dictionary; (3) if this then fails, match the surface form of the input term to stems of source terms in the dictionary; (4) if all else fails, match the stem of the input term to stems of source terms in the dictionary. Oard *et al.* (2000) demonstrated that backoff translation is useful for recovering OOV terms.

Another common technique to mitigate the problems of OOV terms involves query expansion. Query expansion can be done before or after translation (McNamee and Mayfield 2002). In the case of pre-translation query expansion, monolingual retrieval in the query language is performed and the most important terms from top-ranked documents are added to the original query to create an 'expanded' query. The new query is then translated into the document language and used to search the target document collection. The rationale for pre-translation query expansion is that it offers useful terms to be translated into the target language, hence increasing CLIR effectiveness. A similar way of compensating for poor queries is to expand the translated query, which is called post-translation query expansion. Post-translation expansion adds content terms extracted from top-ranked documents back to the query to reduce the effect of inappropriate translations. The results of post-translation query expansion may vary according to the experiment condition (Ballesteros and Croft 1997; McNamee and Mayfield 2002).

Query expansion techniques may be applied to documents where the entire documents are used as if they were queries. Document expansion was first introduced for the retrieval of error-prone automatic speech transcription (Singhal *et al.* 1998) and later applied to CLIR tasks (Levow 2003). The basic idea is to enhance *noisy* translated documents with terms selected from a *clean* document collection. The effectiveness of document expansion for CLIR tasks, however, could be influenced by several factors, such as the topical closeness of the side collection and the number of top documents and top terms used for expansion.

Web mining can also help to identify the translations for OOV terms. Besides obtaining translations for OOV terms as a by-product of creating parallel corpus, it is also possible to specifically looking on the Web for the translations for certain types of noun phrases that refer to people, locations, organizations, etc. This type of phrases, called Named Entities (NEs), is among the most common OOV terms (Demner-Fushman & Oard 2003). There are observations that when new terms, foreign terms, or proper nouns are used in web text of some languages such as Chinese, they are sometimes accompanied by the English translation in the vicinity of the Chinese text, for example 西雅图水手队... Seattle Mariners (Chen *et al.* 2004). Therefore, extraction algorithms have been developed to obtain translations for OOV terms via Web text mining to improve CLIR performance (Chen *et al.* 2004, Zhan and Vines 2004, Wu *et al.* 2008).

11.5 Using Translation Knowledge

The major techniques of refining and using translation knowledge include translation disambiguation, weighting translation alternatives, and using translation probability for estimating query term weight.

11.5.1 Translation disambiguation

The problem of translation ambiguity was briefly discussed in Section 11.2. It is recognised as one of the most important factors to influence CLIR effectiveness. Many techniques have been proposed to resolve this problem, which include either using word statistics computed from corpora or exploring syntactic constraints or other evidence such as dictionary structure. The most effective technique of translation disambiguation based on dictionaries is to translate phases (if covered in the dictionaries) instead of individual words (Ballesteros and Croft 1998; Meng *et al.* 2001; He *et al.* 2003a).

Translation disambiguation can also be performed using corpus-based techniques. For example, term statistics computed from monolingual corpora may also be used to decide translation probability when multiple translation alternatives are present. The simplest techniques use the unigram frequency of each translation in a large target corpus. More effective techniques, however, exploit contextual information, such as word co-occurrence statistics in a target corpus (Gao *et al.* 2000). The underlying hypothesis is that correct translations of two words tend to co-occur more frequently than incorrect translations. In addition, syntactic constraints such as part of speech (POS) may be used to reduce the number of translation candidates (Ballesteros and Croft 1998). Recently, He and Wu (2008) use relevance feedback information from users (i.e., interactive relevance feedback) or without users (i.e., pseudo relevance feedback) to tune query translation closer to users' current search tasks. Their method, which is called Translation Enhancement (TE), uses the documents in relevance feedback and their machine translations to construct a pseudo parallel corpus. Such corpus is then aligned either at sentence level or at word level for extracting translation relationships and probabilities. Their experiments demonstrate that TE improves CLIR retrieval effectiveness, and more importantly, by combining with query expansion, TE helps to make CLIR performance more robust to different query types (He and Wu 2008).

11.5.2 Weighting translation alternatives

Translation disambiguation or translation selection in CLIR systems may still produce more than one translation for each source term, since more than one translation may be appropriate. However, including all translations in the queries or documents will change term statistics such as term frequency (TF), document frequency (DF), and document length. As a result, terms with more translation alternatives may be more influential in query-document matching than terms with fewer alternatives. However, one term having more translations does not necessarily mean it is more important. Therefore, translation alternatives should be weighted so that only truly important translations carry more influence. Techniques for weighting translation alternatives include weighted Boolean model, balanced translation, structured queries, probabilistic structured queries, and language models.

The weighted Boolean model is perhaps the earliest technique for weighting multiple translations (Hull 1997). Original query terms are linked with Boolean conjunctions and the translation alternatives of each query term are linked with Boolean disjunctions. The Boolean conjunction results in disambiguation since the correct translation equivalents of two or more query terms are much more likely to co-occur in the target language than any incorrect corresponding translation equivalents. The Boolean disjunction can suppress the weight of query terms under many translation equivalents. Therefore, translation disambiguation and weighting occur in the same model without using extra resources such as a corpus.

The most intuitive way to mitigate the unbalanced contribution of multiple translation alternatives is to average the term weights (Leek *et al.* 2000; Oard *et al.* 2000). The idea is called 'balanced translation' since the weight of a source term is defined as the arithmetic mean of the weight of each of its translations. In turn, term weight can be computed with standard term weighting schemes. With balanced translation, however, rare translations tend to contribute more to the source term weight than common translations.

A more effective weighting method involves the use of multiple translation alternatives in translating queries, the so-called structured queries (Pirkola 1998). The basic idea of the technique is to treat multiple translation alternatives as if they were all instances of the query term. Specifically, the term frequency (TF) of a query term with regard to a document is computed as the sum of the TF of each of its translation alternatives found in that document, and its document frequency (DF) is computed as the number of documents in which at least one of its translation alternatives appears. The DF computation can be approximated as the sum of the DF of each translation alternative (Kwok 2000).

11.5.3   *Using translation probabilities in term weighting*

It is natural to incorporate translation probabilities into the computation of term weights. Within the vector space model framework, one effective technique developed for using translation probabilities is the 'probabilistic structured query' method (Darwish and Oard 2003). The technique extends structured queries by incorporating a weighting factor, the translation probability, in estimating the TF and DF of a term. One obvious advantage of this method is that it can greatly reduce the effect of low-probability translation alternatives that would otherwise have a big influence on term weighting. Translation probabilities, in particular translation probabilities from document terms to query terms, have been commonly used in the language modelling approaches for CLIR tasks (Hiemstra 2000; Xu and Weischedel 2000; Kraaij 2004).

Translation probabilities in both directions may be used simultaneously for computing term weights in CLIR systems. One method of using translation in both directions merges the ranked list from query translation with the ranked list from document translation, which has proven to be more effective than either one alone (Franz *et al.* 2000; Kang *et al.* 2004). Wang (2005) developed a technique for combining the translation probabilities in both directions with the calculation of term weights. The technique identifies terms that share the same meaning as a set of synonyms (called a synset), and establishes translation knowledge based on the synset for matching a query against a document. Synsets may be obtained from existing resources such as WordNet or derived statistically from parallel corpora. Synsets can also be aligned across languages using resources such as EuroWordNet or statistical techniques. Both the structured queries and probabilistic structured queries have been shown to be special cases of this technique. Cross-language search results based on this technique were statistically indistinguishable from strong monolingual baselines for both French and Chinese documents (Wang 2005).

11.6   Interactivity in CLIR

11.6.1   *Interactive CLIR*

In our discussion of problems and techniques in previous sections, we defined CLIR as a process of matching queries with documents, assuming searchers' information needs can be expressed as a 'static' set of queries. In practice, however, searchers need to be integrated into the CLIR process because it is the searchers who initiate and refine the queries, and it is the searchers who ultimately decide whether the retrieved information is relevant to their needs. Furthermore, CLIR is both an interactive and an iterative process. Figure 11.1 shows a typical CLIR process, which starts when a human searcher initiates a retrieval task to satisfy his/her information need, and ends when the relevant information for meeting such a need has been found or the searcher gives up. CLIR that involves users is generally known as 'interactive CLIR'.

Therefore, the matching between a query and documents is just one stage among several in interactive CLIR (called 'Ranked Retrieval' in Figure 11.1). There are four other stages, all of which are possible interaction points between searchers and the system. Three of these, *Query Formulation*, *Selection*, and *Examination*, exist for all kinds of information retrieval whereas the fourth, *Query Translation*, is unique to interactive CLIR.

Interactions between searchers and retrieval systems are important to the success of the CLIR process. Humans and machines can bring complementary strengths to an interactive CLIR process. Machines are excellent at repetitive tasks that are well specified; humans bring creativity and exceptional pattern recognition capabilities. Properly combining these capabilities can result in a synergy that greatly exceeds the ability of either human or machine alone (He *et al.* 2003a).

We all know that queries come from the searchers. But saying that raises the question of how the searchers learned to formulate the correct query. Searchers often find that, over the course of a

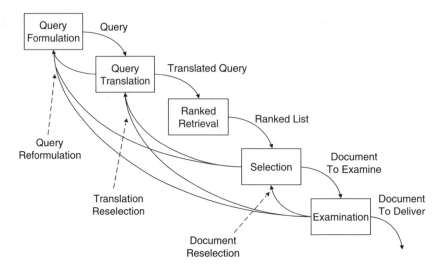

Figure 11.1   Stages in interactive cross-language information retrieval

search session, their understanding of what they actually need changes. Moreover, they may also need to learn to effectively express those information needs. Strategies based on iterative refinement are commonly used in such cases (Marchonini 1995). The success of iterative refinement depends on two types of knowledge: an understanding of why the machine produced the results that were obtained, and an understanding of the ways in which the outcome could be altered. Therefore, searchers can be viewed as seeking to develop and refine three mental models: (1) their understanding of their own information needs; (2) appropriate query terms whose translations might be present in the looked-for documents; and (3) ways of combining these terms to accurately express the needs. The four stages points provide opportunities to develop such mental models.

In interactive systems, searchers can leverage feedback to support refinement of their mental models. The five backward arrows in Figure 11.1 represent such feedback opportunities. These iterative refinement processes make CLIR, like monolingual IR, a process of repeated stages, ranging from query formulation to document examination.

Compared with interactive monolingual IR, supporting interactive CLIR has its unique challenges. For example, document selection often cannot be performed with original documents. Some form of automatic translation has to be applied, which inevitably introduces ambiguities and errors that would impact searchers' accurate understanding of document content. Thus, query refinement in CLIR tasks is more challenging because: (1) searchers' identification of expanding terms is based on automatic translations of documents as mentioned above; and (2) searchers usually cannot control which translation would be selected by the CLIR system for their expanded terms. Therefore, a correct expanded term might not work well when the system mistranslates the term. In the remainder of this section, we will examine techniques to solve these two problems.

11.6.2   Query translation in interactive CLIR

Interactive CLIR systems have close connections with the CLIR search engine inside them. Depending upon the approach that the CLIR search engine takes, interactive CLIR systems can be based on query, document, or interlingual translation methods. Because the query translation approach is simple, flexible, and provides an opportunity for the searchers to review the translations, most interactive CLIR systems use the query translation approach (He *et al.* 2003b; Petrelli *et al.* 2004). Therefore, our discussion will be concentrated on the support for query translation.

Queries can be translated automatically. Kim *et al.* (1999) were among the first to explore automatic query translation in interactive CLIR. Their 'FromTo-CLIR' system allowed searchers to conduct CLIR tasks with a process identical to monolingual IR. The searchers are only concerned with formulating queries, and reading the translations of the returned documents. Automatic translation of queries can also be performed on multi-word expressions, which exhibit markedly less homonymy than single words. López-Ostenero *et al.* (2003) generated all possible translations for the constituent words in noun phrases and then filtered the results using a representative text collection to remove all but the most common rendition. Because the resulting noun phrases always induce a single unique translation, translation ambiguities can often be avoided.

Even though fully automated translation assisted by translation knowledge is an effective approach, it can be difficult for the searchers to understand (and ultimately control) what the system is doing, which often produces unexpected and potentially inscrutable results. The searchers can identify the query terms that are responsible for the poor search results by examining the original returned documents in monolingual IR, but this often fails in CLIR processes because the translations are used rather than the original.

Therefore, researchers have also explored manual translation of queries, or more accurately 'user-assisted query translation' (He *et al.* 2003a,b). The rationale is that the queries are generated by the searchers. They can determine the correct translation if they know the meanings of the translations. By participating in the query translation process, the searchers also can learn about the CLIR system's translation resources and translation abilities, which are useful to understand so that the searchers can control the system's behaviour. However, the keys to the success of user-assisted query translation are: (1) that the searchers can understand the meaning of translation alternatives, and (2) that the searchers can see the effect of the selection.

The meanings of the translations can be presented by manually generated English definitions (Ogden *et al.* 1999) or by machine-generated 'back translations' which are the English terms that share the same translation (Capstick *et al.* 2000). However, manually generated definitions are expensive to obtain, and the back translations are sometimes difficult to understand. He *et al.* (2003a, b) developed the Miracle system, which contains two other cues that are based on 'keyword in context'; (KWIC) using parallel corpus and comparable corpus respectively. The idea is that the searchers can understand the meaning and the usage of a term in a foreign language by examining the meaning and the usage of its English counterpart in the parallel or comparable corpus.

In user-assisted query translations, selecting correct translations could improve results, but omitting a useful translation could just as easily have an adverse effect. Therefore, the searchers should be able to see the effect of their selection/deselection of translations as soon as they perform the action. Thus, the appropriate design is a progressive refinement strategy in which a fully automatic search is first performed and subsequent searches are then refined by interactive deselection of inappropriate translations (He *et al.* 2003a; Petrelli *et al.* 2004; Oard *et al.* 2008). Consequently, interactive CLIR systems should include a user-assisted query translation capability to support iterative query refinement. If searchers make poor choices in query terms or translations of query terms, they can see the effect immediately and learn to better control the system.

11.6.3 Document selection in interactive CLIR

Interactive CLIR does not stop at the point when the retrieval system generates a ranked list of documents. Since no retrieval system, even in a monolingual environment, can return only relevant documents, searchers need the means to examine returned documents and select relevant information from them. As happens with web search engines, the selection process is usually performed on surrogates of the documents, but the searcher does get the chance to examine the full content of the document. However, in CLIR tasks, some forms of translations of the original documents have to be utilised for the searcher to understand the content (Oard *et al.* 2004).

The key to document selection and examination is determining the types of translations that should be used. Translations can be generated manually, or by some automatic methods. Manual translation has the advantage of accuracy and a higher level of quality, but it is too expensive to use when the collection contains thousands of documents.

Research work on cross-language document selection has concentrated on automatic methods. Oard and Resnik (1999) were among the first to study translation methods in cross-language document selection. They conducted usability tests on automatically generated word-by-word gloss English translations of Japanese documents. They found that searchers can categorise the documents using gloss translations, but the performance was inferior to those using human prepared translations. Ogden and Davis (2000) studied the use of machine translation (MT) outputs in cross-language document selection, and found that the quality of the translations from a state-of-the-art MT system (i.e. Systran system) is adequate for cross-language document selection. Wang and Oard's (2001) study, comparing the effectiveness of using gloss translation and machine translation in cross-language document selection, showed results similar to Oard and Resnik's: gloss translation is adequate for document selection, but the performance of document selection when using MT is significantly better than when using gloss translation.

However, MT is still expensive to run, so López-Ostenero and his team worked on using the translations of noun phrases instead of the translations of whole documents in cross-language document selection (López-Ostenero *et al.* 2002; Oard *et al.* 2004). The translation of noun phrases was automatically generated using a comparable corpus, and it was at least one order of magnitude faster than the state-of-the-art commercial MT system. Their results show that using noun phrase translations will allow searchers to obtain results comparable to those using MT output.

11.7   Evaluation of CLIR Systems

The experimental evaluation of the CLIR systems shares many characteristics and considerations with the evaluation of monolingual IR systems. The evaluation can test on the core CLIR retrieval engine or examine the whole interactive CLIR system. However, CLIR evaluation has unique challenges because 'the languages covered by the translation resources must match the languages covered by the evaluation resources' (Oard and Diekema 1998).

11.7.1   Cranfield-based evaluation framework

Like its monolingual counterpart, large scale evaluation of CLIR search engines has been following the Cranfield tradition first done in the Cranfield II experiment (Cleverdon *et al.* 1966). This tradition has the characteristics of viewing static and objective topic similarity as the relevance criteria, and evaluates the effectiveness of the retrieval engine based on the returned ranked lists of documents (Harman 1992). There is a test collection containing fixed document and query sets, with carefully established relevance judgments linking documents to certain queries. However, the language of the documents and that of the queries in a CLIR test collection are different, which makes the relevance judgments on each document and query pair much more difficult. The CLIR test collections may also provide translation resources such as dictionaries and bilingual corpora to further standardize the retrieval process.

CLIR evaluation shares many commonly used measures of the retrieval effectiveness with monolingual IR. This includes precision, which examines the proportion of relevant documents among returned sets of documents; recall, which examines the proportion of relevant documents being returned among all relevant documents; and mean average precision (MAP), which looks at the quality of the ranked lists by examining both precision and recall factors across all topics. Each of these measures has strengths and weakness. An oft-used strategy is to employ several of them during the evaluation. The effectiveness of a CLIR search engine is often compared to a similar monolingual technique on the same collection. Expressed as a percentage of monolingual effectiveness, researchers have established that CLIR typically ranges from 50% for unconstrained dictionary-based query translation to 105% for more sophisticated techniques (Wang 2005).

11.7.2   Evaluations on interactive CLIR

Evaluation can also be applied to interactive CLIR systems. People are more interested in how a CLIR system can help the human user to identify relevant information in the other language. As shown in Figure 11.1, the evaluation of an interactive CLIR system can be modelled by examining how well a CLIR system can support: (1) query formulation and translations, and (2) document selection and examination.

The framework for evaluating interactive CLIR systems can be built on top of the test collections developed for evaluating core CLIR search engines (Oard and Gonzalo 2004). By adopting documents, topics and relevance judgments from the test collections for evaluating core CLIR search engines, it greatly reduces the amount of effort for running the evaluation. However, this poses challenges in terms of experiment design: (1) the design must account for differences in the searcher's knowledge of those topics; (2) the design must account for the fact that the searchers recruited for this evaluation must search on predetermined topics rather than their own. For the past several years, the CLEF interactive track has adopted both within-subject design and training with clear instructions to overcome those problems (Gonzalo and Oard 2003, Oard and Gonzalo 2004).

Another challenge in evaluating interactive CLIR systems is that some form of translations of the returned documents has to be provided. Evaluating interactive CLIR systems often involves human searchers reading and identifying relevant returned documents, where translations of returned documents are used to facilitate these actions. The outputs from a machine translation system are often used to provide the baseline, and other types of translations, such as word-by-word gloss translations (Resnik 1997; Wang and Oard 2001) and phrase translations (López-Ostenero *et al.* 2002), have also been explored in such evaluations.

11.7.3   Current CLIR evaluation frameworks

During the past decade or so, there have been four major frameworks developed to evaluate various aspects of CLIR systems. The Text Retrieval Conference (TREC) CLIR track concentrated on CLIR between English and Chinese/Arabic (e.g. Gey and Oard 2001). The Cross-Language Evaluation Forum (CLEF) examines CLIR across major European languages (e.g. Peters *et al.* 2007). The NII-NACSIS Test Collection for IR Systems (NTCIR) evaluation studies CLIR between English and major East Asian languages (Chinese, Korean, and Japanese) (e.g. Kando and Takaku 2005;), and the Topic Detection and Tracking (TDT) works on CLIR containing spoken documents (Allan 2004). CLEF also contains interactive track that specifically examines the interactive CLIR systems (e.g. Gonzalo and Oard 2003, Oard and Gonzalo 2004).

11.8   Summary and Future Directions

11.8.1   Current achievements in CLIR

After a decade of rapid development, researchers have a much clearer understanding of the challenges of using CLIR systems. Query translation has been the most common approach for CLIR, but document translation and interlingual representation provide valid alternatives for certain applications. The three major challenges in translation-based CLIR are problems related to translation units, translation knowledge, and translation ambiguities. A great number of techniques and models have been proposed, developed, and evaluated in large scale quantitative experimental frameworks. We also have seen several interactive CLIR systems built, and their usefulness in supporting human searchers in CLIR tasks have been examined over various platforms as well.

Probably the most noticeable achievement in CLIR is that cross-language document ranking can often achieve near 100%, or even higher, of the retrieval effectiveness of monolingual document ranking. However, the effectiveness largely depends on the availability of high-quality translation resources, including translation lexicons with good probability estimates, large and domain-relevant bilingual corpus, and so on.

11.8.2   Future directions for CLIR

However, the CLIR process does not start at a predefined static query, nor does it end at a ranked list of documents that the human searchers cannot read, select or use. Although there has been work done in the interactive CLIR area, most challenges related to cross-language information use remain unsolved. Some tools for helping searchers to select translations have been developed, but more work still needs to be done in iterative search processes to better support searchers' development of their mental models about the need, the system, and the collection.

For most low-density languages, acquiring translation knowledge is still the biggest challenge. The quality of CLIR largely depends on the quality of the translation resources, which requires vast amount of text resources on both language sides. The key problems to be resolved include how to acquire a large amount of electronic texts for those languages, and/or how to develop robust CLIR techniques that can rely on translation resources derived from a small amount of bilingual texts.

Our understanding of CLIR for non-text information is still quite limited. Hence, techniques for searching non-text information across languages are still in their infancy. The improvement of CLIR in recent years has made it feasible to combine CLIR with other retrieval tasks. Cross-language spoken document retrieval (e.g. Oard *et al.* 2007, Pecina *et al.* 2008), question answering (e.g. Echihabi *et al.* 2004, Giampiccolo *et al.* 2008), photo retrieval (e.g. Grubinger *et al.* 2008), and medical image retrieval (Clough *et al.* 2006, Müller *et al.* 2008) are just a few examples of combining CLIR with other retrieval tasks.

Finally, CLIR is just one facet of information retrieval, whose ultimate task is to support people finding the information that they want. As we needed to know how to support retrieval beyond language barriers, CLIR became a significant research area. Eventually, both language and modality aspects of information should be made transparent to human searchers. At that time, the functionalities of CLIR will be an integrated component of the ultimate information access systems. Consequently, CLIR research is still far from complete – there are more and more interesting problems waiting to be solved.

11.8.3   Further reading

Although it was published about a decade ago, (Oard and Diekema 1998) can still provide an excellent overview about major issues, problems, and approaches in CLIR. Grefenstette's book *Cross-Language Information Retrieval* (Grefenstette 1998) is an excellent source to the details of a few well known techniques in CLIR developed in the late 1990s. Readers who want to know the latest overview of the CLIR field can look at the discussions in SIGIR 2006 workshop 'New Directions in Multilingual Information Access' and subsequent publication at ACM SIGIR Forum (Gey *et al.* 2006).

As the two active CLIR evaluation frameworks, CLEF and NTCIR publish latest CLIR techniques for various CLIR tasks regularly. Readers who are interested in CL question answers, CL image retrieval, CL geographic information retrieval, etc. should follow the proceedings of CLEF (http://www.clef-campaign.org) and NTICR (http://research.nii.ac.jp/ntcir/) closely. The annual SIGIR and CIKM conferences are also the two places to obtain publications about the latest CLIR techniques. Readers who want to know thorough discussions of certain CLIR techniques can also check on related international journals such as *Information Processing and Management, ACM Transactions on Information Systems, Information Retrieval, Journal of the American Society for Information Science and Technology*, and *ACM Transactions on Asian Information and Language Processing*.

Exercises

11.1 *CLIR search with you as a CLIR system*. Think of a topic that makes sense to search cross language, and pick up a meaningful query. Then using online translation resources such as Google Translate (http://www.google.com/translate_t#), Yahoo Babel Fish (http://babelfish.yahoo.com), or online dictionaries like http://dictionary.reference.com to translate the query into a language that you want to search on. For Google Translate and Yahoo Babel Fish, you can translate the query as a whole, or one word each time, whereas you have to translate each query term individually if you use online dictionaries. Once you have obtained the translated query, perform search on a web search engine such as Google or Yahoo. With the returned search results, you can congratulate yourself that you have successfully acted as a cross-language information retrieval system and complete a search.

11.2 *CLIR with Google Translate*. Either think of a new suitable CLIR search topic or simply reuse the one you adopted in Exercise 11.1. At Google Translate site (http://www.google.com/translate_t#), click on 'Translated Search'. Type your original query into the text box after 'Search for', select appropriate languages for 'My language' and 'Search pages written in', then click on the button 'Translate and Search'. The returned results are the outcome of a cross-language search. Now determine whether you like the results you obtained, and ask yourself which part of the search results were you looking at? Did you look at the original web pages or the translation version? Why does Google not only return the original websites in the foreign language, but also the translated versions? If you do not like the search results, you can change your original query or click on the link 'Not quite right? Edit'. Why do we need this feature? Can it be improved? How?

References

Allan, J., (2004). Introduction to Topic Detection and Tracking, in *Topic Detection and Tracking, Event-based Information Organization*, (eds J. Allan) Kluwer Academic Publishers, pp. 1–16.

Ballesteros, L. and W. B. Croft, (1997). Phrasal Translation and Query Expansion Techniques for Cross-Language Information Retrieval, in *Proceedings of the 20th International ACM SIGIR Conference on Research and Development in Information Retrieval*, New Orleans, Louisiana, pp. 84–91.

Ballesteros, L. and W. B. Croft, (1998). Resolving Ambiguity for Cross-Language Retrieval, in *Proceedings of the 21st Annual International ACM SIGIR Conference on Research and Development in Information Retrieval*, Melbourne, Australia, pp. 64–71.

Brown, P., S. Della Pietra, V. Della Pietra, and R. Mercer, (1993). The Mathematics of Statistical Machine Translation: Parameter Estimation. *Computational Linguistics*, **19**(2), 263–311.

Buckley, C., M. Mitra, J. Walz, and C. Cardie, (2000). Using Clustering and Super Concepts Within SMART: TREC 6. *Information Processing and Management*, **36**(1), 109–31.

Capstick, J., A. K. Diagne, G. Erbach, H. Uszkoreit, A. Leisenberg, and M. Leisenberg, (2000). A System for Supporting Cross-Lingual Information Retrieval. *Information Processing and Management*, **36**(2), 275–89.

Carbonell, J., Y. Yang, R. Frederking, R. Brown, Y. Geng, and D. Lee, (1997). Translingual Information Retrieval: A Comparative Evaluation, in *Proceedings of the 15th International Joint Conference on Artificial Intelligence*, Morgan Kaufmann, pp. 708–15.

Chen, A., (2002). Multilingual Information Retrieval using English and Chinese Queries. in *Evaluation of Cross-Language Information Retrieval Systems: Second Workshop of the Cross-Language Evaluation Forum, CLEF 2001 Darmstadt, Germany, September 3–4, 2001*, Springer, Berlin, pp. 373–81.

Chen, A. and F. Gey, (2002). Building an Arabic Stemmer for Information Retrieval, in *Proceedings of Text REtrieval Conference (TREC-2002)*, NIST Gaithersburg, MD.

Cheng, P.-J., J.-W. Teng, R.-C. Chen, J.-H. Wang, W.-H. Lu, and L.-F. Chien, (2004). Translating Unknown Queries with Web Corpora for Cross-Language Information Retrieval, in *Proceedings of the 27th Annual International ACM SIGIR Conference on Research and Development in Information Retrieval*, Sheffield, United Kingdom ACM Press, pp. 146–53

Cleverdon, C. W., J. Mills, and E. M. Keen, (1966). Factors Determining the Performance of Indexing Systems, Vol. **1**: Design, Vol. **2**: Test Results, Cranfield England.

Clough, P., H. Muller, T. Deselaers, M. Grubinger, T. M. Lehmann, A. Hanbury, and W. Hersh, (2007). The Cross Language Image Retrieval Track: ImageCLEF 2006, in *Evaluation of Multilingual and Multi-modal Information Retrieval: 7th Workshop of the Cross-Language Evaluation Forum, CLEF 2006, Alicante, Spain, September 20–22, 2006, Revised Selected Papers*. (Eds. Peters, C., Clough, P., Gey, F. C., Karlgren, J., Magnini, B., Oard, D. W., de Rijke, M., Stempfhuber, M.). pp. 579–94.

Croft, W. B., (1995). What Do People Want from Information Retrieval? *D-Lib Magazine*, November 1995.

Darwish, K. and D. W. Oard, (2002). Term Selection for Searching Printed Arabic, in *Proceedings of the 24th Annual International ACM-SIGIR Conference (SIGIR 2002)*, Tampere, Finland, pp. 261–8.

Darwish, K. and D. W. Oard, (2003). Probabilistic Structured Query Methods, in *Proceedings of the 21st Annual 26th International ACM SIGIR Conference on Research and Development in Information Retrieval*, Toronto, Canada, pp. 338–44.

Demner-Fushman, D. and D. W. Oard, (2003). The Effect of Bilingual Term List Size on Dictionary-Based Cross-Language Information Retrieval, in *36th Annual Hawaii International Conference on System Sciences (HICSS'03) - Track 4*, Hawaii.

Echihabi, A., D. W. Oard, D. Marcu, and U. Hermjakob, (2004). Cross-Language Question Answering at the USC Information Sciences Institute, in *Comparative Evaluation of Multilingual Information Access Systems: 4th Workshop of the Cross-Language Evaluation Forum, CLEF 2003, Trondheim, Norway, August 21-22, 2003, Revised Selected Papers*. Springer: Berlin. pp. 514–22.

Franz, M., J. S. McCarley, and R. T. Ward, (2000). Ad Hoc, Cross-Language and Spoken Document Retrieval at IBM, in *Proceedings of Text REtrieval Conference (TREC-8)* NIST Gaithersburg, MD, pp. 391–8.

Gale, W. and K. Church, (1991). A Program for Aligning Sentences in Bilingual Corpora, in *Proceedings of ACL 1991*, Berkeley, CA, pp. 177–84.

Gao, J., J.-Y. Nie, J. Zhang, E. Xun, Y. Su, M. Zhou, and C. Huang, (2000). TREC-9 CLIR Experiments at MSRCN. in *the Ninth Text REtrieval Conference (TREC-9)*. NIST Gaithersburg, MD, pp. 343–54.

Gey, F. and A. Chen, (1997). Phrase Discovery for English and Cross-Language Retrieval at TREC 6, in *the Proceedings of the 6th Text Retrieval Conference (TREC 6)* Gaithersburg, MD: NIST Gaithersburg, MD, pp. 637–48.

Gey, F. C., N. Kando, C.-Y. Lin, and C. Peters, (2006). New Directions in Multilingual Information Access: SIGIR2006 Workshop Report. *ACM SIGIR Forum*, **40**(2).

Gey, F. C. and D. W. Oard, (2001). The TREC-2001 Cross-Language Information Retrieval Track: Searching Arabic Using English, French or Arabic Queries, in *the Tenth Text REtrieval Conference (TREC 2001)* NIST Gaithersburg, MD. pp. 16–25.

Giampiccolo, D., Forner, P., Penas, A., Ayache, C., Cristea, D., Jijkoun, V., Osenova, P., Rocha, P., Sacaleanu, B., and Sutcliffe, R., (2008). Overview of the CLEF 2007 Multilingual Question Answering Track, in *Advances in Multilingual and Multimodal Information Retrieval: 8th Workshop of the Cross-Language Evaluation Forum, CLEF 2007, Budapest, Hungary, September 19-21, 2007, Revised Selected Papers*. Springer, Berlin. pp. 200–36.

Gonzalo, J. and D. W. Oard, (2003). The CLEF 2002 Interactive Track, in *Advances in Cross-Language Information Retrieval: Third Workshop of the Cross-Language Evaluation Forum, CLEF 2002 Rome, Italy, September 19-20, 2002 Revised Papers*. Springer, Berlin. pp. 372–82.

Gonzalo, J., F. Verdejo, and I. Chugur, (1999). Using EuroWordNet in a Concept-Based Approach to Cross-Language Text Retrieval. *Applied Artificial Intelligence Special Issue on Multilinguality in the Software Industry: the AI contribution*. **13**, 647–78.

Grefenstette, G., (1998). The Problem of Cross-Language Information Retrieval, in *Cross-Language Information Retrieval*, ed. G. Grefenstette Kluwer Academic Publishers, pp. 1–10.

Grubinger, M., Clough, P., Hanbury, A., and Müller, H., (2008). Overview of the ImageCLEFphoto 2007 Photographic Retrieval Task. in *Advances in Multilingual and Multimodal Information Retrieval: 8th Workshop of the Cross-Language Evaluation Forum, CLEF 2007, Budapest, Hungary, September 19-21, 2007, Revised Selected Papers*. Springer, Berlin. pp. 433–44.

Harman, D., (1992). Overview of the First TREC. In *The First Text REtrieval Conference TREC1*. 1–21.

He, D., D. W. Oard, J. Wang, J. Luo, D. Demner-Fushman, K. Darwish, P. Resnik, S. Khudanpur, M. Nossal, M. Subotin, and A. Leuski, (2003a). Making MIRACLEs: Interactive Translingual Search for Cebuano and Hindi. *ACM Transactions on Asian Language Information Processing*, **2**(3), 219–44.

He, D., J. Wang, D. W. Oard, and M. Nossal, (2003b). Comparing User-Assisted and Automatic Query Translation, in *Advances in Cross-Language Information Retrieval: Third Workshop of the Cross-Language Evaluation Forum, CLEF 2002 Rome, Italy, September 19-20, 2002 Revised Papers*. Springer, Berlin. pp. 400–15.

He, D. and D. Wu. Translation Enhancement: a New Relevance Feedback Method for Cross-Language Information Retrieval, (2008). in *Proceeding of the 17th ACM conference on Information and knowledge management*, ACM: Napa Valley, California, USA. pp. 1203–17.

Hiemstra, D., (2000). Using Language Models for Information Retrieval, Centre for Telematics and Information Technology, University of Twente. Ph.D. Thesis.

Hlava, M., R. Hainebach, G. Belonogov, and B. Kuznetsov, (1997). Cross-Language Retrieval – English/Russian/French, in *AAAI Spring Symposium on Cross-Language Text and Speech Retrieval*, Palo Alto, CA. pp. 63–83.

Huang, X., F. Peng, D. Schuurmans, N. Cercone, and S. Robertson, (2003). Applying Machine Learning to Text Segmentation for Information Retrieval. *Information Retrieval*, **6**(3/4), 333–62.

Hull, D. A., (1997). Using Structured Queries for Disambiguation in Cross-Language Information Retrieval. in *AAAI Symposium on Cross-Language Text and Speech Retrieval*, pp. 84–98.

Kando, N. and M. Takaku, (2005). *Proceedings of the Fifth NTCIR Workshop Meeting on Evaluation of Information Access Technologies: Information Retrieval, Question Answering and Cross-Lingual Information Access*. Tokyo, Japan.

Kang, I.-S., S.-H. Na, and J.-H. Lee, (2004). POSTECH at NTCIR-4: CJKE Monolingual and Korean-related Cross-Language Retrieval Experiments. in *Working Notes of the 4th NTCIR Workshop*. Toyko, Japan.

Kim, T., C. Sim, S. Yuh, H. Jung, Y. Kim, S. Choi, D. Park, and K. S. Choi, (1999). FromTo-CLIRTM: Web-Based Natural Language Interface for Cross-Language Information Retrieval. *Information Processing and Management*, **35**(4), 559–86.

Knight, K. and J. Graehl, (1997). Machine Transliteration, in *Proceedings of the Thirty-Fifth Annual Meeting of the Association for Computational Linguistics and Eighth Conference of the European Chapter of the Association for Computational Linguistics*, (eds. W. Wahlster Somerset), New Jersey: Association for Computational Linguistics, pp. 128–35.

Kolak, O., W. J. Byrne, and P. Resnik, (2003). A Generative Probabilistic OCR Model for NLP Applications, in *Proceedings for HLT-NAACL 2003*. Edmonton, Canada. pp. 55–62.

Kraaij, W., (2004). Variations on Language Modeling on Information Retrieval, University of Twente. PhD thesis.

Kwok, K. L., (2000). Improving English and Chinese ad-hoc Retrieval: A Tipster Text Phase 3 Project. *Information Retrieval*, **3**(4), 313–38.

Leek, T., H. Jin, S. Sista, and R. Schwartz, (2000). The BBN Crosslingual Topic Detection and Tracking System, in *Working Notes of the Third Topic Detection and Tracking Workshop*, Gaithersburg, MD.

Levow, G. A., (2003). Issues in pre- and post-translation document expansion: untranslatable cognates and missegmented words. in *Proceedings of the Sixth International Workshop on Information Retrieval with Asian Languages*, Sappro, Japan. pp. 77–83.

Levow, G.-A. and D. W. Oard, (2000). Translingual Topic Tracking: Applying Lessons from the MEI Project, in *Proceedings of the TDT Workshop 2000*, Gaithersburg, MD.

Littman, M., S. Dumias, and T. Landauer, (1998). Automatic Cross-Language Information Retrieval Using Latent Semantic Indexing. in *Cross-Language Information Retrieval*. (eds. G. Grefenstette) Kluwer Academic Publishers. Norwell, MA. USA. pp. 51–62

López-Ostenero, F., J. Gonzalo, A. Penas, and F. Verdejo, (2002). Noun Phrase Translation for Cross-Language Document Selection, in *Evaluation of Cross-Language Information Retrieval Systems: Second Workshop of the Cross-Language Evaluation Forum, CLEF 2001 Darmstadt, Germany, September 3–4, 2001*, Springer, Berlin, pp. 1639–50.

López-Ostenero, F., J. Gonzalo, A. Penas, and F. Verdejo, (2003). Interactive Cross-Language Searching: Phrases are Better Than Terms for Query Formulation and Refinement, in *Advances in Cross-Language Information Retrieval: Third Workshop of the Cross-Language Evaluation Forum, CLEF 2002 Rome, Italy, September 19-20, 2002 Revised Papers*. Springer, Berlin, pp. 416–29.

Loukachevitch, N. and B. Dobrov, (2002). Cross-Language Information Retrieval Based on Multilingual Thesauri Specially Created for Automatic Text Processing, in *Proceedings of Workshop on Cross-Language Information Retrieval: A Research Road Map, a workshop of ACM SIGIR 2002*, Berkeley, CA.

Ma, H., B. Karagol-Ayan, D. Doermann, D. Oard, and J. Wang, (2003). Parsing and Tagging of Bilingual Dictionaries. *Traitement Automatique Des Langues*, **44**(2), 125–50.

Marchonini, G., (1995). *Information Seeking in Electronic Environments*, Cambridge, England: Cambridge University Press.

McNamee, P. and J. Mayfield, (2002). Comparing Cross-Language Query Expansion Techniques by Degrading Translation Resources. in *Proceedings of the 25th Annual International ACM SIGIR Conference on Research and Development in Information Retrieval*, Tampere, Finland, pp. 159–66.

Meng, H., B. Chen, E. Grams, S. Khudanpur, W.-K. Lo, G. A. Levow, D. W. Oard, P. Schone, K. Tang, H.-M. Wang, and J. Wang, (2001). Mandarin-English Information (MEI): Investigating Translingual Speech Retrieval, in *Technical Report, Summer 2001 Workshop, Center for Language and Speech Processing, Johns Hopkins University*. pp. 23–30.

Monz, C., (2000). Computational Semantics and Information Retrieval, in *Proceedings of the 2nd Workshop on Inference in Computational Semantics (ICoS-2)*, (eds. J. Bos and M. Kohlhase), pp. 1–5.

Müller, H., Deselaers, T., Kim, E., Kalpathy–Cramer, J., Deserno, T. M., and Hersh, W., (2008). Overview of the ImageCLEF 2007 Medical Retrieval and Annotation Tasks, in *Advances in Multilingual and Multimodal Information Retrieval: 8th Workshop of the Cross-Language Evaluation Forum, CLEF 2007, Budapest, Hungary, September 19-21, 2007, Revised Selected Papers*. Springer, Berlin. pp. 472–91.

Nie, J.-Y., M. Simard, P. Isabelle, and R. Durand, (1999). Cross-Language Information Retrieval based on Parallel Texts and Automatic Mining of Parallel Texts from the Web. in *Proceedings of the 22nd Annual International ACM SIGIR Conference on Research and Development in Information Retrieval*, Berkeley, CA, pp. 74–81.

Oard, D. W., (1998). A Comparative Study of Query and Document Translation for Cross-Language Information Retrieval, in *Proceedings of the Third Conference of the Association for Machine Translation in the Americas (AMTA)* Philadelphia, PA. pp. 472–83.

Oard, D. W. and A. R. Diekema, (1998). Cross-Language Information Retrieval. *Annual Review of Information Science and Technology*, **33**, 223–56.

Oard, D. W. and J. Gonzalo, (2004). The CLEF 2003 Interactive Track, in *Comparative Evaluation of Multilingual Information Access Systems: 4th Workshop of the Cross-Language Evaluation Forum, CLEF 2003, Trondheim, Norway, August 21–22, 2003, Revised Selected Papers*. Springer, Berlin, pp. 425–34.

Oard, D. W., J. Gonzalo, M. Sanderson, F. López-Ostenero, and J. Wang, (2004). Interactive Cross-Language Document Selection. *Information Retrieval*, **7**(1–2), 205–28.

Oard, D. W., D. He, and J. Wang, (2008). User Assisted Query Translation for Interactive Cross-Language Information Retrieval. *Information Processing and Management*, **44**(1), 181–211.

Oard, D. W., G.-A. Levow, and C. I. Cabezas, (2000). CLEF Experiments at Maryland: Statistical Stemming and Backoff Translation, in *Cross-Language Information Retrieval and Evaluation: Workshop of Cross-Language Evaluation Forum, CLEF 2000*, (eds. C. Peters), pp. 176–87.

Oard, D. W., C. Peters, M. Ruiz, R. Frederking, J. Klavans, and P. Sheridan, (1999). Multilingual Information Discovery and AccesS (MIDAS) A Joint ACM DL'99/ACM SIGIR'99 Workshop. *D-Lib Magazine*, **5**(10).

Oard, D. W. and P. Resnik, (1999). Support for Interactive Document Selection in Cross-Language Information Retrieval. *Information Processing and Management*, **35**(3), 365–82.

Oard, D. W., J. Wang, G. J. F. Jones, R. W. White, W. Pecina, D. Soergel, and X. Huang, (2007). Overview of the CLEF-2006 Cross-Language Speech Retrieval Track, in *Evaluation of Multilingual and Multi-modal Information Retrieval: 7th Workshop of the Cross-Language Evaluation Forum, CLEF 2006, Alicante, Spain, September 20-22, 2006, Revised Selected Papers*. Springer, Berlin. pp. 744–58.

Och, F. J., (2003). Minimum Error Rate Training for Statistical Machine Translation, in *The Proceedings of the 41st Annual Meeting of the Association for Computational Linguistics (ACL2003)* Sapporo, Japan. pp. 160–7.

Och, F. J. and H. Ney, (2000). Improved Statistical Alignment Models. in *Proceedings of the 38th Annual Conference of the Association for Computational Linguistics*, Hong Kong, pp. 440–7.

Ogden, W., J. Cowie, M. Davis, Y. Ludovik, S. Nirenburg, H. Molina-Salgado, and S. N., (1999). Keizai: An Interactive Cross-Language Text Retrieval System, in *Machine Translation Summit VII, Workshop on Machine Translation for Cross Language Information Retrieval*.

Ogden, W. C. and M. Davis, (2000). Improving Cross-Language Text Retrieval With Human Interactions, in *Proceedings of the 33rd Hawaii International Conference on System Sciences*. pp. 3044.

Pecina, P., Hoffmannová, P., Jones, G. J. F., Zhang, Y., and Oard, D. W., (2008). Overview of the CLEF-2007 Cross Language Speech Retrieval Track. in *Advances in Multilingual and Multimodal Information Retrieval: 8th Workshop of the Cross-Language Evaluation Forum, CLEF 2007, Budapest, Hungary, September 19–21, 2007, Revised Selected Papers*. Springer, Berlin. pp. 674–86.

Peters, C., Clough, P., Gey, F. C., Karlgren, J., Magnini, B. Oard, D. W. deRijke, M., and Stempfhuber, M. (eds.) (2007). *Evaluation of Multilingual and Multi-modal Information Retrieval, 7th Workshop of the Cross-Language Evaluation Forum, CLEF 2006*. Springer Berlin/Heidelberg.

Petrelli, D., M. Beaulieu, M. Sanderson, G. Demetriou, P. Herring, and P. Hansen, (2004). Observing Users, Designing Clarity. *Journal for American Society of Information Science and Technology*, **55**(10), 923–34.

Pirkola, A., (1998). The Effects of Query Structure and Dictionary Setups in Dictionary-Based Cross-Language Information Retrieval. in *Proceedings of the 21st Annual International ACM SIGIR Conference on Research and Development in Information Retrieval*, Melbourne, Australia, pp. 55–63.

Pirkola, A., J. Toivonen, H. Keskustalo, K. Visala, and K. Jarvelin, (2003). Fuzzy Translation of Cross-Lingual Spelling Variants, in *Proceedings of the 21st Annual International ACM SIGIR Conference on Research and Development in Information Retrieval (SIGIR 2003)*, Toronto, Canada, pp. 345–52.

Porter, M., (1980). An algorithm for suffix stripping. *Program*, **14**(3), 130–7.

Qu, Y., G. Grefenstette, and D. A. Evans, (2003). Automatic Transliteration for Japanese-to-English Text Retrieval. in *Proceedings of the 26th Annual International ACM SIGIR Conference on Research and Development in Information Retrieval*, Toronto, Canada, pp. 353–60.

Resnik, P., (1997). Evaluating Multilingual Gisting of Web Pages, in *the AAAI Symposium on Natural Language Processing for the World Wide Web*, Stanford, CA.

Resnik, P., (1998). Parallel Strands: A Preliminary Investigation into Mining the Web for Bilingual Text. in *Proceedings of the Third Conference of the Association for Machine Translation in the Americas on Machine Translation and the Information Soup*, pp. 72–82.

Resnik, P., M. B. Olsen, and M. Diab, (1999). The Bible as a Parallel Corpus: Annotating the 'Book of 2000 Tongues'. *Computers and the Humanities*, **33**(1–2), 129–53.

Singhal, A., J. Choi, D. Hindle, and F. Pereira, (1998). ATT at TREC-7. in *the Seventh Text REtrieval Conference*, Gaithersburg, MD, pp. 239–52.

Vinokourov, V., J. Shawe-Taylor, and N. Cristianini, (2002). Finding Language-Independent Semantic Representation of Text Using Kernel Canonical Correlation Analysis, in *the proceedings of ICML-2002 Workshop on Development of Representations*.

Vossen, P., (1998). *EuroWordNet: A Multilingual Database with Lexical Semantic Networks*, Dordrecht: Kluwer Academic Publishers.

Wang, J., (2005). Matching Meanings for Cross-Language Information Retrieval, University of Maryland at College Park. PhD Thesis.

Wang, J. and D. W. Oard, (2001). iCLEF 2001 at Maryland: Comparing Word-for-Word Gloss and MT, in *Evaluation of Cross-Language Information Retrieval Systems: Second Workshop of the Cross-Language Evaluation Forum, CLEF 2001*, Darmstadt, Germany, pp. 336–54.

Wang, J. and D. W. Oard, (2006). Combining Bidirectional Translation and Synonymy for Cross-language Information Retrieval, in *Proceedings of the ACM SIGIR 2006*, Seattle, WA, pp. 202–9.

Wu, D., D. He, H. Ji, and R. Grishman, (2008). The Effects of High Quality Translations of Named Entities in Cross-Language Information Exploration. In *The 2008 IEEE International Conference on Natural Language Processing and Knowledge Engineering*. Beijing, China, 2008, pp. 443–450.

Xu, J. and R. Weischedel, (2000). TREC-9 Cross-Lingual Retrieval at BBN. in *The Nineth Text REtrieval Conference*. Gaithersburg, MD, pp. 106–15.

Zhang, Y. and P. Vines, (2004). Using the Web for Automated Translation Extraction in Cross-Language Information Retrieval. in *Proceedings of the 27th Annual International ACM SIGIR Conference on Research and Development in Information Retrieval*, Sheffield, UK, pp. 162–9.

12

Performance Issues in Parallel Computing for Information Retrieval

Andrew MacFarlane

12.1 Introduction

The use of performance models when considering the deployment of parallel computing to information retrieval (IR) applications is a much neglected area. During the late 1980s and early 1990s there was a great deal of research into parallel computing for IR, but because of various factors interest has faded and very little research is now being done (MacFarlane 2000). Much of the work was: (a) empirically based and did not try to produce models of performance in order to obtain some kind of theoretical underpinning for the research; (b) only tackled one issue (typically search); and (c) only looked at one method for the distribution of data to nodes in a parallel machine. The PhD thesis written by the author (MacFarlane 2000) attempted to tackle these issues and succeeded with issue (c), but a great deal still needs to be addressed in (a) and (b). In this chapter we will briefly review the relevant literature, present a model of performance developed in MacFarlane (2000) and outline areas of further research that the author regards as being fruitful. The chapter is organised as follows. Section 12.2 justifies the use of parallel computing to solve IR problems. Section 12.3 reviews previous work in modelling parallel computing for IR problems. In Section 12.4 we define a number of IR tasks and then describe various distribution methods for inverted files in Section 12.5 which could be used improve run time performance on these tasks when using parallelism. We define a task as being a specific aspect of an IR system that has its own functionality, but which is designed to solve a particular information seeking problem. Using these defined tasks and distribution methods, a synthetic model of performance is outlined in Section 12.6, which is in turn examined using empirical evidence in Section 12.7. Finally we draw conclusions from the material and propose areas for further research.

12.2 Why Parallel IR?

There is a limit to the gains in run time performance which can be achieved algorithmically (that is, by improved software design) and while great strides have been made in the power of computer hardware

Information Retrieval: Searching in the 21st Century   edited by A. Göker & J. Davies
© 2009 John Wiley & Sons, Ltd

(particularly CPUs) there will always be an absolute limit. This limit, combined with increasing amounts of information being made available, particularly on the Internet, is putting increasing strain on sequential processing systems. The only way forward for many computationally intensive tasks is to use parallelism. There are a number of tasks in IR which do require the use of parallelism (a more detailed description of these tasks is given in Section 12.5). There is increasing interest in using much larger collections for experimentation: for example, the introduction of a Terabyte track within the TREC series of conferences (Clarke *et al*. 2005) with a sizeable test collection, that may warrant the deployment of some kind of parallelism to speed up searching.

At the most fundamental level, parallel computing offers performance improvements by enabling the execution of two or more tasks simultaneously, thereby reducing the overall run time to complete the task. Task in this context is defined as units of sequential computation, according to a given level of granularity, which can be split up and treated independently. The key challenge of course is to execute in parallel only those tasks which are independent of each other, do not require significant data exchange and would not therefore require a communication overhead. It should be noted that there is a limit to the application of parallelism; i.e. we cannot take every line of code from a program and execute it in parallel – all programs have segments which have to be treated sequentially and cannot be split any further into independent segments. The fundamental problems in parallelism are therefore to find the sequential sections in a given algorithm, to parallelise them, and to ensure that any communication overhead due to data exchange between these sequential segments does not negate the benefits gained by applying parallelism.

It should be noted that the use of parallelism for IR caused some controversy in the late 1980s with criticism from Stone (1987) and Salton and Buckley (1988). The basis of this criticism was some work done on the massively parallel Connection Machine which used a signature file method (a fixed length surrogate is used to represent each document). Of the criticisms, Stone's was more useful in that he put forward an alternative parallel method based on inverted files – a far more efficient method for search which indexes documents by words (Harman *et al*. 1992). Stone's ideas have proved to be very influential and most parallel IR systems use inverted files. From the literature, Rasmussen (1992) puts forward a number of reasons for applying parallelism to IR problems:

- *Response times*: there may be situations (the web is a good example of this) where many users require access to the same document collection. The purpose of parallelism would be to reduce contention between queries, thereby increasing the system response time and query throughput.
- *Very large databases*: the time to process queries increases linearly with the document collection, e.g. a query on a 10 gigabyte collection takes 10 times longer than a query on a 1 gigabyte collection. The purpose of parallelism is to scale up the methods used to handle much larger collections.
- *Superior algorithms*: there are some models such as the extended Boolean models (Fox *et al*. 1992) which have the potential to offer improved retrieval effectiveness at the cost of extra computation. The role of parallel computing is to make these algorithms useable in a realistic search environment.
- *Search cost:* Stanfill *et al*. (1989) showed that the resources needed to search a database approach a level of cost effectiveness, given the assumption that search time is linear with collection size and resource costs are static. The role of parallelism is to make the deployment of hardware economically effective (web search engines use this factor to very good effect).

In this chapter we concentrate on addressing the issue of *response times*. The reasons for deploying parallelism given above are related to queries, but the issue applies as much to other tasks in IR such as indexing and inverted file maintenance. Given this, the critical issue in applying parallelism to IR algorithms is the distribution of inverted file data to nodes in a parallel computer. In terms of the problems in applying parallelism stated above, we would want to ensure that our algorithm which uses inverted files could utilise either the whole inverted file or some segment of it, without having to transfer to much document, word list or other data.

12.3   Review of Previous Work

There have been two major reviews of parallel computing for IR, namely Rasmussen (1992) and MacFarlane *et al.* (1997). For the most part these reviews concentrated on the practical implementation of IR systems, such as the architectures and algorithms used and models implemented. The Rasmussen (1992) review did have a section on evaluation performance using such metrics as Speedup, but does not tackle the issue of modelling performance. Performance modelling in this context means a predictive model that can be used to estimate how a given algorithm will behave under a given set of conditions. Of particular concern is the lack of recent interest in doing research in parallel IR: it is seen as a 'solved problem'. However there has been success in applying parallelism to real world IR problems: the most significant example is that of the Internet search engines (Pedersen and Magne Risvik 2004). There is therefore good reason to produce predictive models of performance.

   A few attempts haven been made to produce such models, but they have limitations. Those models that provide a general performance overview of IR do not deal with the problem of distribution (Cardenas 1975; Fedorowicz 1987; Wolfram 1992a; 1992b). Distribution in this context is the method of allocation of index data (i.e. word and document records) to nodes in a parallel computer. Much of the work described in the literature on the subject, which does look at distribution, either tackles one task (Jeong and Omiecinski 1995; Tomasic and Garcia-Molina 1993a; 1993b) or one aspect of a task, such as the consideration of only one distribution method (Ribeiro-Neto *et al.* 1999; Hawking 1996). In an attempt to address the issue of modelling performance, MacFarlane (2000) produced a synthetic model which tackles the issue of distribution of index data on a number of different IR tasks. It is this model which is discussed in this chapter. It should be noted that we draw a distinction between synthetic models that only predict the relative difference between two algorithms and analytical models which actually try to predict real performance of the two algorithms. Attempts by the author to produce an analytical model of performance failed due to the complexity of trying to model real performance on a disparate number of tasks: this is the main reason that the issue has not be tackled satisfactorily in the literature.

   In order to set the scene for our attempt at solving this problem we describe the issue of distributing inverted file data and define some tasks that we attempt to model using our techniques.

12.4   Distribution Methods for Inverted File Data

We look at four distribution methods; *On-the-fly* distribution, *Replication* and two types of *Partitioning*. It should be noted that some data distribution methods are invalid for some tasks (this issue is tackled when the tasks are defined below). In order to explain the differences between these distributions we use an abstract task, which is defined as follows. We have some inverted file data D, together with some work W to be done on D (W can be any IR task e.g. index a document or do a search). When W is applied to D, we get a result R. Any or all of the three variables D, W and R may be subdivided or partitioned in some way; W', D' and R' will denote some such partition or subset, which may nevertheless be the whole of the original variable in some cases. We define some algorithmic steps on these variables that are common to all distribution methods.

- A central node sends W'' plus any data needed to a number of *i* identical processor nodes.
- The processor nodes apply W' to D' to produce results R'.
- The results R' are sent back to the central node which prepares the final result R using all R' results.

   It may help the reader to think of an inverted file (or more formally D) as a matrix with the rows made up of term references and the columns made up of references to documents, see Figure 12.1.

   In this example we have *m* terms in the collection, which invert *n* documents from the indexed text. Relations between terms and documents are signified by a **cross** (x). The matrix will be very sparse, and some relations between documents and terms or *vice versa* will be richer than others. We will use this matrix representation to show how data is distributed in the methods to be examined in this chapter.

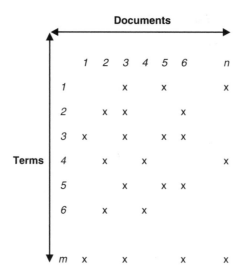

Figure 12.1 Example matrix

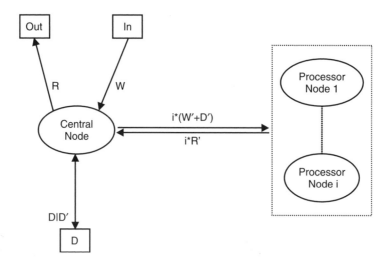

Figure 12.2 Example of abstract task on On-the-fly distribution

12.4.1 On-the-fly distribution

We define *On-the-fly* distribution as the distribution of part of the data as it is required by a given task. It is a dynamic data distribution method, all the others are static. The method is very flexible in that we can distribute the whole matrix D or any D′ which could be the whole or part of either a row or column. The inverted file data is held in one location. Consider the example in Figure 12.2. Note that the ellipse is a node, a box is a disk, and the dashed line box is the universe of our processor nodes. The arrows signify the direction of data exchange between nodes and or disks. Our abstract task uses the following algorithmic steps. The work W is input and the data associated with W (D′) is lifted from the inverted file and is packed up with a subset of W (W′). Each of the *i* processor nodes gets its own W′ and D′, and processes the data, producing R′. All results from *i* processor nodes (R′) are sent back to the central node which prepares the final result R. The significant disadvantage with this method is that a large amount of data must be distributed before computation can be done. The communication may swamp any useful work, and this makes the distribution method impractical for many IR tasks.

12.4.2 Inverted file replication

Replication is the duplication of inverted file data on local disks of a parallel computer. Each node in the system has access to all the data on a local disk, therefore the need to transmit large amounts of data is greatly reduced. The advantage of replication as against *On-the-fly* distribution is that the high communication cost is much reduced without loss of flexibility. The disadvantage is that space and maintenance costs are considerably higher than any of the other distribution methods discussed here. Consider the example in Figure 12.3.

The key issue here is that all nodes have access to matrix D locally. In our abstract task we send each processor node W' together with some scheme for partitioning the matrix as required by the given computation or task. We then produce R' for each processor node and return the result to the central node to compute the final R as would be done with *On-the-fly* distribution. A further advantage in having the whole matrix D available to the node is that load can be re-balanced by exchanging subsets of W' between the processor nodes, without having to communicate any aspect of D.

12.4.3 Inverted file partitioning

Partitioning is the fragmentation of inverted file data over local disks in a parallel computer (see Figure 12.4). In the example given in Figure 12.4 each node has access to its own subset of D, D', which can only be accessed by that node. In terms of the abstract task, each node manipulates a subset of D', D'' in order to service work W or W'. The node services W or W', depending on the task and partitioning type. The advantage of partitioning is that the space costs are lower than *Replication*, but it is a static distribution method and is therefore not as flexible as the *On-the-fly* distribution method. The process of distribution is also much more complex than inverted file *Replication*.

There are two main inverted file *Partitioning* methods (Jeong and Omiecinski 1995): by term identifier (*TermId*) and by document identifier (*DocId*). These partitioning methods are orthogonal to each other. With *DocId* partitioning the terms for a single document are placed on one disk, therefore postings for the same term may be held on multiple disks (see Figure 12.5).

We assume for arguments sake that we have three partitions in our example. In the example documents 1 to $i - 1$ are given to node 1, documents i to $j - 1$ are given to node 2 and documents j to n are given to node 3. The demarcation of the partitions is signified by the dotted line. With respect

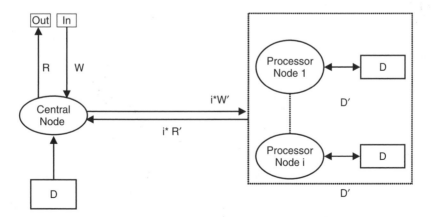

Figure 12.3 Example of abstract task on inverted file replication

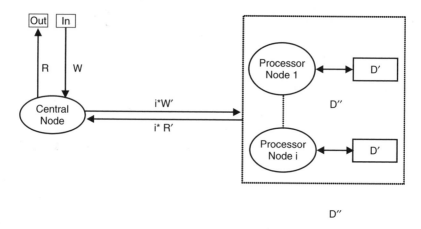

Figure 12.4   Example of abstract task on inverted file partitioning

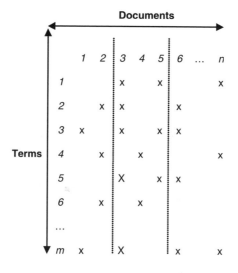

Figure 12.5   Example *DocId* matrix partitioning

to our abstract task, we need to distribute W (if W is a query) to all partitions as all may have data for any or all of the terms in W. With *TermId* partitioning however, all postings for a given term are on one disk, therefore postings for the same document may be on multiple disks (see Figure 12.6). In the example, terms 1 to $i - 1$ are given to node 1, terms i to $j - 1$ are given to node 2 and terms j to n are given to node 3. With respect to our abstract task and our specific example given above on *DocId*, nodes get their own unique subset of work W′ as each node has its set of unique terms.

12.5   Tasks in Information Retrieval

To recap, we define a task as being a specific aspect of an IR system that has its own functionality. We do not attempt to examine every task in IR, as the field is large. Thesaurus construction, clustering and hypertext creation tasks are among the notable exceptions that we do not investigate. We largely concentrate on what we regard as the main tasks in IR such as indexing and search: we define a main task as one with which an IR system using inverted files would be unable to function if such did not

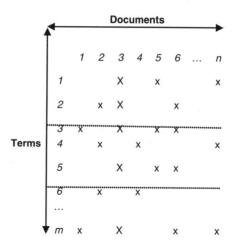

Figure 12.6   Example *TermId* matrix partitioning

exist. The other (non-main) tasks studied can be built using the core tasks and extended as required. For example index update can be built from search and index functionality, while routing/filtering and passage retrieval can be built from search functionality. Each of the tasks to be studied in this chapter is defined below.

12.5.1   The indexing task

The indexing task is the process of taking raw text and building an index over that text using some criteria such as removal of stop-words and stemming. The process of indexing is one of the most computationally intensive aspects of IR requiring vast CPU, memory and disk resources (but is done only once and incremented thereafter). It is therefore a prime candidate for the application of parallelism. We consider *Inverted file Partitioning* only for this task. Parallelism on creating Inverted files could be applied to *On-the-fly* distribution and Replication, by deploying one of the *Partitioning* methods and merging the results together – we therefore do not consider these methods.

12.5.2   The probabilistic search task

The probabilistic search task is the process of servicing queries on inverted files to produce a ranked list of documents using a term weighting scheme such as that derived by Robertson and Sparck Jones (1976). Query processing in this context is very fast, which is why inverted files in some form have become the dominant indexing technology in IR. However, it is still possible to increase the speed of query processing further using parallelism, and by doing so increase the system throughput. With respect to distribution methods, *Replication* is not a suitable method for probabilistic search (with the possible exception of concurrent query service) while *On-the-fly* would restrict efficiency due to the extra communication overhead. It is likely that this extra overhead would outweigh any gain made by parallelism. Our discussion on the probabilistic search task is restricted to *Partitioning* methods only.

12.5.3   The passage retrieval task

The passage retrieval task described here is one used in Okapi experiments conducted within the TREC conference framework, in particular Okapi at TREC-3 (Robertson *et al.* 1995). We classify passage retrieval as being the retrieval of part of a document that is most likely to be of interest to a user, given a query. This algorithm takes an atom of text, say a paragraph, and iterates through the atoms of a document to a given maximum size. Passages that do not have query terms at either the start or end text atoms are ignored. Each passage is assigned a weight (using probabilistic model techniques), and the best-weighted passage used for display to the user, as a surrogate for the whole document or in relevance feedback. This passage processing technique is very computationally intensive and benefits from parallelism (MacFarlane *et al.* 2004). We study *Partitioning* methods only for the same reasons as given for the probabilistic search task.

12.5.4   The routing/filtering task

The idea behind information filtering is to disseminate incoming documents to users who require them. Users have long-term information needs that may be satisfied by newly published documents. A method for information filtering is to take the documents that have been marked relevant by the user in the past and apply a relevance feedback mechanism to obtain a set of terms that can be applied to the new documents. We use TREC definitions of routing and filtering Harman (1996). In routing we provide the user with the top *n* documents, while in filtering we make a binary decision on which *n* documents will be presented to the user. There are a number of filtering techniques: batch filtering that takes *n* documents as a batch and adaptive filtering where documents are considered one at a time, with the possibility of feedback after each one. In Robertson *et al.* (1995) it was stated that an alternative to some term ranking methods described would be to 'evaluate every possible combination of terms on a training set and use some performance evaluation measure to determine which combination is best'. This is a combinatorial optimisation problem in term selection. A number of Hill Climbers (an optimisation heuristic) have been implemented in Okapi at TREC-4 (Robertson *et al.* 1996) to solve this problem. The method we concentrate on in this chapter is the 'Find Best' algorithm which is a steepest ascent Hill Climber – the best term is chosen from the terms in the evaluation set iteratively until some stopping criterion has been reached, e.g. no further improvement can be made. In each iteration various operations are available for term selection: add only, remove only and add/remove. Query terms can also be re-weighted any number of times. We consider two operations only: add only with no re-weighting; add only re-weighting individual terms no more than twice. It is this term selection task in routing/filtering that we attempt to model here. We examine all the data distribution techniques theoretically for this task.

12.5.5   The index update task

The index update task consists of a number of different aspects. We consider the issue of transaction processing, where a transaction is either a document to be inserted or a probabilistic search request. We define index update as data to be periodically added to the index when a buffer with document insertions has exceeded some memory limit. This requires some form of index reorganisation which must be done concurrently with transaction processing to prevent delays: we assume that transaction processing cannot be suspended and specify a requirement that queries should be serviced as soon as possible. It should be noted that we consider insertions only, not deletions or modifications. To do otherwise would complicate the modelling process further, and in any case most text collections are archival in nature (the web being a notable exception). Index maintenance is a computationally intensive activity and we study the modelling of the task in order to investigate the viability of parallelism in a transaction processing context. We study partitioning methods only for the same reasons as given for the probabilistic search task.

12.6   A Synthetic Model of Performance for Parallel Information Retrieval

In this section we briefly outline synthetic models of performance in order to compare distribution methods for all tasks under consideration in the chapter (a detailed description of the actual models themselves can be found in MacFarlane 2000). However, due to practicalities we do not study absolute performance of the tasks (for reasons stated above), and our emphasis is restricted to the derivation of synthetic models that can only be used for comparative purposes. Not having to address the issue of absolute performance simplifies the process of modelling greatly. While our primary aim is to produce models that are strong enough to compare distribution methods and make choices between them, we also try to look beyond this simple requirement. We would like to define models that are good enough to predict the relative difference between data distribution schemes. We would also like to be able to make generic statements about parallel IR performance beyond the algorithms and architectures which we examine in this chapter: this may be difficult to do, given the range of systems described in the literature (see Rasmussen 1992; MacFarlane *et al.* 1997). These issues will be examined later on in the chapter by reviewing the empirical results gathered. The algorithms and methods modelled in this chapter are those described in MacFarlane *et al.* (1999). We have a number of general variables for the models that are declared in Table 12.1.

The format of the models is functional. This allows us to specify equations and reuse them in other defined equations. This functional model makes it easier to reuse various elements of a given model in order to study different type of methods not under consideration in this chapter such as query processing optimisation and compression (further functions can be defined and added as necessary). We attempt to make our models as generic as possible. All functions return a single figure in abstract time. The functions only take variables as arguments: we do not specify higher order functions. Lookup values declared in the form $x[y]$ (e.g. $LI[P]$) are not recorded as parameters, but as global variables. The scope rules for any declared variable are the normal ones found in most programming languages: variables declared locally take precedence over global ones. Sequential and parallel models are declared for all tasks. Simplifying assumptions for each of the models is declared in the relevant sections.

We make a number of assumptions on the general variables that impact (with varying degrees) on the synthetic models. We assume a low-latency network in order to simplify the modelling of communication (otherwise we would have to break down T_{comm} using a $T_{\text{comm}}[x]$ format). For I/O we do allow two forms as blocks of data can be either static or dynamic in size. The $T_{\text{i/o}}$ form of the variable can be used if fixed size blocks are transferred, and it is safe to assume that the balance between transfer and seek time is constant per query term. We use the $T_{\text{i/o}}[x]$ where variable sized blocks are transferred and the balance between seek and transfer time must be an integral part of the modelling process. For seek time we assume that an I/O request entails a single disk head movement. We assume an accumulated increase in load imbalance (variable $LI[P]$), at a rate of 0.015 for all synthetic models. It is difficult to know what the load balance will be for the parallel version of a particular task, without running a program and measuring the imbalance. We take this approach to provide a reasonable level of load imbalance for a given parallel machine size. For a given model we assume that the same parallel machine is used, that is the communication, CPU and I/O costs are identical across nodes in the machine: we do not address the issue of heterogeneous parallelism in the models (e.g. nodes with different architectures).

What follows in Section 12.7 is a brief description of the model given for each task described in Section 12.5, and the results using that model are examined and compared with empirical data. Details of the models and how they were constructed and derived can be found on a website from MacFarlane (2000).

Table 12.1   General variables for the synthetic models

T_{cpu} :	CPU time (for some operation)
$T_{\text{i/o}}[x]$:	I/O time - Components : $1T_{\text{seek}} + xT_{\text{trans}}$
	T_{seek} : Time to seek for I/O
	T_{trans} : Time to transfer data for I/O
T_{comm} :	Communication time
T_{t} :	Unit time
P :	No of nodes in a parallel machine
$LI[P]$:	Load imbalance estimate at P processors

12.7   Empirical Examination of Synthetic Model

Our purpose in this section is twofold, to compare the theoretical results in order to show which distribution method is appropriate for each task, and then to compare the theoretical results with empirical results from our implementation of the tasks in order to see how well the models can distinguish different distribution methods. In each section we describe the evidence used to instantiate the theoretical models, in order to produce them. All diagrams declare the theoretical results on abstract unit time as against the number of processors P.

12.7.1   Comparative results using indexing models

The evidence we used to develop the theoretical models is from the BASE1 collection Hawking *et al.* (1999): this collection consists of 187 000 documents with an average document length of around 465 words. We use the following values in the models: T_{cpu} of 0.01, $T_{i/o}$ of 0.015 and T_{comm} of 1 – these were chosen to reflect the approximate balance between the different aspects. We assume the transfer to and from disk is with fixed sized blocks. With communication time we assume a fast network, given the amount of data to be transferred between nodes. We display two theoretical models, one for the *TermId* partitioning method and one for the *DocId* partitioning method.

From Figure 12.7 it can be seen that there is an advantage in theory in using *DocId* partitioning over the *TermId* method for indexing in that the models for the former predict better performance over the latter on all parallel machine sizes. It should be noted that the model predicts a narrowing of the gap between partitioning methods with increasing machine size (we will examine this issue later in this section). If we assumed a much lower bandwidth network, there would be a clear difference between the builds and *TermId* would not compare well with the *DocId* methods.

When comparing the performance of both partitioning methods with our empirical results MacFarlane *et al.* (2005), the synthetic model for indexing was able to predict that *DocId* builds are faster than *TermId* builds. Our empirical results show definitively that the *DocId* partitioning scheme is a superior method in terms of speed, than the *TermId* method MacFarlane *et al.* (2005) i.e. the gap between the methods was more pronounced on the actual runs. The clear reason for this was the extra communication costs implied by the *TermId* method, even using the assumption of a fast network. The extra communication costs are largely due to an extra merging process needed for *TermId* indexing. Documents are distributed to processors in *TermId* for a parsing phase, which meant that intermediate results must be exchanged, requiring $N(N - 1)$ data sets to be transmitted over the network. Problems in the modelling of communication in the synthetic model on this merging process meant that a widening gap in performance between the two partitioning methods for increasing parallel

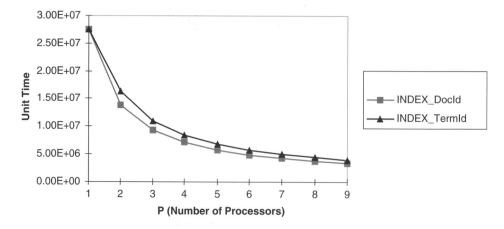

Figure 12.7   Synthetic indexing performance on 1–9 processors

machine size was not anticipated (the model actually predicted a narrowing of the gap in run time between the two partitioning methods). This failure in modelling is discussed in more detail in the conclusion.

12.7.2   Comparative results using search models

The BASE1 collection was again used as source of evidence to develop the theoretical search models. The same values for T_{cpu} and T_{comm} where used as in the indexing models, but assumed in the search case that variable sized blocks would be transferred from disk and therefore split $T_{i/o}$ into T_{trans} and T_{seek} using the values 0.015 and 0.1 respectively. These are the values we assumed for all the theoretical search models discussed below. We assumed a query size of 2.5 terms for the models as users tend to submit just over two terms per query (Silverstein *et al.* 1999). The results of applying these values shown in Figure 12.8 demonstrate that in theory the *DocId* partitioning method (SEARCH$_{docid}$) would perform better than the *TermId* method using either a parallel sort for results ranking (SEARCH$_{termid2}$) or a sequential sort (SEARCH$_{termid1}$). The comparative results also predict that the *TermId* method with parallel sort (SEARCH$_{termid2}$) will outperform the algorithm with a sequential sort (SEARCH$_{termid1}$) by a substantial amount: the synthetic model predicts that a sequential sort will be a bottleneck. The prediction on the *TermId* partitioning method with a sequential sort is particularly bleak, with little or no advantage to be gained from parallelism.

Our empirical results MacFarlane *et al.* (2000) show that runs on the *DocId* partitioning scheme outperform a *TermId* scheme, whether or not a parallel sort is implemented. The prediction in the synthetic model that a sequential sort would be a bottleneck is confirmed by empirical results (MacFarlane *et al.* 2000). The synthetic model is also able to predict the relative performance difference between the partitioning methods to a great extent. However our empirical results show that *TermId* with parallel sort performance is nearer to *TermId* with sequential sort, whereas our synthetic model predicted that search on *TermId* with parallel sort would be nearer to *DocId*. The implemented parallel sort required that the final merged results be distributed to processors and the sorted data retrieved when each processor has completed its sort: as with indexing this communication was not dealt with correctly by the synthetic model.

12.7.3   Comparative results using passage retrieval models

We used the same values for the models as per the search models in Section 12.7.2, but needed some further assumptions to make on the average number of text segments to inspect per document

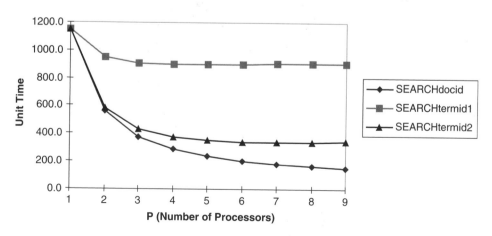

Figure 12.8   Comparative results for search models on 1–9 processors

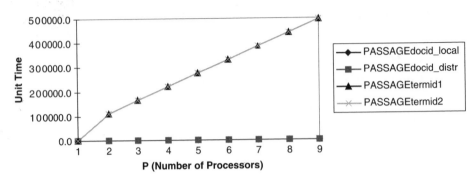

Figure 12.9   Synthetic passage retrieval model on 1–9 processors

and the total number of documents on which to do passage retrieval: we chose the values of 11 and 1000 respectively. Figure 12.9 shows the results on the theoretical passage retrieval models for both partitioning methods. There is clearly a significant problem with using *TermId* partitioning with the passage retrieval algorithm we model. The predicted comparative performance is so poor that the *DocId* models appear to be near zero. Note that the values on the graph are so close in Figure 12.9 that *TermId* models cannot be distinguished. It is clear from the model that predicted communication costs would increase the run time of any parallel program using such a partitioning method. Given the problems with modelling communication in both indexing and search models, the costs for passage retrieval are likely to a significant underestimate. As a result we chose not to run any practical experiments using this partitioning method, and therefore concentrated on the *DocId* partition method alone using both the *local* and *distributed* document allocation methods. With the *local* method, the local top 1000/P documents are used for passage processing, whereas in the *distributed* method the global top 1000 documents are processed.

Figure 12.10 shows synthetic model results for *DocId* partitioning only. With both types of passage retrieval using *DocId* the prediction is time reduction with increasing numbers of processors, but the local method (marked PASSAGE$_{docid\_local}$) shows slightly better theoretical results than the distributed method (marked PASSAGE$_{docid\_distr}$). There is a little extra communication for the *distributed* method, where the master process needs to identify the top 1000 documents, and inform the search processes as to which of their documents are to be examined for passages. Also, there is no guarantee that passages will be equally distributed in the *distributed* method, and this has the potential to affect load

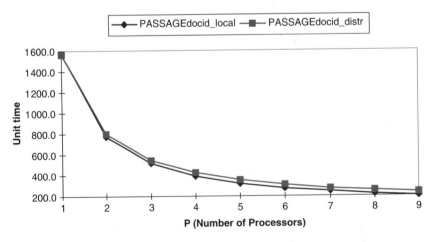

Figure 12.10   Synthetic passage retrieval models – *DocId* only

balance. This load balancing issue is very difficult to model, as the distribution of documents to which passage processing is to be applied, cannot be determined prior to search.

The empirical results show that our synthetic model correctly predicted that the *local* passage processing method would outperform the *distributed* version. However, the model was unable to predict the relative difference due to the super linear speedup performance for the former and the erratic nature of the performance in the latter on short queries (MacFarlane *et al.* 2004). The remarkable performance of the *local* passage processing method was somewhat hard to fathom at first, as the number of passages processed was around the same as the *distributed* method. However, because individual text atoms can vary in size considerably, it is clear that the local method was processing a different set of documents which were less computationally costly to process than the *distributed* method. The variable performance of the *distributed* method was due to the different numbers of documents processed at each machine size, which tended to vary quite considerably and non-deterministically. The factors found by empirical experiment added another unforeseen level of complexity to the modelling process which cannot be dealt with currently, namely modelling the effect of text atom size on passage processing.

12.7.4   Comparative results using term selection models

For our modelling of the term selection we assumed that the optimisation process would be done on 300 terms with a maximum of 100 iterations. The values chosen for these variables were used in the empirical experiments (MacFarlane *et al.* 2003). We modelled only one of the Hill Climbers under consideration, namely 'Find Best'. Figure 12.11 shows the comparison between models on all distribution and partitioning methods with the number of processors P set between 10 and 100.

The theoretical models predict that the best performing distribution scheme overall would be the *replication* method. On smaller processor sets, the *DocId* partitioning method shows better theoretical results, but predicts that the performance would deteriorate substantially due to the restriction on the level of parallelism in that distribution method: the communication/computation balance would be skewed with increasing numbers of processors. We therefore did not implement the *DocId* partitioning method, as it was clear that other distribution methods where more likely to succeed when larger parallel machines were used. The prediction with respect to *TermId* partitioning and *On-the-fly* distribution is that there is little difference between them particularly with large processor sets.

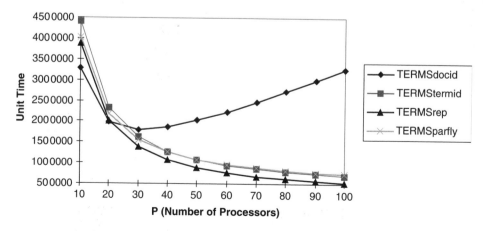

Figure 12.11   Term selection model results using large processor set and iteration sizes

The synthetic model was successfully able predict that *Replication* is a superior method to *On-the-fly* distribution, but not that the latter would perform so poorly (MacFarlane 2000). The 'Find Best' algorithm show near linear speedup on the *Replication* method. However, the results for the *On-the-fly* distribution method actually show a slowdown with more processors than just using one processor: the performance gets worse with more processors. This is in spite of the fact that we used a 50 Mb/s network or on a large scale Fujitsu parallel computing (with 100 processors). The bottleneck at the synchronisation point, where data must be exchanged between iterations, proved to be the stumbling block for that method.

We can make a further statement on the synthetic models, given the evidence found with *On-the-fly* distribution. The *TermId* partitioning method would require more communication at the synchronisation point than *On-the-fly* distribution. There is no guarantee that query terms will be distributed evenly with the *TermId* method, and this has the potential to affect load balance detrimentally. The *TermId* partitioning method is therefore not a viable method, and as a result we did not implement this scheme for our experiments.

12.7.5   Comparative results using index update model

In the index update model we assumed that for every 10 searches, there would be 1 update to process. Figure 12.12 shows the prediction using these values and compares the theoretical performance between both types of partitioning methods where transactions are affected by a concurrent index update (models are labelled with the suffix RO) and normal transaction processing without contention for resources. What we mean by concurrent index update is that updates held in a temporary buffer are merged with the inverted file, while transactions are serviced concurrently.

The models predict that *DocId* partitioning will outperform *TermId* whether or not there is contention for resources due to index update. As you would expect, where there is contention for resources the predicted performance is worse than in non-contention models. The model of *TermId* partitioning in the presence of an index update predicts deterioration in performance with increasing numbers of processors: with *DocId* the prediction is that performance will remain constant after a certain number of processors is reached.

Our synthetic model for the index update task is strong enough to distinguish between the partitioning methods (MacFarlane *et al.* 2007). Problems highlighted in our probabilistic search experiments (see Section 12.7.2) impose severe restrictions on transaction processing when the *TermId* method is used, which are difficult to solve within our experimental context. These problems (most notably the sort aspect of search) had an impact on the relative difference between the two partitioning methods during transaction processing, which the synthetic model was not able to deal with very well. The

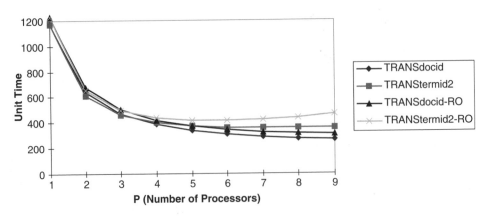

Figure 12.12   Comparison between partitioning methods in presence and absence of index update

synthetic model correctly predicted that the performance of transactions serviced on *TermId* indexes during an index reorganisation would deteriorate (although the empirical results were relatively worse because of a list size problem described in MacFarlane *et al.* 2007). The synthetic model also correctly predicted that transaction performance on *DocId* partitioning indexes would be constant after a given number of processors are reached.

12.8   Summary and Further Research

The synthetic models outlined in this chapter predict that for most tasks the *DocId* partitioning method would be the better performing data distribution scheme of those studied. For the index, probabilistic search, passage retrieval and index update tasks the prediction is unambiguous. For the passage retrieval task in particular it is very clear that the *TermId* partitioning method is simply not viable: we did not therefore implement this partitioning method for the passage processing task (MacFarlane *et al.* 2000). The theoretical evidence on the routing/filtering task is more complicated and therefore needs more discussion. On small numbers of processors the prediction is that *DocId* partitioning would be the best data distribution scheme, but on larger parallel machines the performance would deteriorate due to excessive communication. Due to the restrictions on inter-set parallelism, the proposed method for *DocId* partitioning does not show the same promise as intra-set parallelism usable on the other distribution methods. Of the other three, the prediction is that *replication* would be the best performing method and that performance with *on-the-fly* distribution would be about the same with *TermId* partitioning. We therefore concentrated our efforts on *replication* and *on-the-fly* distribution schemes, partly because of the theoretical comparison and partly because such methods are easier to manage. Only one indexing run needed to be initiated in order to undertake those experiments, whereas with *TermId* partitioning a new indexing run would be required for every processor added.

Further work on the synthetic modelling technique used in this chapter is merited both in terms of strengthening the actual model and extending it for use in other tasks not studied here. While the models were able to correctly predict that one distribution scheme was superior to the others, given the empirical evidence, they were not able to predict the relative difference between the models. This was largely because the synthetic models were not able to model communication completely. The key aspect to concentrate on initially therefore will be the modelling of communication, given that it was such a significant problem in our models. The challenges in being able to do this successfully should not be underestimated. The model will not only have to cope with interactions between two processors, it will also have to model the pattern of communication throughout the whole system (one of the reasons for our simplified modelling of communication was to avoid this complexity). A particular problem with passage retrieval is the ability to model the effect of the text atom size on passage processing: this non-deterministic factor is very hard to model. An interesting and worthwhile piece of research would be to extend the functional modelling and use formal techniques to prove various aspects of it: this research may well provide insights that would not be obtained otherwise. When many problems have been solved by using synthetic modelling, and more of an understanding of the theory of performance of parallelism in IR is obtained, it may well be possible to derive an analytical model of performance in order to model real performance. Hopefully, this analytical model would be able to predict the actual performance in a given IR task, more accurately then we can at present.

More details of the research presented in this chapter can be found in the authors PhD thesis (MacFarlane 2000). General reviews for research in parallel computing and information retrieval can be found in Rasmussen (1992) and MacFarlane *et al.* (1997). A technical overview of the area can also be found in Brown (1999) and a general overview of the area of parallel computing can be found in Grama *et al.* (2003).

Exercises

12.1 What difficulties prevent a realistic analytical model being derived for parallel computing as applied to information retrieval tasks?

12.2 With regard to the synthetic model, what are the limitations preventing the synthetic model from actually predicting the relative difference between different types of inverted file distribution methods?

References

Brown, E. (1999). Parallel and Distributed IR, In: Baeza-Yates and Ribeiro-Neto (eds), *Modern Information Retrieval*, Addison-Wesley, 229–256.

Cardenas, A. F. (1975). Analysis and performance of inverted data base structures. *Communications of the ACM*, **18**(5) 253–263.

Clarke, C., Craswell, N. and Soboroff, I. (2005). Overview of the TREC 2004 Terabyte Track, in *Proceedings of the 11th Text REtrieval Conference, (TREC 2004)*, (eds E. Voorhees and L. Buckland), NIST Special Publication 500-261, Gaithersburg, U.S.A, [available on: http://trec.nist.gov - visited 15 March 2007].

Fedorowicz, J. (1987). Database performance evaluation in an indexed file environment. *ACM Transactions on Database Systems*, **12**(1) 85–110.

Fox, E., Betrabet, S., Koushik, M. and Lee, W. (1992). Extended Boolean models. In *Information Retrieval, Data Structures and Algorithms*. (eds W. B. Frakes, and R. Baeza-Yates), Prentice-Hall, New Jersey, pp. 393–418.

Grama, A., Gupta, A, Karypis, G. and Kumar, V. (2003). *Introduction to Parallel Computing*, 2nd Edition, Addison-Wesley.

Harman, D. K. Fox, E. Baeza-Yates, R and Lee, W. (1992). Inverted Files, in *Information Retrieval, Data Structures and Algorithms*. (eds W. B. Frakes, and R. Baeza-Yates), Prentice-Hall, New Jersey, pp. 28–43.

Harman, D. K. (1996). Overview of the fourth text retrieval conference (TREC-4), in *Proceedings of the 4th Text Retrieval Conference (TREC-4)*, (ed D. K. Harman), NIST Special Publication 500-236, Gaithersburg, U.S.A, pp. 1–24.

Hawking, D. (1996). Document retrieval performance on parallel systems, in *Proceedings of the 1996 International Conference on Parallel and Distributed Processing Techniques and Applications*, (ed H. R. Arabnial), CSREA, Athens, pp. 1354–1365.

Hawking, D., Craswell, N. and Thistlewaite, P. (1999). Overview of TREC-7 very large collection track, in *Proceedings of the 7th Text Retrieval Conference (TREC-7)*, (eds E. M. Voorhees and D. K. Harman), NIST Special Publication 500-242, Gaithersburg, USA, pp. 257–268.

Jeong, B. and Omiecinski, E. (1995). Inverted file partitioning schemes in multiple disk systems. *IEEE Transactions on Parallel and Distributed Systems*, **6**(2) 142–153.

MacFarlane, A. Robertson, S. E. and McCann, J. A. (1997). Parallel computing in information retrieval – an updated review. *Journal of Documentation*, **53**(3) 274–315.

MacFarlane, A., McCann, J. A. and Robertson, S. E. (1999). PLIERS: a parallel information retrieval system using MPI, in *Proceedings of the 6th European PVM/MPI Users' Group Meeting, Barcelona*, *Lecture Notes in Computer Science 1697*, (eds J. Dongarra, E. Luque, and T. Margalef), Springer-Verlag, Berlin, pp. 317–324.

MacFarlane, A. (2000). *Distributed Inverted files and performance: a study of parallelism and data distribution methods in IR*, PhD Thesis, City University London, August 2000. [Available on: http://www.soi.city.ac.uk/~andym/PHD/: visited 15 March 2007]

MacFarlane, A., McCann, J. A. and Robertson, S. E. (2000). Parallel search using partitioned inverted files, in *Proceedings of String Processing and Information Retrieval (SPIRE 2000)*, (ed P. De La Fuente), IEEE Computer Society Press, Los Alamitos, pp. 209–220.

MacFarlane, A. Robertson, S. E. and McCann, J. A. (2003). Parallel computing for term selection in routing/filtering, in *Advances in Information Retrieval, Proceedings of the 25th European Conference on IR Research (ECIR 2003)*, *Lecture Notes in Computer Science 2633*, (ed F. Sebastiani), Springer-Verlag, Berlin, pp. 537–545.

MacFarlane, A. Robertson, S. E. and McCann, J. A. (2004). Parallel computing for Passage Retrieval. *ASLIB Proceedings: New Information Perspectives*, **56**(4), 201–211.

MacFarlane, A., McCann, J. A. and Robertson, S. E. (2005). Parallel methods for the generation of partitioned inverted files. *ASLIB Proceedings: New Information Perspectives*, **57**(5) 434–459.

MacFarlane, A., McCann, J. A. and Robertson, S. E. (2007). Parallel methods for the update of partitioned inverted files. *ASLIB Proceedings: New Information Perspectives*, **59**(4/5) 367–396.

Pedersen, J. and Magne Risvik, K. (2004). Web Search Tutorial, *The 27th Annual International SIGR Conference on Research and Development in Information Retrieval – SIGIR 2004*.

Rasmussen, E. (1992). Parallel information processing. in *Annual Review of Information Science and Technology (Volume 27)*, (ed ME. Williams), ARIST, pp. 99–130.

Ribeiro-Neto, B., Moura, E. S., Neubert, M. S. and Ziviani, N. (1999). Efficient distributed algorithms to build inverted files, in *Proceedings of the 22nd International Conference on the Research and Development in Information Retrieval (SIGIR'99)*, (ed M. Hearst, F. Gey, and R. Tong), ACM Press, New York, pp. 105–112.

Robertson, S. E. and Sparck Jones, K. (1976). Relevance weighting of search terms. *Journal of the American Society for Information Science*, May-June, 129–145.

Robertson, S. E., Walker, S., Jones, S., Hancock-Beaulieu, M. M. and Gatford, M. (1995). Okapi at TREC-3, in *Proceedings of the 3rd Text Retrieval Conference (TREC-3)*, (ed D. K. Harman), NIST Special Publication 500-226, Gaithersburg, USA, pp. 109–126.

Robertson, S. E., Walker, S., Jones, S., Hancock-Beaulieu, M. M., Gatford, M. and Payne, A. (1996). Okapi at TREC-4, in *Proceedings of the Fourth Text Retrieval Conference (TREC-4)*, (ed D. K. Harman), NIST Special Publication 500-236, Gaithersburg, USA, pp. 73–96.

Salton, G. and Buckley, C. (1988). Parallel text search methods. *Communications of the ACM*, **31**(2) 202–215.

Silverstein, C., Henzinger, M., Marais, H and Moricz, M. (1999). Analysis of a very large web search engine log. *SIGIR Forum*, **33**(1), 6–12.

Stanfill, C., Thau, R. and Waltz, D. (1989). A parallel Indexed algorithm for Information Retrieval. In *Proceedings of the 12th annual conference on research and development in Information Retrieval (SIGIR'89)*, (eds N. J., Belkin, and C. J., van Rijsbergen), ACM Press, New York, pp. 88–97.

Stone, H. S. (1987). Parallel querying of large database: a case study. *IEEE Computer*, **20**(10) 11–21.

Tomasic, A. and Garcia-Molina, H. (1993a). Performance of inverted indices in shared-nothing distributed text document information retrieval systems in *Proceedings of the 2nd International Conference on Parallel and Distributed Information Systems*, IEEE Computer Society Press, Los Alomitos, pp. 8–17.

Tomasic, A. and Garcia-Molina, H. (1993b). Caching and database scaling in distributed shared-nothing information retrieval systems in *Proceedings of the 1993 ACM SIGMOD International Conference on Management of Data*, (eds P. Buneman, and S. Jajoida), ACM Press, New York, pp. 129–138.

Wolfram, D. (1992a). Applying informetric characteristics of databases for IR system file design, part i: informetric models. *Information Processing and Management*, **28**(1) 121–133.

Wolfram, D. (1992b). Applying informetric characteristics of databases for IR system file design, part ii: simulation comparisons. *Information Processing and Management*, **28**(1) 135–151.

Solutions to Exercises

Chapter 1 – Information Retrieval Models
Djoerd Hiemstra

1.1(c) The Venn diagrams of Figure 1.2 show exactly 8 disjoint subsets of documents, including the area around the diagram. Whatever the final result of a Boolean query, each subset is either selected or not, so in total $2^8 = 256$ subsets can be defined.

1.2(b) Vector spaces are metric spaces, i.e., a set of objects equipped with a distance, where the distance from $\vec{q}$ to $\vec{d}$ is the same as from $\vec{d}$ to $\vec{q}$, so if we change $\vec{q}$ for $\vec{d}$ and $\vec{d}$ for $\vec{q}$, the distance should remain the same. In practice however, many practical term weighting algorithms do not use the same weights for queries and documents. In such a case, the similarity might not be equal to 0.08. One might argue that such a model is *not* a vector space model, though.

1.3(c) If we add a single document, the document frequencies of terms that occur in the added document will increase by 1. Furthermore, the number of documents N changes, which affects the *idf*s of all other terms.

1.4(b) Assuming that the *idf* is calculated using the number of documents N as shown in Equation (1.5), all weights need to change, as is the case in Exercise 3.

1.5(a) For each term, the probabilistic model considers either presence of the term in the document, or absence of the term in the document. So, per term, there are two cases, hence $2^3 = 8$ different scores in the case of three query terms.

1.6(c) As said above, the model only considers presence or absence, but in the case of presence it does not consider the number of occurrences of the term in the document. If the term is present in both D and E, then they will be assigned the exact same score. If the system needs to provide a total ranking, then the implementation has to determine which document is ranked first.

1.7(b) Without smoothing, the probability is a simple fraction of the number of occurrences, divided by the total number of terms in the document. Of course, language models do not need an additional term weighting algorithm.

1.8(c) If $\lambda = 0$, the model does not use smoothing. Without smoothing, terms that do not occur in the document are assigned zero probability. The score of a document is determined by multiplying the probabilities of the single terms. If one of them is zero, the final score of the document is zero.

Information Retrieval: Searching in the 21st Century   edited by A. Göker & J. Davies
© 2009 John Wiley & Sons, Ltd

Chapter 2 – User-centred Evaluation of Information Retrieval Systems
Pia Borlund

2.1–2.4 These are self-reflective. Ask yourself the following questions:

- How well do the search engines do against each other in terms of precision?
- Are there any differences in effectiveness between the search engines? Are there any differences at different cut-off points, e.g., p@5 and p@10?
- In particular, how well do the meta-search engines do compared with the search engines?
- In terms of sources and quality, which systems produce the better results?
- Do diagnostic measures such as Link Broken, etc. have an impact on precision results?
- How do natural language queries perform compared with Boolean queries?
- How does any advert in search results affect the precision values? What is the impact of the position of any explicit adverts in the search results?
- How well does binary relevance (the relevance categories of relevant and non-relevant) reflect satisfaction of the information need represented in the simulated work task situation?
- How many different criteria did you apply when judging relevance?
- How can situational relevance be judged and measured best?

Chapter 3 – Multimedia Resource Discovery
Stefan Rüger

3.1 Search types

It is relatively easy to come up with a usage scenario for each of the matrix elements in Figure 3.1: for example, the image input speech output matrix element might be 'given an X-ray image of a patient's chest, retrieve dictaphone documents with a relevant spoken description of a matching diagnosis'. However, creating satisfying retrieval solutions is highly non-trivial and the main subject of the multimedia information retrieval discipline.

In essence, Figure 3.1 lists cross-modal retrieval scenarios. The left-hand column describes situations where the search is by text, for example, using a web-search-engine-type interface such as the one in Figure 3.4 for TV news. Users have grown accustomed to this search method, and it can easily be implemented as long as the document repositories (be it text, video, images, speech, music, sketches, etc.) have textual metadata associated with their individual entries. Museum catalogues would be good examples of sources of these metadata. Depending on the document types in the repository at hand it may be more or less easy to extract a surrogate text of the original document. This is almost trivial when the repository consists of text documents. One would strip these from formatting instructions and, for example, stem words in order to remove variations brought about by grammar rules, to arrive at a 'bag of words' that can be indexed. The audio track of video or speech documents could undergo automated speech recognition; one would hope that redundancy of word repetitions in the original document somehow alleviates the effect of transcription errors.

In the case of 'query-by-example' modes, where the query is of the same type as the document in the repository, one immediate approach would be relying on a type of content-based or fingerprinting retrieval. The trick here is to be able to extract features that are specific to the document and invariant under perceptually or otherwise irrelevant transformations of the query or the document: it should not matter whether spoken queries are issued by a man or a woman, whether an image as query input is scaled, has undergone a high or low image compression, whether a hummed query is slightly higher or lower in pitch, etc. This requirement is non-trivial: e.g. for medical image retrieval the features should pick up only medically relevant aspects of the imaging, and this normally requires much domain expertise and arguably a fair amount of signal processing skills to achieve that.

In the case of retrieving speech documents against spoken queries the common best practice appears to be *not* to use low-level feature matching of the processed speech signal, but instead transcribe both queries and speech documents to text. In contrast, approaches to query by humming – a thriving challenge of the growing music information retrieval community – most commonly deploy low-level features for the matching process. Currently, it is not clear what the best approach for image search by image similarity would be: low-level feature matching or automated annotation. In the first case common research questions are which features and which similarities to use, while the second case requires an appropriate choice of symbolic vocabulary and suitable models, typically from machine learning, to facilitate the automated annotation.

The level of desired similarity or sameness is also an important factor in deciding which techniques are more likely to succeed. In order to locate the exact performance of, say, a music piece that you are listening to on the radio, one would likely employ a signal-based fingerprinting technique that is invariant to distortions brought about by the broadcasting and recording (e.g. invariant to the background noise in a car when recording the music sample with a mobile phone) but still sensitive to the qualities of the particular performance. If on the other hand, one was interested more generally in versions of the same melody or lyrics by possibly different artists performed within possibly even different genres (a pop version of a jazz song, say) then a more symbolic or coarse feature extraction process may be more appropriate.

Cross-modal retrieval where queries and documents are of *different* type clearly require some sort of translation, which in the simplest case could be a textual annotation of either using a discrete set of vocabulary. The cross-modal translation is the most involved, and least explored, of all the cases in Figure 3.1.

3.2 Colour histograms

(a) The first thing to note is that $r/256$ is smaller than 1, as r runs from 0 to 255; hence, $i_r := \lfloor n_r r/256 \rfloor$ is an integer that takes on values from 0 (for $r = 0$) to $n_r - 1$. Note that the floor function $\lfloor x \rfloor$ returns the largest integer smaller than or equal to x. Similarly, $i_g := \lfloor n_g g/256 \rfloor \in \{0, \ldots, n_g - 1\}$ and $i_b := \lfloor n_b b/256 \rfloor \in \{0, \ldots, n_b - 1\}$. The 3-D bins in colour space correspond to a 3-D array that has the dimensionality $\text{bin}[n_r][n_g][n_b]$. i_r, i_g and i_b are valid indices for this 3-D array. A compiler might assign the location $i_r n_g n_b + i_g n_b + i_b$ in an equivalent one-dimensional array; this assumes that array indices start from 0 and that the last index of a multidimensional array addresses neighbouring locations in memory as is the case, e.g. in the programming language C++.

(b) Of the 64 different 3-D bins (numbered $0, \ldots, 63$) in the above scheme only 8 bins are occupied, as there are only 8 different colours used in the example picture: bin 0 (black), 3 (blue), 12 (green), 15 (cyan), 48 (red), 51 (magenta), 60 (yellow) and 63 (white). The following figure visualises this 3-D colour histogram:

The area of the circles in the occupied bins indicate how much this bin is occupied; the colour corresponds to average colour that this bin summarises.

In contrast, the three 1-D histograms are identical: The values $0, \ldots, 255$ are subdivided into 22 different bins, and bin 0 and bin 21 are the only ones that are occupied in each of the r, g and b histograms.

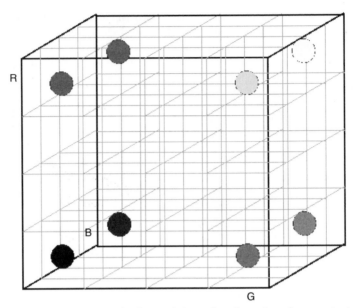

(c) The 3-D colour histogram has retained more information about the colour use in the original image than the three 1-D histgrams dispite the fact that there are more bins in the three 1-D histograms than in the 3-D histogram (66 vs 64). The 3d histgram captures the *joint distribution* of the r, g and b values and clearly recognises that 1/8th of the pixels are black, blue, green, cyan, red, magenta, yellow and white, respectively. The three 1-D histograms have only retained the *marginal distribution* of the r, g and b values. As such images with a completely different colour distribution can have the same marginal distributions. In fact, a black and white image with half the pixels black and the other half of the pixels white would have the same marginal distributions. Marginalisation of multidimensional distributions always loses information.

(d) Normalising histograms makes them scaling invariant, i.e. larger and smaller versions of the same image result in the same histogram.

3.3 Image search

Figure 3.2 can be modified to cater for the colour-and-texture-based image search engine at hand. The query would be an example image containing curtain fabrics of similar colour and texture. The feature extraction process would deploy colour histograms, for example, the 3-D colour histogram from the previous exercise and a texture feature vector. Both vectors could be concatenated into a singe vector representation. In the end the query image will have a certain representation as a point in feature space and so will every single image in the database.

The images whose representations are closest to the representation of the query are ranked top by this process. In fact, all images in the database will be sorted by their distance in feature space to the query image. Thus the feature representation and distance notion in feature space are key aspects of the workings of this type of content-based search engine. A very simple distance of the query feature vector f^q to an arbitrary image feature vector f^i is the Manhatten distance

$$d(f^q, f^i) = \sum_{j=1}^{n} |f_j^q - f_j^i|,$$

which is simply the sum of the absolute differences of the respective n components of the feature vector.

Chapter 4 – Image Users' Needs and Searching Behaviour
Stina Westman

4.1 At the time of writing, the all-time popular tags in Flickr (http://www.flickr.com/photos/tags/) included the following tags A–C: animals, architecture, art, Australia, autumn, baby, band, Barcelona, beach, Berlin, bird, birthday, black, blackandwhite, blue, bw, California, cameraphone, camping, Canada, Canon, car, cat, Chicago, China, Christmas, church, city, clouds, color, concert, cute. Within this sample you can find tags referring to the following image attributes (example tags in parenthesis):

- Non-visual information (cameraphone, Canon);
- Global syntax (blackandwhite, black, color);
- Generic objects (animals, baby, church);
- Generic scenes (beach, city, camping);
- Specific scenes (Australia, Barcelona, California);
- Abstract concepts (birthday, Christmas, cute);

The usefulness of these tags for retrieval would depend on the needs of the searcher and the match of the tag(s) to the image(s) sought after. This sample highlights the ambiguity of some tags, as 'blue' is used to describe both color values and emotional state. On a related note, both 'blackandwhite' and 'bw' have been used to tag black and white images. It also raises the issue of whether tags should be in singular ('animal' and 'animals' both produced more than a million hits). The distinction between object and scene may also be difficult: 'church' refers to both the building and the indoors location.

Chapter 5 – Web Information Retrieval
Nick Craswell and David Hawking

5.1 The minimal answer, as indicated here, should show a queue and some mechanism for only adding unseen URLs to the queue. Another diagram is in (Heydon and Najork 1999). Answers could also indicate options such as URL policy filters, politeness mechanisms and parallelisation of the crawl. A depth-first crawler would use a last-in-first-out (LIFO) stack, rather than first-in-first-out (FIFO). More complex data structures are needed to support focused crawling or variable-frequency revisit policies.

5.2 Readers must trust their own observations.

5.3 Real users may be performing navigational, informational or transactional search. They have a rich information need, but describe their query in a few words. Based on this short query, a relevance judge under lab conditions might pick a very unlikely interpretation of the query. For example, judging the query 'amazon' according to the criterion: 'The user is searching for documents about the history of Amazon and the factors that led to its success in the online arena.' Some users might want this, but the most likely intent is actually that the user wants to go shopping. If the relevance judgments systematically disagree with real usage, then the evaluation will reward systems that fail to satisfy real users.

In summary, click data comes from real users, whereas judgments are made in lab conditions. Clicks are made in the context of a list (the probability of click depends on what other documents appear), whereas judgments are usually independent of a list. Clicks are affected by summaries, whereas judgments are usually made on document full text. Clicks are shallow because most users only look at a few documents, whereas judgments can be made on a large number of documents for each query. It may be necessary to combine click information over many users to get a reliable indication of relevance and even then a bad document may be highly clicked, because clicks are only an implicit indication of relevance, whereas relevance judgments are explicit.

Finally, evaluations based on independent document-level judgments alone, cannot penalise systems which retrieve overlapping or duplicate documents.

Chapter 6 – Mobile Search
David Mountain, Hans Myrhaug and Ayşe Göker

6.1(a) Designing the mobile website

- An analysis of the existing website with focus on website navigation and display. The existing menus would be rendered either from HTML, server pages, or scripts. Menus defined as HTML-links will be easiest to transfer to mobile phones, whilst the use of server pages and scripts may generate undesired presentation or the lack of it. Multiple display areas are generally difficult to present at the same time to a user on a mobile phone.

- Large images, animated interactive elements, large fonts, film trailers and audio, and advertisement areas will be potentially difficult to transfer without changes. In addition to this tables with multiple columns and the use of frames will make presentation and visibility of content on mobile phones very difficult.

- A menu structure needs to be specified that can reach all content with few clicks. This must be accompanied with a description of which menu items are similar and which are different from the original design. There might be new categories in the design, some categories may have been merged, whereas other categories may have been omitted because the content would not make sense on a mobile phone. Menu navigation on mobile phones require few clicks in order not to confuse the consumer, and the designed menu can therefore be either organised as a flat list or hierarchically with the possibility to navigate back and forward between the levels. Any banners, categories, and/or title headers would be quite natural to use for naming new menu items.

- Sequences of web pages can typically be identified by following where the eye is reading to solve certain tasks, or by clicking on the existing menu items to see what information appears on the screen after the clicks. Making a storyboard of the sequence is a good way to express this. In addition to the storyboard one should mention which information is actually similar to the old one, which information has been left out, how the start of each storyboard is reached from the menu, and if all pages in the sequences provide an opportunity to reach to the menu if one needs to click through all the stages before the menu appears again.

- Most content on mobile phones at present tends to be text or static images. Images need to be generally smaller on a mobile screen size in terms of pixels than for a laptop – both in width and height. Since laptop screens generally have a wide layout and mobile screens mostly have a tall orientation, square or tall images would be a better design choice than images with a wide orientation. Hence, 140x140 pixels could be a good choice for both presentation and fast delivery given the design constraints above. Tables mostly present texts and numbers, but tables are generally difficult to present on a small mobile phone. Tables should therefore generally be omitted and the use of plain text, lists, or web pages that represent the same information as within the single rows or single columns of the existing website may be a good choice here.

- Banners and product categories should generally be made small and compact, especially in terms of height in pixels. The smaller the banners and categories are, the better it will be in terms of less consumed space on the mobile screen and faster delivery time. The transfer of text instead of image graphics is always faster too due to significantly smaller file sizes for text only.

- One or more product description pages in HTML should be proposed according to the new navigation menu. If images were used in the original product description, a compressed and smaller image would be a natural part of the new design too as a means to communicate the new design.

- A discussion that reveals that one needs to choose whether one should develop and maintain two separate content repositories, one for laptops and one for mobile phones, or if one would be able to create and publish web pages that are 100% reusable across mobiles and laptops should be the core of this summary. The conclusion should be that the introduction of a mobile website will inevitably affect the content creation and publishing process of the existing website designed for laptop use, therefore careful design considerations would be of high value to save content creation and publishing efforts.

6.1(b) The mobile search and spatial filters

- Local search aims to make information more relevant at the spatial level by filtering retrieved information based upon spatial criteria. The aim is to improve precision at the expense of recall: the set of results is much smaller but the information that remains is relevant both to an individual's query (based, for example, on a key word search) and their spatial location, or spatial behaviour.
- Some services can work effectively with low resolution information. Weather reports and local news items tend to be relevant to a region: this type of information does not change from one street to the next. Some services, however, will require very high resolution information: a "buddy finder" service built on a social networking platform-which promises to tell you when your trusted friends are nearby and navigate you to a rendezvous point-will need to know both your location, and that of your friends, to within a few metres.
- For services which aim to help an individual visit, the physical location associated with one or more of their results (for example finding local restaurants, then going to one of them), the 'accessibility' filter is the most appropriate. This can be hard to calculate, hence the 'nearby' filter is often used in its place. Services designed for people in transit (for example those reporting traffic incidents) can benefit from the 'likely future location' filter. Those wishing to engage people with their visible surroundings (for example, a mobile guide in an area of outstanding natural beauty) would benefit from a 'visibility' filter.

6.1(c) Ways to enable the mobile search

- The existing website may have a search page or search box available from every web page, or the search may be available for instance by choosing from a menu. In the first case, there are no extra clicks in order to reach the search facility, however in the latter one there might be up to several clicks before one can formulate a query. The consumer may have to click several places in the search page before the search is performed, or click to navigate to the search page(s). In general, fewer clicks are better than many clicks, and keying text may take considerably longer time than ticking choices. There might be several parameters to formulate a query with, and some of these can be single or multiple choice or check lists. Some of the query parameters will be more important to use, and the more alternatives, the more the consumer may spend time to formulate queries. A table or list that indicates the importance of each of the available query parameters, together with a justification of the importance is needed to provide a good answer. A screen shot might also help to illustrate the alternative options.
- A search facility with only one query parameter is likely to be using a traditional IR method, whereas if there is an SQL database, there is likely to be several search options that together can help reduce the number of results. Free text search can be supported by query operators, and this might reveal something about the underlying IR model in use. This should be reflected upon.
- A new design in HTML needs to be made and shown in a web browser. The best is to show the design on a mobile phone so one can compare it with the existing search page for laptops. A description of the similarities and differences is needed along with justifications for each of these. A table of pros and cons for multi-step versus single-step search pages should also be provided. Hint: the more available parameters to formulate a query, the more time and effort it may take the consumer to formulate each query, however, the relevance of the retrieved results might become higher the more effort the consumer spends to formulate

each query. In terms of navigation and presentation it should be pointed out that it is desirable in general to reduce the query formulation time and effort as much as possible.

- A personal settings page will enable the user to effectively formulate persistent queries and filters for information, which may lead to increased and faster use of the search facility. The design should be in HTML and shown in a browser – preferably on a mobile phone. A good design may include the query parameters identified above, or at least the important ones. Such a design can maximise the reuse of the existing search facility, and the search page should be one page away or less from any page.

- Introduction of a mobile website means more content that needs to be indexed. The mobile search page should only retrieve web pages designed for mobile phone use, and the existing search page for laptop use should only retrieve web pages designed for laptop use. Thus, if the same IR search index is used to store and retrieve relevant pages respectively for laptop use and mobile phone use, then one needs a method to separate the web pages from each other so that laptop pages are not relevant for mobile phones and vice versa. This will be a problem if one indexes mobile web pages together with laptop web pages. If the content creation and publishing process implies two separate web repositories, i.e. one for mobile and one for laptop, then the IR search index would be better off stored separately. In any case, any changes of published web pages would have to be reflected in the IR search index – regardless of whether the content is for mobile or laptop use. A good answer here reveals a clear understanding of these aspects along with a brief description of how the search facility works when product descriptions are respectively added, updated, or removed.

Chapter 7 – Context and Information Retrieval
Ayşe Göker, Hans Myrhaug and Ralf Bierig

7.1 The answer should include a list of the queries and an analysis of different interpretations. This may be based on a variety of aspects such as: differences in interest, geographic location, cultural background, time and so on. The answer should try to map differing contextual features on to the five dimensions or sub-contexts mentioned in the chapter (Social, Task, Personal, Environment, and Spatiotemporal.

7.2(a) Existing context information for publishing and search

    i. To help structure your answers, make a table called Existing context information with the following column labels: Name, Description, Is predefined category (tag), Is menu selection, Is text input. These are for menu input, text input, etc. Describe each input field as row in the table.

    ii. Publish five images in the web service, but remember to save all of the context information you specify for each image as five separate text paragraphs with the file name as the header for each of them, because you will use these terms below for query formulation.

    iii. Search for the five images by formulating queries from the context information you saved in the five text paragraphs, for example. Describe your observation for each query.

    iv. Describe if and how the existing context information of the table is being used in the search index, and conclude which of the context information attributes are in use, not in use, and if an image provider can contribute to the term vocabulary with new categories (tags) or not. You may find a range of possible context attributes discussed in Section 7.4.1.2.

    v. Summarise if it is possible to provide free text only, predefined categories (tags) only, or a combination of both free text and predefined categories (tags) as context information when publishing images. Conclude which context information is certainly in use, and how you think this context information is included in the search index. Again, it may be best to organise and represent this in the table you have made.

7.2(b) Improved context information for publishing and search

You need to create a context information model suitable for both image publishing and search, and describe it according to the figures 'User Context', and 'Context and Content' earlier in the chapter:

    i. Make a table called Current Context Attributes with the column labels: Name, Value type, Value set/range, Context type. Fill in the Name of the identified attribute; Value type as one of: String, Number, or Boolean; Value set/range as one of: Predefined, or Free; Context type as one of: Social, Task, Personal (Mental or Physiological), Environment, or Spatiotemporal.

    ii. Whether the information is more context or more content, try anyway to propose a new context attribute for the information. Revise if any of the identified attributes capture the same context information, and merge any similar attributes into one attribute.

    iii. Make a similar Competitor's Context Attributes table for a competitor's website on the Web. You may design or sketch this too as a start.

    iv. Propose an Improved Context Attributes table. This may be by joining the Current Context Attributes table with the Competitor's Context Attribute table, and by merging duplicate attributes. Alternatively, you may decide to drop some attributes in the competitor's case.

    v. In drawing the UML (Unified Modelling Language) class diagram ensure it represents the User Context with attributes as part of the different sub-contexts. Look at the Improved Context Attributes table for this. Create an XML-schema (eXtensible Mark-up Language) representing the UML class diagram. Hint: Convert the different Context classes into Context elements in XML – attributes can become either XML-elements or attributes of an XML-element.

    vi. Draw or sketch a new web page for publishing images, discuss how context information represented by your context XML-schema, can be used to update and maintain the search index.

    vii. In discussing the two approaches imagine a repository of 10 000 images and some search terms from above. Analyse in terms of result list length, precision, and recall (if you can think of ways to estimate recall). Consider which alternative you think will solve the problem of low relevance – expressed by consumers as: 'it gives random results', and 'the results are poor'. Consider other scenarios/situations that may help decide between the two.

Chapter 8 – Text Categorisation and Genre in Information Retrieval
Stuart Watt

8.1 You should consider the argument that text categorisation involves a qualitative decision. Consider the reason, need, and purpose of this decision for your chosen scenario.

Categorisation is part of a wider context of use. The main points can be summarised as follows:

1. Categorisation is intended to make a set of documents/objects easier to manage in some way.
2. Effective management of documents/objects depends on its intended purpose, for both sender and recipient.
3. The intended purpose of a document/object is reflected in its structure and layout, as well as in its use of language.

For example, for a photograph categorisation scenario involving your personal collection consider why you tend to retrieve or wish to retrieve photographs (e.g., to share with others, browse yourself). Consider how you are currently storing them (e.g., chronologically, event based, people based, a combination, random) and its effect on your retrieval. Consider how you have retrieved your photos in the past and how you think this may be in the future. Now, imagine you have a photograph collection but it is for work, say a news agency. How might some of these issues be different? For example, consider the role of captions, the circumstances in which you may need to retrieve various images in the future, and so on.

8.2 Remember that genres emerge through social interaction in communities, and reflect the needs and aims of those communities, so there is no universally common set of structures. Academic papers, for example, typically have a genre consisting of an abstract, introduction, conclusions, and references; in particular sub-fields, the middle sections may also be structured. Other common textual genres include business letters, memos, academic papers, blogs, and so on. The actual topic of the text is separate from the genre: blogs can be written on everything from politics to cooking, but in each case, blogs have an easily-recognised structure (date, headline, text, comments), with some standardised use of terms ('permalink', 'tag').

So, for your personal collection consider which genres might be possible and which you would prefer. For example, you may prefer to have a blog on your favourite films. Bear in mind that the purpose of genre is to guide information consumers what to do with the text – it is an aid to decision-making. In the case of a news story, the purpose of the headline is to grab people's attention; in an academic paper, the abstract aids filtering. If your personal collection includes a database of some reference books, then consider the use of book abstract, for example. In books, you may even keep academic reference books at work (where you are more likely to need it for reference, or share it with peers), and fiction at home (where you are more likely to want it recreationally). This is a simple classification that reflects the different purposes for each category of book.

Chapter 9 – Semantic Search
John Davies, Alistair Duke and Atanas Kiryakov

9.1 Two difficulties are index term synonymy and query term polysemy. The use of semantic annotation of documents, that is annotation based on the concepts in the documents, not simply the character strings used to represent the concepts, can overcome these difficulties by modelling the relationship between the meanings of terms.

9.2 From the ontology, it is known that London is part of the UK and that XYZ is a company. Thus the relevance of the document terms to the query terms can be inferred.

Chapter 10 – The Role of Natural Language Processing in Information Retrieval: Searching for Meaning and Structure
Tony Russell-Rose and Mark Stevenson

10.1 It is likely that one of the possible meanings of the polysemous term will dominate the search results and other meanings may not appear at all. In addition, the queries may retrieve unexpected web pages, for example where the terms are used as a name or abbreviation.

10.2 Performance of the search engine is likely to vary widely for a range of questions. Whether the user prefers formulating queries as natural language questions or using search terms is likely to be influenced by several factors, including query type, experience of using search engines and personal preference.

Chapter 11 – Cross-language Information Retrieval
Daqing He and Jianqiang Wang

11.1 During this exercise, readers will probably notice two things. First, unless they understand the document side language, it will not be easy for them to judge the relevance of the returned documents. This is exactly why we stated in this chapter that a batch model CLIR system is useful only for examining search algorithms, whereas a true interactive query translation-based CLIR system needs to translate the returned documents back to the query language side. The second insight would be that translation is very important in CLIR. If the quality of the translation is poor, the search results will be affected. Just by looking at the different translation qualities obtained from Google Translate, Yahoo Babel Fish or the online dictionary, readers can see this effect.

11.2 Through this exercise, readers should have more understanding of interactive query translation-based CLIR systems, and how translation is used twice in such processes. The readers should also notice that they spent more time looking at the translated returned documents for relevance judgment rather than at the original returned documents. The 'Not quite right? Edit' is unique for query translation-based CLIR. Here, to change the search results readers can not only modify the query terms, but can also modify the translations of the query terms. Readers will also notice one problem with the handling of translations in the current version of Google Translate. There is no way to put in multiple translations for a query term and at the same time indicate their relative importance, just like the situation in a MRD with translation probabilities. Maybe one day, Google Translate will provide that capability.

Chapter 12 – Performance Issues in Parallel Computing for Information Retrieval
Andrew MacFarlane

12.1 The first problem is modelling a task, which involves a significant complexity problem in its own right. Adding more variables such as more tasks and different distribution methods makes the problem very difficult if not impossible.

12.2 The problem is the limitations of modelling communication between machines in a parallel computer. This is due to the non-deterministic nature of the interaction between processes completing a given task, and is still and outstanding research issue in parallelism.

Index